STP 1302

High Temperature and Environmental Effects on Polymeric Composites: 2nd Volume

Thomas S. Gates and Abdul-Hamid Zureick, Editors

ASTM Publication Code Number (PCN):
04-013020-33

ASTM
100 Barr Harbor Drive
West Conshohocken, PA 19428-2959
Printed in the U.S.A.

Library of Congress Cataloging-in-Publication Data

Nontraditional methods of sensing stress, strain, and damage in materials and structures / George F. Lucas and David A. Stubbs, editors.
p. cm. — (STP ; 1318)
Includes bibliographical references and index.
ISBN 0-8031-2403-1
1. Strains and stresses—Measurement. 2. Detectors. I. Lucas, George F., 1951– . II. Stubbs, David A., 1958– . III. Series: ASTM special technical publication ; 1318.
TA417.6.N64 1997
620.1′123′0287—dc21 97-18165
CIP

Photocopy Rights

Peer Review Policy

Each paper published in this volume was evaluated by three peer reviewers. The authors addressed all of the reviewers' comments to the satisfaction of both the technical editor(s) and the ASTM Committee on Publications.

The quality of the papers in this publication reflects not only the obvious efforts of the authors and the technical editor(s), but also the work of these peer reviewers. The ASTM Committee on Publications acknowledges with appreciation their dedication and their contribution of time and effort on behalf of ASTM.

Printed in Fredricksburg, VA
August 1997

Foreword

This publication, *High Temperature and Environmental Effects on Polymeric Composites: 2nd Volume,* contains papers presented at the symposium of the same name held in Norfolk, Virginia on 13 November 1995. The symposium was sponsored by ASTM Committee D-30 on High Modulus Fibers and Their Composites. Dr. Thomas S. Gates of NASA Langley Research Center in Hampton, Virginia and Dr. Abdul-Hamid Zureick of the Georgia Institute of Technology in Atlanta, Georgia presided as symposium chairmen and are editors of the resulting publication.

Contents

Overview

The second symposium on High Temperature and Environmental Effects on Polymeric Composites was organized to address current research in a specialized and important area that is attracting international attention. The topics covered in the symposium represent state-of-the-art research from branches of materials science and mechanics of materials that range across the engineering disciplines from aerospace to civil. This diversity is due not only to the wide use of polymeric composites in today's world but also the tendency of manufacturers and designers to continue to expand the design envelope of these advanced materials. It is expected that this volume will provide information of use to those involved in the structural use of advanced polymeric composites for infrastructure, marine applications, and aerospace vehicles. Results are presented for research in: durability analysis procedures, moisture effects, time-dependent and fatigue behavior, long-term aging and degradation studies. Many of the papers combine analytical and experimental results while several of them rely on experimental studies to advance their research results.

Analytical models that predict long-term performance must account for the synergy and interaction between environmental factors such as load and temperature. Initially these models may be developed to provide a qualitative assessment and as they mature their predictive capabilities can be considered more quantitative and accurate. McManus and Cunningham proposed such a model to assess composite durability. Factors considered in their model included weight loss, shrinkage, and mechanical property degradation. Maier and Friedrich suggested a thermomechanical treatment procedure to improve the mechanical performance. The mechanisms causing this improvement were discussed and linked to the high permanent deformation induced in the matrix of the composite. Micromechanics, at the fiber-matrix level, were used to help describe this behavior. A methodology for the prediction of remaining strength and life in polymeric composites subjected to cyclic and thermomechanical conditions was presented by Case, Plunkett, and Reifsnider. Their model, which simulated damage accumulation and failure modes was based on micromechanical and ply-level models of strength that combined the material properties, geometry, and arrangement of the constituents to estimate global behavior.

Moisture in an elevated temperature environment can have a pronounced influence on the physical and mechanical properties of polymeric composites. VanLandingham, Eduljee, and Gillespie examined the effects of moisture at various temperatures and relative humidity on the glass transition temperature, fracture toughness, and diffusion kinetics. These properties were measured experimentally for a graphite/thermoplastic composite. Pratt and Bradley measured the effects of absorbed moisture on interfacial strength of a graphite/thermoset composite by using a single fiber fragmentation test. Changes in interfacial strength were associated with the changes in residual compressive stresses at the fiber/matrix interface. Choqueuse et al. measured the absorption kinetics and the evolution of properties for five glass fiber polymer composites immersed in distilled water for 2 years under elevated temperature and pressure conditions. They attempted correlation between the quantity of water absorbed and the loss of mechanical properties and assessed the usefulness of absorption curves for material selection. Zaffaroni used a thermodynamics formulation to determine the 2-dimensional diffusion in hybrid composites. The model was compared to the analytical Fick's law solution and experimental measurements.

Time-dependent behavior such as fatigue and viscoelasticity can have significant dependence on environmental conditions. The interaction between creep and fatigue at elevated temperatures and the effects of this interaction on Mode I delamination was investigated by Uematsu, Kitamure, and Ohtani. Crack propagation rates were measured and correlated with the stress intensity factor for two different polymeric composite materials. Shokreih, Eilers, and Lessard used a three-rail shear test to characterize the in-plane shear behavior of a graphite/epoxy tested using room-temperature fatigue and static loading. Fatigue life, residual strength, and residual stiffness were measured for both notched and unnotched specimens. Physical aging, a thermo-reversible effect, will occur in all polymers whereas chemical aging, a thermo-irreversible effect, may only occur in those polymers that have an incomplete state of cure. Elevated temperature is known to accelerate these effects. Kamvouris et al. investigated the use of time-temperature superposition to quantify the effects of physical and chemical aging on the stress relaxation behavior of high-temperature polyimide thermoset polymers. Both vertical (stiffness) and horizontal (time) shifts were measured.

Parvatareddy et al. used a fracture mechanics-based approach to evaluate the durability of a titanium/adhesive system. Testing was accomplished with wedge and double cantilever beam specimens. Aging was carried out at elevated temperatures with three different air pressures as well as a set of specimens subjected to several organic solvents. Strain energy release rates, static, and fatigue properties were measured. Bowles, Roberts, and Kamvouris measured the compressive properties of fabric reinforced composites after long-term isothermal themo-oxidative aging. Experimental properties included failure modes and compressive strength. The degradation mechanisms associated with the aging process were noted and correlated with specimen thickness and residual properties.

Many experimental and analytical techniques used to assess durability and long-term performance require data on the physical changes in the polymer matrix and fiber-matrix interface. The research by Bank and Puterman provided information on the physical changes associated with surface degradation of glass fiber pultruded rods embedded in concrete castings and subjected to standard and aqueous environments. Measurement techniques in this study included optical microscopy and scanning electron microscope. Prian et al. used thermogravimetric analysis (TGA) and weight loss to assess the degradation of a large number of polymeric composites and then correlated these results with changes in mechanical strength. The test materials were exposed to neutral, basic, and acidic media at elevated temperatures for extended periods. The results were examined for potential accelerating factors. The effects of a space environment at low earth orbit on polymeric composites was investigated by Shively et al. using laboratory test apparatus. The test environment consisted of simultaneous fluxes of hyperthermal neutral oxygen atoms, energetic electrons, vacuum ultraviolet light, and hypervelocity-particle impacts.

As evidenced by the papers presented in this volume, there is a large breadth of potential research problems in the area of high-temperature and environmental effects of polymeric composites with applications ranging from civil infrastructure to supersonic aircraft. Despite this diversity, the common threads that bind these areas together are the need for an understanding of degradation mechanisms, formulation of predictive models, and the development of accurate experimental techniques.

Dr. Thomas S. Gates

NASA Langley Research Center,
Hampton, Virginia.

Dr. Abdul-Hamid Zureick

Georgia Institute of Technology
Atlanta, Georgia.

Hugh L. McManus[1] *and Ronan A. Cunningham*[1]

Materials and Mechanics Analyses of Durability Tests for High-Temperature Polymer Matrix Composites

REFERENCE: McManus, H. L. and Cunningham, R. A., **"Materials and Mechanics Analyses of Durability Tests for High-Temperature Polymer Matrix Composites,"** *High Temperature and Environmental Effects on Polymeric Composites: 2nd Volume, ASTM STP 1302,* Thomas S. Gates and Abdul-Hamid Zureick, Eds., American Society for Testing and Materials, 1997, pp. 1–17.

ABSTRACT: Models of the environmentally-induced degradation of composite materials and the effect of this degradation on composite laminate behavior are presented. Although the models are not sufficiently mature for predictive design use, they provide insights into the mechanisms of degradation of composite materials. The models are used to study test methods frequently used to assess composite durability, including weight loss, shrinkage, and mechanical property degradation testing. It is found that even simple tests of material undergoing environmentally induced degradation are difficult to interpret. In particular, the spatial non-uniformity of degradation converts originally simple specimens into complex structures with non-uniform chemical and mechanical states. Consideration of this complexity is vital to the full understanding of the results of such tests.

KEY WORDS: polymer matrix composites, durability, environmental effects, aging, degradation, thermochemical modeling, micromechanics, shrinkage, microcracking, test methods

Composite materials are increasingly used in demanding structural applications exposed to harsh environmental conditions. The durability of such materials is a major concern, potentially limiting both the integrity of the structures and their useful lifetime. High temperatures, thermal cycling, and exposure to moisture, oils, and solvents all accelerate the degradation of composites. Hence, application of composites to structures such as turbine engine components and high-speed aircraft airframes may be ultimately limited by durability issues.

The present work is an attempt at developing an overall understanding of the mechanisms involved in composite degradation using a material modeling approach. A coupled model incorporating diffusion of oxygen into the material, degradation chemistry, matrix shrinkage and property loss, resulting effects on the ply and laminate level, and structural behavior of the degraded laminate is presented.

The model is used to replicate the results of a number of tests commonly used to study degradation effects in composite materials. Weight loss from rectangular prism specimens,

[1] Class of 1943 assistant professor and research assistant, respectively, Massachusetts Institute of Technology, Department of Aeronautics and Astronautics, Technology Lab for Advanced Composites, 77 Massachusetts Ave., Cambridge, MA 02139.

the shrinkage and cracking of such specimens, cracking of laminates, and the degradation of measured properties from bending tests are all simulated as functions of time, temperature, and environmental conditions. The results are correlated with test results. They are also explored in detail using parametric studies to elucidate the mechanisms responsible for the behaviors observed.

Background

Extensive testing has been carried out to characterize the performance of polymer matrix composites at high temperatures. This paper will concentrate on a commonly used material system, the NASA-Lewis developed PMR polyimide material. Other materials, notably other polyimide thermosets, and thermoplastic materials such as PEEK, have also been included in aging studies. Many of the phenomena observed are similar to those reported for PMR materials. Recent development efforts in the PMR material system are discussed in Ref *1*. Most testing has concentrated on isothermal aging, that is, the effects of continuous soaking at a single high temperature [*2–12*].

Experimental studies of the effects of aging at elevated temperatures can be divided into two categories, aging in inert atmospheres and aging in air. They can also be divided by the nature of the material tested. Various investigations have considered neat resin [*7,13*], bare fibers [*6*], and composite materials [*6,8,12,14,15*]. Only limited correlation exists between individual studies.

Neat resin studies provide insight into major composite degradation mechanisms. The observed effects of long-term aging on neat resin include specimen weight loss, specimen shrinkage, the formation of a distinct surface layer, development of surface microcracks, and the degradation of mechanical properties. Weight loss appears to be the result of two mechanisms—thermal and oxidative. Thermal weight loss occurs within the first 100 to 200 h of aging. It is a bulk mechanism, that is, it depends only on specimen volume, and it does not alter the physical nature of the polymer significantly, that is, there is no evidence of it causing an altered layer on the surface, voids or cracks [*11*].

Oxidative weight loss occurs throughout the aging periods observed. Over-extended aging times, a distinctive thin layer forms on the exposed surfaces and progresses inward. The surface layer exhibits a different chemical composition than the unreacted core material [*7,12*]. Voids develop within the surface layer and increase in size and density over time. These voids act as starter points for microcracks that grow from the exposed surface. Similarity of observed surface layer growth rates and weight loss rates [*7*] suggests that the weight loss occurs primarily in the degraded surface layer. Shrinkage of the polymer is also observed. It results in dimensional changes that are sensitive to specimen geometry.

Unidirectional composites demonstrate similar degradation mechanisms to the neat resin, although the weight loss is less severe. Weight loss appears to be dependent upon the matrix volume fraction [*14*]. Unidirectional composites also exhibit different rates of growth of surface degradation layers in each of the three principal directions, axial, transverse, and through-thickness. Highly enhanced degradation occurs along the fiber-matrix interface, accelerating growth of surface layers in the axial direction. Growth of surface layers in the transverse and through-thickness directions is slower than in the axial direction [*16*], except when resin-rich molded surfaces are present [*17*]. In these cases, the growth rate is initially enhanced, but as time progresses, the growth rate returns to the rate seen on surfaces without extra resin.

Laminated composites, which have plies at different angles, exhibit a more rapid weight loss than unidirectional composites. The weight loss at longer aging times in cross-plied composites appears to depend only on the specimen volume [*18*]. Again, a surface layer forms (approximately one ply thick) and microcracks develop at the surface. However, microcracks also form in the interior of the composite. Tensile stresses develop in the matrix due to residual thermal stresses from curing and degradation-induced shrinkage, leading to cracking of the matrix throughout the laminate. The cracks in the laminate enhance oxidation of the composite by providing additional paths for air to penetrate the material. This can lead to a vicious circle, with oxidation promoting cracking, allowing more oxidation, and so on [*18*].

Attempts have been made to analyze and model various aspects of this problem. Mass loss rates have been empirically fitted to Arrhenius rate curves [*9,19*]. This approach provides a simple means to model the stability of many different systems, but it is only useful for comparative purposes. More sophisticated models have combined modeling of the diffusion of oxygen into the material with chemical reaction rate equations to predict mass loss and the growth of degraded layers. In many such cases an effective diffusion coefficient model is used [*8,17*]. In these cases an apparent diffusivity is found based on the experimental weight loss curve for a composite. Cracking of the degraded layers has been predicted by modeling the degraded layers as layers with different material properties in a finite element model [*4*]. Aging-induced changes in viscoelastic properties have been incorporated into standard viscoelastic analysis techniques [*5*].

More general analysis techniques are also available that can be applied to some aspects of this problem. Here, only those used during this work are noted. Micromechanical models allow the prediction of composite properties and behavior given the matrix and fiber properties [*20*]. The effects of temperature and moisture on composite properties can also be predicted [*21,22*]. Existing semi-empirical analysis methods for composite durability [*23*] could be applied to this problem, although success is unlikely as these methods do not account for the unique physics observed in the high-temperature degradation process.

Problem Statement

Given material properties (including chemical reaction constants, gas diffusivities, and mechanical properties), geometry (including composite volume fractions and lay-up), and the environment (temperature and surface environment, as functions of time), a method is needed to predict the extent of oxygen diffusion into the material, the chemical state of the matrix, matrix shrinkage and property losses, and resulting ply and laminate properties and behavior. In this paper, coupled oxygen diffusion and chemical state models will be developed and used in conjunction with earlier models of shrinkage, property loss, and ply and laminate properties [*24*] to develop such a method.

Model

Approach

A model of the behavior of degrading composites is developed from basic engineering principles. Emphasis is placed on modeling all relevant phenomena and their possible interactions; therefore, the modeling of each phenomenon is kept as simple as possible. The model consists of the following submodels:

- Fickian model of diffusion of oxygen from the environment
- Arrhenius model of degradation chemistry
- Empirical matrix shrinkage and mechanical property loss model
- Micromechanical model of effects at the ply level

- Laminated plate model

The first two models will be presented and their coupled implementation discussed. The next three models, referred to here as the mechanical models, are fully described in Ref *24*. The critical features of these models, and the coupling between the diffusion and chemistry models and the mechanical models, will be summarized here.

A flat plate of thickness $2h$ is considered. Its width and length are assumed to be much greater than its thickness. It is composed of unidirectional composite plies of thickness t, aligned to a reference axis at angle θ. Uniform in-plane loads N and moments M (stress resultants in the notation of Jones [*25*]) are applied at the edges. One or both faces are exposed to an oxygen-containing atmosphere at time-varying temperature T and pressure P. It as assumed that the thermal transient time is small compared to that of all degradation mechanisms, so the temperature throughout the plate is assumed to be T.

Diffusion Model

It is assumed that substances from the environment diffuse into the material according to the following modified Fickian diffusion law.

$$\frac{\partial c_s}{\partial t} = \frac{\partial}{\partial x_i}\left(D_{ij}^s \frac{\partial c_+}{\partial x_j}\right) - r_s \tag{1}$$

where c_s is the relative concentration of substance s, D_{ij}^s is the non-isotropic tensor of diffusivity of substance s through the composite material, and r_s is the rate of consumption of substance s by reactions. The calculation of r_s is discussed below. The relative concentration c_s is defined such that when $c_s = 1$, the material is locally saturated with substance s, while $c_s = 0$ indicates its absence. The diffusivity constants D_{ij}^s are strongly dependent on temperature. Typically [*26*]

$$D_{ij}^s = D_o e^{-C/T} \tag{2}$$

where D_o and C are constants. Generally, a different D_o and C are possible for each component of D_{ij} and for each substance s.

Degradation Chemistry Model

Simple Arrhenius reactions are assumed. The reactions may be temperature-dependent only (representing thermal breakdown of material) in which case the reaction rates are described by

$$\frac{\partial \alpha_r}{\partial t} = k_r(1 - \alpha_r)^{p_r} \exp\left(\frac{-E_r}{RT}\right) \tag{3}$$

where α_r is a measure of the completeness of reaction r, and k_r, p_r, and E_r are the rate constant, order, and activation energy of the reaction, respectively, and R is the gas constant. The metric α_r is taken to be equal to zero at the beginning of the reaction, and one at the end.

Oxidative reactions (or, more generally, reactions requiring any substance s diffusing from the environment) are modeled using the following rate law, where q_r is the order of the concentration dependence.

$$\frac{\partial \alpha_r}{\partial t} = k_r(1 - \alpha_r)^{p_r} c_s^{q_r} \exp\left(\frac{-E_r}{RT}\right) \tag{4}$$

Implementation of Coupled Diffusion/Chemical Model

The diffusion and chemistry models are implemented using a 1-D computer code. Oxygen concentration and degradation due to a single oxidative reaction are predicted as functions of time and distance z from the surface. These quantities are assumed to vary continuously through the thickness, so a large number of computational nodes are used. From Eq 1, assuming that changes in D_{zz}^{ox} with z are small

$$\frac{\partial c_{ox}}{\partial t} = D_{zz}^{ox} \frac{\partial^2 c_{ox}}{\partial z^2} - r_{ox} \tag{5}$$

and from Eq 4

$$\frac{\partial \alpha_{ox}}{\partial t} = k_{ox}(1 - \alpha_{ox})^{p_{ox}} c_{ox}^{q_{ox}} \exp\left(\frac{-E_{ox}}{RT}\right) \tag{6}$$

The consumption rate of oxygen r_{ox} is found from the reaction rate

$$r_{ox} = R_{ox} \frac{\partial \alpha_{ox}}{\partial t} \tag{7}$$

where R_{ox} is a constant dependent on the chemistry of the reaction. Equations 5 through 7 can be seen to be coupled due to the presence of the consumption term r_{ox} in Eq 5 and the concentration term $c_{ox}^{q_{ox}}$ in Eq 6. The reduction in solid mass due to reactions is also dependent on the chemistry of the reactions. It can be expressed as

$$m = m_o - m_r\alpha_{ox} \tag{8}$$

Here, m, m_o, and m_r represent the mass per unit *original* volume, the original mass per unit volume, and the change in mass per unit original volume when the reaction is complete, respectively. Note m is not necessarily equal to the material density, as the material may shrink during degradation, changing the volume. If the material does not shrink, m equals the local material density ρ. The total mass lost from a specimen of width W, length L, and thickness $2h$ is then

$$\Delta M = 2WL \int_0^h (m_o - m)\, dz \tag{9}$$

Equations 5 through 9 are solved numerically using an explicit finite difference scheme to predict oxygen concentrations, reaction rates, reaction states, and mass loss as functions of time and position, and total mass loss as a function of time. Initial conditions are $c_{ox} = 0$ and $\alpha_{ox} = 0$ everywhere. Boundary conditions are $c_{ox} = 1$ at the surface ($z = 0$), and no mass flux ($\partial c_{ox}/\partial z = 0$) at the midplane of the plate ($z = 0$).

Matrix Shrinkage and Mechanical Property Loss Model

The mechanical modeling of the degradation effects is performed on a ply-by-ply basis. Thus, the progress of the oxidation reaction α_{ox}, which is a continuous function of z, is averaged across each ply. The result is the matrix degradation metric b. Taking the definition of b found in Ref *24* and averaging across a given play k

$$b_k = \frac{1}{t_k} \int_{z_k^b}^{z_k^t} \frac{m_o - m}{m_o} dz \tag{10}$$

where t_k is the ply thickness and z_k^t and z_k^b are the positions of the top and bottom of ply k, respectively. Substituting Eq 8 into Eq 10 yields

$$b_k = \frac{m_r}{t_k m_o} \int_{z_k^b}^{z_k^t} \alpha_{ox}\, dz \tag{11}$$

showing that b_k is simply an expression of the average chemical degradation in the ply.

Matrix properties are assumed to degrade as a (empirical) function of b_k. Matrix degradation also causes an increase in void content ϕ found from

$$\phi = \phi_o + Db \tag{12}$$

where ϕ_o is the initial void content and D is an experimentally determined constant. Mass loss not arising from the formation of voids is assumed to cause shrinkage, hence

$$\varepsilon_b^* = -\frac{(1 - D)b}{3} \tag{13}$$

The shrinkage strain resembles a thermal strain. In a uniformly shrinking body free of restraint it will not induce stress, but in a non-uniformly shrinking body and/or one in which deformations are restrained, it can induce stresses.

Ply Level Model

The property changes in, and shrinkage of, the matrix cause property changes in, and shrinkage of, unidirectional composite plies. The effects are calculated for each ply k in the laminate. The ply properties are calculated by assuming that the fiber properties do not degrade and the fibers do not shrink. Micromechanical techniques are used to determine the changes in ply properties due to changes in the matrix properties [*24*].

The shrinkage of the matrix also causes shrinkage of the plies. A modified set of micromechanical relations are used to calculate ply deformations due to matrix shrinkage. These relations, combined with Eq 13, yield relations between ply shrinkage and the degradation metric b of the form

$$\varepsilon_{11}^* = \chi_{11} b \qquad \varepsilon_{22}^* = \chi_{22} b \tag{14}$$

where ε_{11}^* and ε_{22}^* are ply shrinkage strains in the longitudinal and transverse directions, and the χ's are shrinkage strain coefficients, derived in Ref *24*.

The in-plane stress-strain behavior of the degraded ply can now be described by

$$\boldsymbol{\sigma} = \mathbf{Q}(\boldsymbol{\varepsilon} - \boldsymbol{\varepsilon}^T - \boldsymbol{\varepsilon}^H - \boldsymbol{\varepsilon}^*) \quad (15)$$

or equivalently

$$\boldsymbol{\sigma} = \mathbf{Q}(\varepsilon - \alpha\Delta T - \boldsymbol{\beta}\Delta M - \boldsymbol{\chi} b) \quad (16)$$

where σ is the ply in-plane stress vector, and ε, ε^T, ε^H, and ε^* are the corresponding vectors of strain, thermal strain, hygral strain, and shrinkage strain, and α, β, and χ are the thermal, hygral, and shrinkage coefficient vectors, respectively. The matrix Q is the reduced stiffness matrix, calculated using the degraded ply material properties.

Laminated Plate Model

Equation 16 can be used in place of the conventional stress-strain relations to derive a modified classical laminate plate theory to predict the stresses and strains in, and deformations of, a laminated plate. Ply property degradations affect laminate properties, and ply shrinkage can cause deformations of, and stresses in, the laminate. Failure criteria can be applied to determine which plies fail, and in what mode. Details are given in Ref *24*.

Implementation of Mechanical Models

The matrix degradation, ply, and laminate models are implemented as modules of the ICAN computer code. A complete description of this code is given in Ref *20*; the new modules are described in Ref *24*. The user specifies laminate geometry, loadings, temperatures, moisture contents, and degradation states. The user can also modify a material data base file containing constituent material properties and the constants relating degradation to material property changes. The output is a complete description of the laminate, including laminate properties and stresses and strains in all plies.

Results

In this section, results from three studies are presented. Mass loss from a finite size specimen is considered first. A fit to existing data is used to find approximate material constants, then a parametric study is used to explore the relative importance of diffusion and reaction rates to the degradations. The second study considers the shrinkage of a neat resin specimen of the type used in the mass loss studies. Finally, the effect of non-uniform degradation on the results of mechanical tests are considered.

Mass Loss Tests

As reported by Bowles et al. [*7*], 75- by 6.4-mm samples of neat resin, either 6.4- or 2.5-mm thick, were isothermally exposed, and weight loss, shrinkage, and the depth of a visible surface layer measured. The tests are analytically simulated using the coupled diffusion and chemical reaction models. The constants needed to use the model were estimated from available literature. Bowles [*6*] reported the activation energy for the mass loss degradation mechanism to be 128 kJ/mol. Other studies have been carried out using thermogravimetric

TABLE 1—*Empirical/model chemical reaction data.*

	E_{ox}	$E_{ox}/R(K)$	$\log k_{ox}(s^{-1})$	$D_o(mm^2/s^{-1})$	$C(K)$
Starting point data	128	15 395	7.81	16.1	5 690
Model fit	128	15 395	6.81	4.31×10^4	18 000

analyses to measure both activation energies and pre-exponential constants. Kiefer et al. [*19*] performed tests on thermoplastic polyimides. The powdered samples used in these tests gave approximately the same activation energy (121 kJ/mol) as in Ref *7*. For this reason it was decided to use the pre-exponential rate constant from this data as a starting point for determining the actual constants. Limited information on the gas diffusivity of polymer matrix composites was given in Ref *8*. Water diffusivity is more widely studied. Values of D_o and C, which fall in a relatively narrow range, were reported in Ref *26*. These were used as a starting point for further calculations.

Preliminary values in Table 1 were taken as a starting point, and then fit to existing mass loss data [*7*] was used to establish constants. The fit achieved is shown in Fig. 1. The resulting fit constants are also given in Table 1. These constants should not be considered useful for predictive calculations. Instead, they were used as a baseline for parametric calculations. No information on the quantities of oxygen absorbed into polymer surfaces was available, so the concentration c_{ox} was left as a relative, dimensionless parameter. No information was available on the consumption rate R_{ox}, so it was set to a nominal value of 0.01. The initial density m_o was set to 1300 kg/m^3, and the mass loss per unit volume during the reaction m_r was taken to be 234 kg/m^3, based on the data from Ref *7*.

To compare the relative importance of the diffusion and reaction mechanisms, a non-dimensional parameter ζ is defined as

$$\zeta = \frac{k_{ox} h^2 e^{-E_{ox}/RT}}{D_{zz}^{ox}} \tag{17}$$

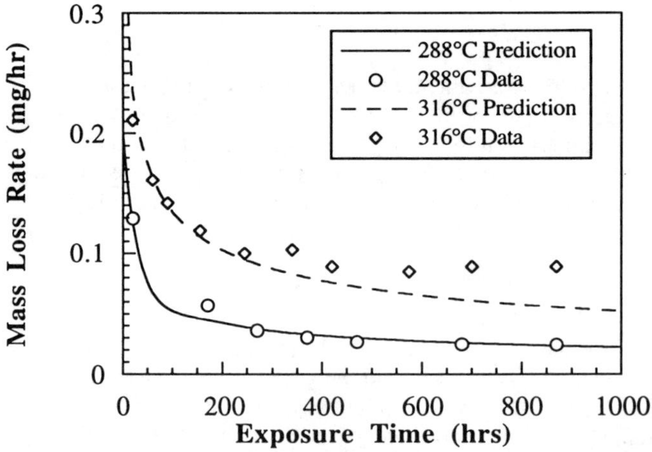

FIG. 1—*Comparison of predicted mass loss rate to data of Bowles at 288 and 316°C.*

TABLE 2—*Parametric study values.*

ζ	log $k_{ox}(s^{-1})$	log D_{zz}^{ox} (mm^2/s^{-1})
0.1	4.48	−6.27
10	5.13	−7.59
100	5.51	−8.21
1 000	5.95	−8.76
10 000	6.54	−9.18
20 000	6.74	−9.27
24 090	6.81	−9.30

The parameter ζ is the ratio of the rate determining terms of the reaction and diffusion processes, normalized by the specimen dimension h. Large values of ζ indicate that the mass loss will be limited by diffusion; small values indicate that the mass loss will be limited by the reaction speed. The speed of these processes may vary by many orders of magnitude depending on conditions such as temperature; hence, a wide range of values of ζ may be encountered. If the consumption of the diffusing substance (r_k in Eq 5) is not too large, ζ can be shown to fully characterize the shape of the mass loss curve.

A parametric study was carried out. The 75- by 6.4- by 2.5-mm specimens of Bowles [7] were considered. Mass loss calculations were repeated for each set of D_{zz}^{ox} and k_{ox} values given in Table 2. These values were normalized such that the total mass loss from a specimen (ΔM in Eq 9) at 1000 h was held constant at 37 mg in each set of calculations. The results are shown in Figs. 2 and 3. Figure 2 shows that the shape of the mass loss versus time curve depends on the parameter ζ. The effect of ζ can be seen even more clearly in Fig. 3, which shows the mass loss rate. This figure includes the data of Bowles. It appears to confirm that in the case studied by Bowles, the reaction is limited by diffusion. It was found that for small values of r_{ox}, ζ entirely controls the shape of this curve. The value of ζ may be kept the same by simply increasing or decreasing both D and k in the same proportion. The effect

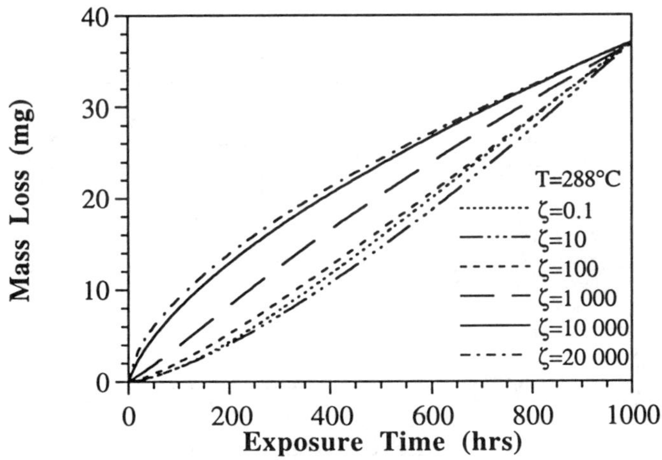

FIG. 2—*Mass loss as a function of time for different values of the dimensionless parameter ζ at 288°C.*

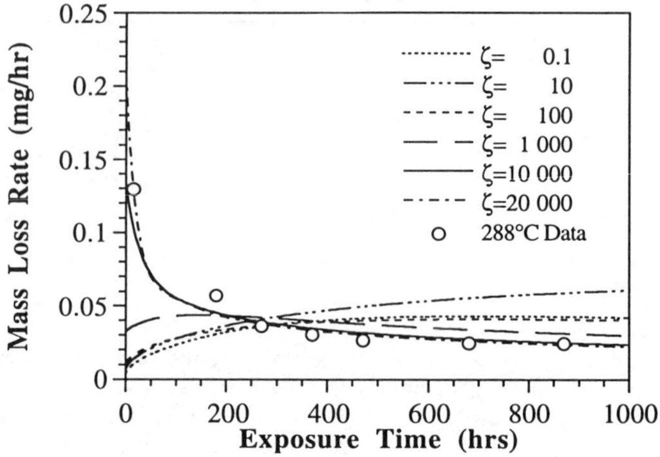

FIG. 3—*Mass loss rate as a function of time for different values of the dimensionless parameter* ζ.

this has is to speed up or slow down the overall reaction, but it does not change the curve shape.

The internal distributions of absorbed oxygen c_{o_x} and oxidation reaction progress α are shown in Figs. 4 and 5 for the case of $\zeta = 24\ 090$, the value which matched the mass loss data of Bowles. It is seen that degradation proceeds in from the surface, paced by the availability of oxygen.

Figures 6 and 7 show the absorbed oxygen concentration and reaction progress in the case where $\zeta = 0.1$. Although not entirely uniform, degradation proceeds throughout the sample. By studying a complete set of such figures, it was determined that for ζ values below 1000,

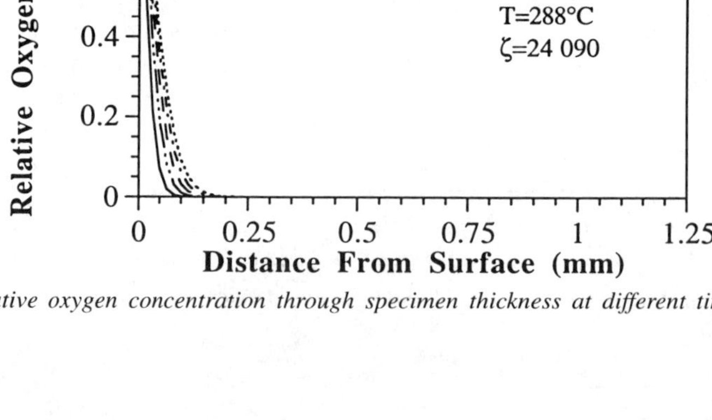

FIG. 4—*Relative oxygen concentration through specimen thickness at different times at 288°C for* $\zeta = 24\ 090$.

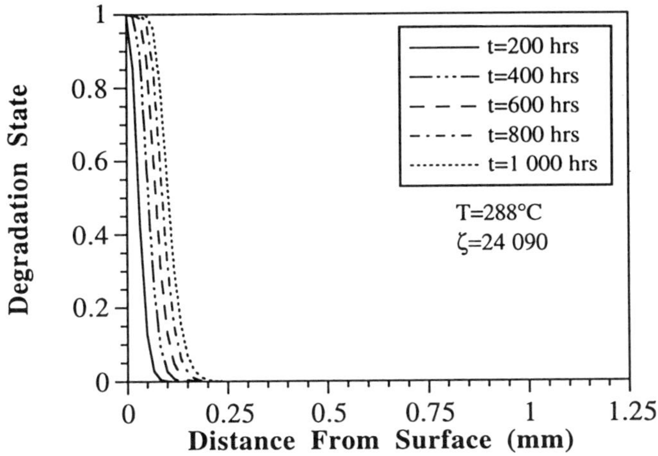

FIG. 5—*Degradation state through specimen thickness at different times at 288°C for* ζ = *24 090.*

the degradation process was reaction-rate limited; for values from 1000 to 5000, it was dependent on both reaction rate and oxygen diffusion, and at values above 5000, it was limited by the availability of diffused oxygen.

Bowles also recorded the progression of the observed degraded layer. Figure 8 compares Bowles' fit to his empirical data and the predictions of the model. It was assumed that when $\alpha = 0.25$ at a point in the specimen, the degradation is visible at that point. The good agreement increases confidence that the model is capturing the correct phenomenon.

To parametrically study the effects of different temperatures, the dimensionless parameter ζ is reconsidered. To consider the temperature dependence of D_{zz}^{ox}, Eq 2 is substituted into Eq 17, which yields

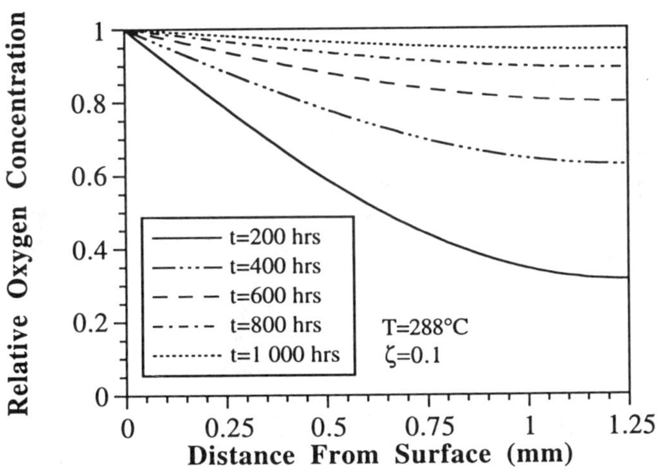

FIG. 6—*Relative oxygen concentration through specimen thickness at different times at 288°C for* ζ = *0.1.*

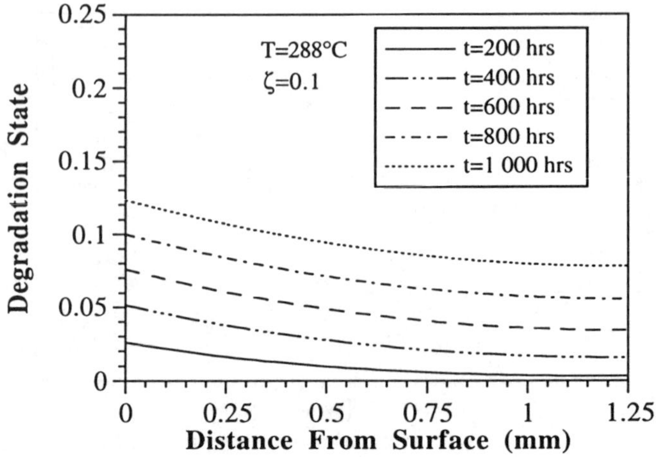

FIG. 7—*Degradation state through specimen thickness at different times at 288°C for* $\zeta = 0.1$.

$$\zeta = \frac{k_{ox}h^2}{D_o} e^{-(E_{ox}/R - C)/T} \tag{18}$$

A parametric study was carried out starting with the nominal values from the last line in Table 2, and varying temperature. The cases studied are summarized in Table 3. At different temperatures, the reactions proceed at very different rates. To allow a comparison of the different cases, the time scale was normalized by t_o, defined as the time necessary to reach 37 mg mass loss in each case. The value of t_o for each case is shown in Table 3. The resulting consolidated plot is shown in Fig. 9. The waviness of the curve at T = 300 K is a numerical artifact, due to the extremely long times required (Table 3). The mass loss curve

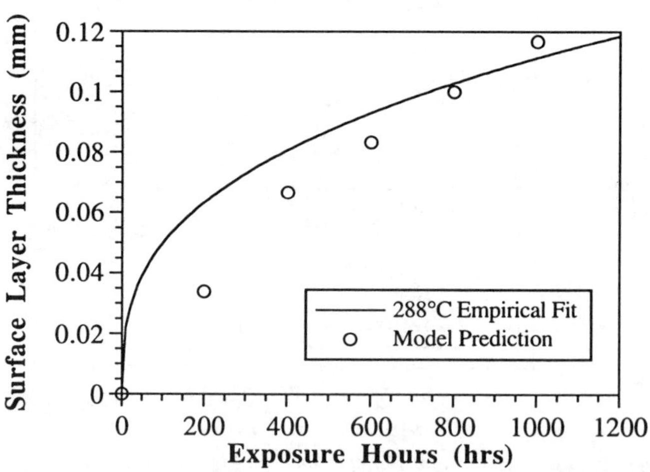

FIG. 8—*Comparison of predicted surface layer thickness to empirical data fit of Bowles at 288°C for* ζ = *24 090.*

TABLE 3—*ζ and normalization factor* t_o *as functions of temperature.*

$T(K)$	ζ	t_o, h
300	1.37×10^6	5.65×10^{14}
400	1.56×10^5	2.45×10^8
500	39 155	4.17×10^4
561	24 090	1.00×10^3
600	17 814	1.35×10^2
700	9 580	2.30

shapes are very dependent on temperature. As noted before, the mass loss curve shape is determined by ζ, which also indicates the relative importance of diffusion and reaction rate mechanism. Table 3 shows a wide range of ζ values. Over practical ranges of testing and use, ζ appears to be very high, indicating diffusion dominated behavior. At higher (accelerated testing) temperatures, ζ is reduced but is still high enough to indicate that the degradation mechanism remains diffusion dominated.

Shrinkage

The specimens of Bowles were modeled using the ply and laminate behavior models to predict their shrinkage along their long axes. They were modeled as 3-layer laminates, two degraded layers and a central core that is not degraded. The thicknesses of the degraded layers were modified to account for the finite width of the specimens. This model is accurate for axial stress and shrinkage predictions, which depend solely on the relative areas of the degraded and undegraded material. Because of the changing degradation layer thickness, a new model was constructed for each temperature and exposure time.

The diffusion/chemical reaction model was used to define the degradation state of the specimen. The degraded layer thickness as a function of time was taken from predictions

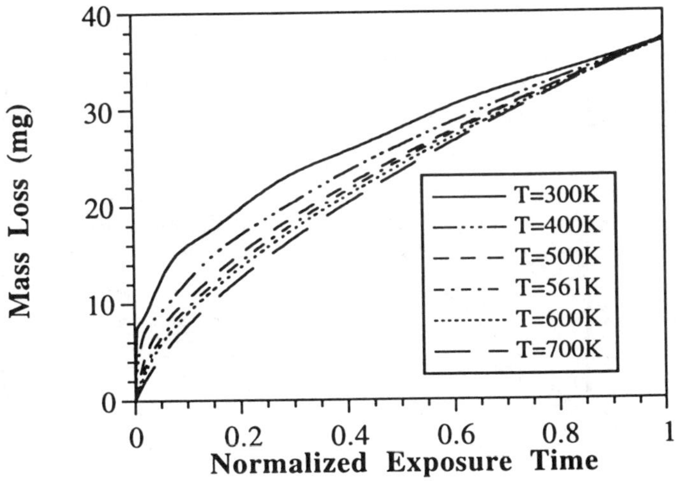

FIG. 9—*Comparison of mass loss curves at different temperatures for fixed set of reaction and diffusion constants.*

shown in Fig. 8. The ply degradation metric b for the degraded plies was then calculated using Eq 11. The ply model was used to find the response of the degraded layers, and the laminate model to find the shrinkage of the entire specimen.

The shrinkage of the specimen was predicted assuming no mechanical loads or residual thermal stresses. It was assumed that $D = 0.1$, based on the observation of moderate void formation in the degraded layer. The results are compared to the measured results in Fig. 10. As can be seen, the correlation is quite good. The results are essentially identical to the results shown in Ref *24*. In that case, the matrix degradation state was determined empirically, directly from Bowles' weight loss and degradation layer thickness data.

The results provide insight into the mechanisms of specimen shrinkage. The mass loss in the surface layer results in shrinkage, which is constrained by the unshrinking, undegraded core material. The results is a net shrinkage of the entire specimen, and stresses in both the degraded layers and the core. The core stress are compressive, and the stresses in the degraded layer are tensile. These stresses reach sufficient magnitude to cause failure in the degraded layer. Although insufficient data was provided in Ref *7* for a quantitative correlation of this effect, the failure in the surface layers predicted by the code qualitatively agrees with the observed surface layer cracking.

Bend Test for Mechanical Properties

Bending tests can be used to measure the strength and stiffness of simple specimens under a variety of conditions. However, the results must be carefully interpreted before material properties can be accurately extracted from these tests. Here, the models are used to illustrate the difficulty of extracting mechanical properties even from simple bend tests on degraded

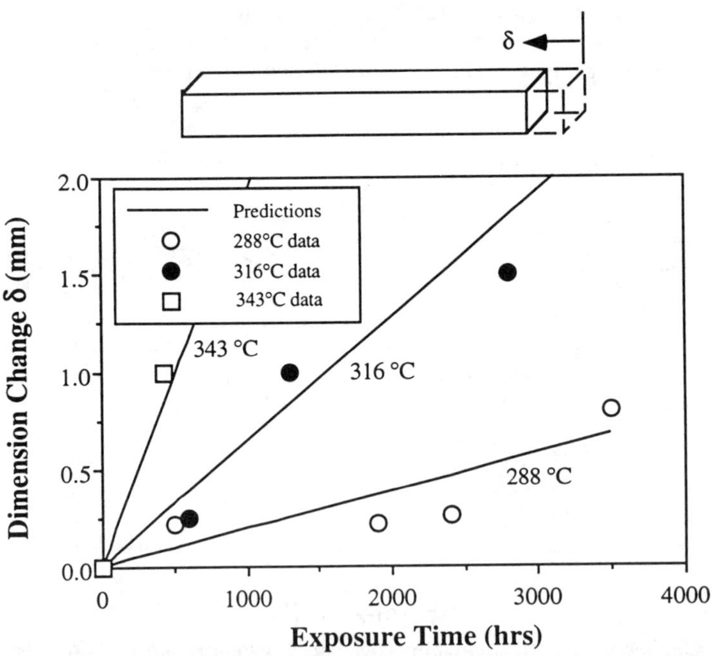

FIG. 10—*Comparison of predicted longitudinal dimension changes to data of Bowles.*

specimens. A 16-ply unidirectional laminate tested in Ref *2* is analyzed. The models were used to predict degradation in the surface plies after 150 h at 371°C, the ply response to this degradation, and the laminate response to degradation and bending loads. This study is more fully described in Ref *24*. The predicted axial stress distribution in the laminate under the bending load that caused failure in the specimens of Ref *7* is shown in Fig. 11, along with the stresses predicted for a uniform material under the same load. The apparent step-wise variation in the stresses is an artifact of the code's calculation of a single stress for each ply and is not important. However, the top and bottom plies have stresses that differ significantly from those predicted for a uniform material due both to shrinkage and changes in stiffness. These effects complicate accurate interpretation of bending test data. An overly simple data reduction technique (for example, use of the stresses calculated assuming uniform material) results in a very inaccurate estimate of the failure stress of the degraded material.

Conclusions

The models developed can replicate the observed physics of degrading polymer matrix composite materials. Mass loss, shrinkage property changes, cracking and failure, and their dependencies on temperature, time, and specimen geometry can be calculated with a set of simple models.

The coupled diffusion and chemical reaction models provide insight into the mechanisms of degradation. A dimensionless parameter can be used to assess the relative importance of diffusion and chemical reaction rates in controlling the behavior of a degrading specimen. This parameter is dependent on material properties, specimen geometry, and test (or service) temperature. The behavior of neat PMR-15 resin specimens under accelerated test conditions

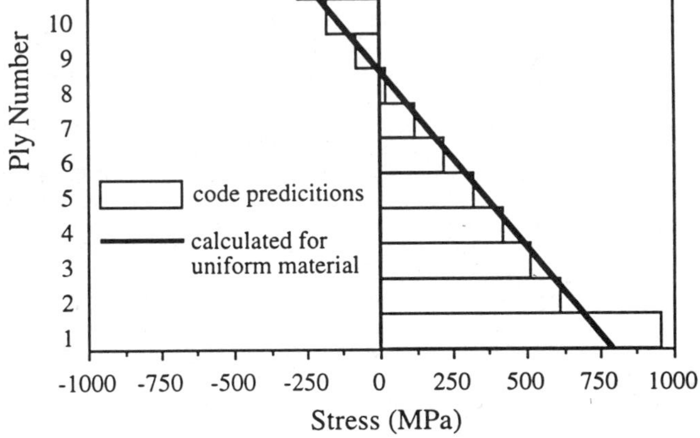

FIG. 11—*Comparison of ply stresses predicted by code to those calculated assuming uniform material at time of failure for specimens aged 150 h.*

seems to be governed by diffusion, and appears to become increasingly dominated by diffusion at lower (service) temperatures.

The predictions of the diffusion and chemical models can be used as input to models of degrading and shrinking ply and laminate behavior. These models can be used to analyze the behavior of mechanical test specimens. It is found that even simple tests such as neat resin sample shrinkage measurements are in fact measurements of complex behavior. Severely degraded surface layers shrink causing a lesser degree of overall specimen shrinkage, and setting up internal stresses that ultimately cause the surface layers to crack. Mechanically loaded specimens such as bending test specimens will behave in even more complex ways, with the mechanical loads, material changes, and shrinkage-induced deformations and internal stresses all interacting. This complex behavior must be taken into account when interpreting the results of such tests.

It appears that relatively simple models of the relevant physical phenomenon can be combined to successfully replicate the complex observed behavior of degrading composite laminates. At their current state of maturity, these models provide qualitative insights into the mechanisms of degradation, and can also aid in interpreting test results and designing accelerated tests. To achieve quantitative, predictive capability, the models need both better material properties as inputs and more extensive verification of their ability to predict laminate behavior.

Acknowledgments

This work was supported by National Science Foundation Young Investigator Award MSS-9257612 and NASA Lewis Grant NAG3-1760, under the direction of Dr. Christos C. Chamis.

References

[1] Meador, M. A., Cavano, P. J., and Malaric, D. C., "High Temperature Polymer Matrix Composites for Extreme Environments," Proceedings of the Sixth Annual ASM/ESD Advanced Composites Conference, Detroit, MI, Oct. 1990, pp. 529–539.

[2] Vannucci, R. D. and Cifani, D., "700°F Properties of Autoclave Cured PMR-II Composites," Proceedings of the 20th International SAMPE Conference, Sept. 1988, pp. 562–575.

[3] Pederson, C. L., Gillespie, J. W. Jr., McCullogh, R. L., Rothschilds, R. J., and Stanek, S. L., "The Effect of Isothermal Aging on Transverse Crack Development in Carbon Fiber Reinforced Cross-Ply Laminates," CCM Report 93-42, University of Delaware Center for Composite Materials, Newark, DE, Nov. 1993.

[4] Pederson, C. L., "The Effect of Temperature on Transverse Cracking in High Performance Composites," CCM Report 92-28, University of Delaware Center for Composite Materials, Newark, DE, June 1992.

[5] Gates, T. S. and Brinson, L. C., "Acceleration of Aging in Graphite/Bismaleimide and Graphite/Thermoplastic Composites," Proceedings of the 36th AIAA/ASME/ASCE/AHS/ASC Structures, Structural Dynamics, and Materials Conference, Hilton Head, SC, April 1994, pp. 2096–2105.

[6] Bowles, K. J., "Thermo-Oxidative Stability Studies of PMR-15 Polymer Matrix Composites Reinforced with Various Continuous Fibers," *SAMPE Quarterly*, Vol. 21, No. 4, 1990, pp. 6–13.

[7] Bowles, K. J., Jayne, D., and Leonhardt, T. A., "Isothermal Aging Effects on PMR-15 Resin," *SAMPE Quarterly*, Vol. 24, No. 2, 1993, pp. 3–9.

[8] Hipp, R. C., Harmon, D. M., and McClellan, P. S., "Accelerated Aging and Methodology Development for Polymeric Composite Material Systems," McDonnell Douglas Co. preprint, 1994.

[9] Grayson, M. A. and Fry, C. G., "On the Use of a Kinetic Map to Compare the Thermal Stability of Polymeric Materials Undergoing Weight Loss," McDonnell Douglas Co. preprint, 1994.

[10] Bowles, K. J. and Nowak, G., "Thermo-Oxidative Stability Studies of Celion 6000/PMR-15 Unidirectional Composites, PMR-15, and Celion 6000 Fiber," *Journal of Composite Materials*, Vol. 22, 1988, pp. 966–985.

[*11*] Bowles, K. J., "A Thermally Modified Polymer Matrix Composite Material with Structural Integrity to 371°C," Proceedings of the 20th International SAMPE Conference, Sept. 1988, pp. 552–561, also published as NASA TM-100922.

[*12*] Martin, R. H., Siochi, E. J., and Gates, T. S., "Isothermal Aging of IM7/8320 and IM7/5260," Proceedings of the American Society for Composites 7th Technical Conference on Composite Materials, University Park, PA, Oct. 1992, pp. 207–217.

[*13*] Alston, W. B., Gluyas, R. E., and Snyder, W. J., "Cyclopentadiene Evolution During Pyrolysis-Gas Chromatography of PMR Polyimides," Proceedings of the Fourth International Conference on Polyimides, Ellenville, NY, 30 Oct.-1 Nov. 1991.

[*14*] Alston, W. B., "Resin/Fiber Thermo-Oxidative Interactions in PMR Polyimide/Graphite Composites," Proceedings of Twenty-Fourth National SAMPE Symposium, San Francisco, CA, May 1979.

[*15*] Alston, W. B., "Characterization of PMR-15 Polyimide Resin Composition in Thermo-Oxidatively Exposed Graphite Fiber Composites," Proceedings of the Twelfth National SAMPE Technical Conference, Seattle, WA, October 1980.

[*16*] Madhukar, M S., Bowles, K. J., and Papadopolous, D. S., "Thermo-Oxidative Stability of Graphite/PMR-15 Composites: Effect of Fiber Surface Modification on Composite Shear Properties," NASA TM-4608.

[*17*] Nam, J. D. and Seferis, J. C., "Anisotropic Thermo-Oxidative Stability of Carbon Fiber Reinforced Polymeric Composites," *SAMPE Quarterly*, Vol. 24, No. 1, Oct. 1992, pp. 10–18.

[*18*] Bowles, K. J., Jayne, D., Leonhardt, T. A., and Bors, D., "Thermal Stability Relationships Between PMR-15 Resin and Its Composites," *Journal of Advanced Materials*, Vol. 26, No. 1, Oct. 1994, pp. 23–32.

[*19*] Kiefer, R., Yue, J. J., and Hinkley, J. A., "Kinetic Mapping of Oxidative Weight Loss in Polyimide Composites," *Journal of Advanced Materials*, Vol. 26, No. 3, April 1995, pp. 55–59.

[*20*] Murthy, P. L. N., Ginty, C. A., and Sanfeliz, J. G., "Second Generation Integrated Composite Analyzer (ICAN) Computer Code," NASA TP-3290, Jan. 1993.

[*21*] Chamis, C. C., "Simplified Composite Micromechanics Equations for Hygral, Thermal, and Mechanical Properties," *SAMPE Quarterly*, Vol. 15, April 1984, pp. 14–23.

[*22*] Chamis, C. C., "Simplified Composite Micromechanics for Predicting Microstresses," NASA TM-87295.

[*23*] Chamis, C. C. and Ginty, C. A., "Fiber Composite Structural Durability and Damage Tolerance: Simplified Predictive Methods," NASA TM-1000179, April 1987.

[*24*] McManus, H. L. and Chamis, C. C., "Stress and Damage in Polymer Matrix Composite Materials Due to Material Degradation at High Temperatures," NASA Technical Memorandum 4682, January 1996.

[*25*] Jones, R. M., *Mechanics of Composite Materials*, Hemisphere, New York, 1975.

[*26*] Tsai, S. W., *Composites Design*, 4th Edition, Think Composites, Ohio, 1988.

Georg Maier[1] *and Klaus Friedrich*[2]

Mechanisms of Property Improvements of Polymer Composites by Thermomechanical Treatment

REFERENCE: Maier, G. and Friedrich, K., **"Mechanisms of Property Improvements of Polymer Composites by Thermomechanical Treatment,"** *High Temperature and Environmental Effects on Polymeric Composites: 2nd Volume, ASTM STP 1302,* Thomas S. Gates and Abdul-Hamid Zureick, Eds., American Society for Testing and Materials, 1997, pp. 18–34.

ABSTRACT: The damage behavior of polymeric composites can be improved considerably by thermomechanical treatment. In this paper the relevant mechanisms causing this improvement are discussed in detail. It includes the physical and mechanical properties of the pure matrix before and after the thermomechanical treatment, as well as micromechanical mechanisms. The experimental results show that a thermomechanical treatment causes a high amount of permanent deformation of the matrix but no essential changes in its mechanical properties. The micromechanical calculations indicate an almost constant absolute value of permanent deformation of matrix across the entire fiber radius and consequently no changes of local stresses. Therefore, the treatment causes only a change in fiber distance, which results in a permanent deformation of transverse plies of the multidirectional laminate. Additional requirements for the mechanical properties of fibers and matrix in order to achieve a successful treatment are suggested from this approach.

KEY WORDS: thermomechanical treatment, polymer matrix composites, mechanisms, transverse crack formation, property improvements, glass transition, epoxy, carbon fiber, frozen in stresses, residual stresses, micromechanics

In a previous paper [*1*], thermomechanical procedures for improving the mechanical properties of thermoset composite laminates were described. They are based on three treatment steps carried out with a multidirectional reinforced laminate, namely heating up to treatment temperature (Step 1), loading (Step 2), and cooling down under load (Step 3). The treatment can be applied to completely cured laminates (Treatment A) and not completely cured laminates (Treatment B). In the first case, the treatment loading (Step 2) is performed above the glass transition temperature for only a short period of time (5 min). In the second case, the laminate is stored under load (Step 2) at curing or post-curing temperature until complete curing has occurred. After this last mentioned treatment, Procedure A with an enhanced load is added.

With both techniques, substantial increases in elongation at the onset of typical damage development like transverse cracking and delamination under tensile loading (up to 150%

[1] Head, Technology of Plastic and Composites Section, Institute for Material Research and Testing of German Armed Forces (WIM-GAF), Landshuterstraβe 70, D-85435 Erding, Germany.

[2] University professor and research directory, Institute for Composite Materials Ltd. (IVW), University of Kaiserslautern, Erwin Schrödinger Str. 58, D-67663 Kaiserslautern, Germany.

relative changes), to some extent in compression strength as well as in the fatigue resistance of the laminates (up to 300% changes of load cycles to failure), were obtained [*1*]. These property improvements are based on the principle of introducing internal stresses into the laminate by permanent deformation of the matrix.

Previously, an exact description of the mechanisms responsible for the property changes could not be given [*1*]. It was not determined if the effect was caused by (1) changes of transverse fiber distance, (2) changes of stress distribution around or between the individual fibers, (3) an anisotropy effect of the matrix, or (4) a combination of all of these possible cases. To solve this open question, research on a neat matrix material and an analytical micromechanical study on unidirectional laminates were performed [*2*]. In this paper a summary and discussion of the extensive results of this study is given.

Fundamentals of Thermomechanical Treatments of Polymers

Physical Effects

Polymeric matrices in composites exhibit a glass transition during heating. When the glass transition is exceeded the material changes from an energy-elastic behavior to an entropy-elastic behavior, which causes an extreme softening and a ductility increase of the polymer. A deformation of the matrix at temperatures above the glass transition temperature can be frozen in by a subsequent cooling down to room temperature. One possibility of intensifying this effect in a thermosetting matrix is to cure the matrix only partially by reducing the usual curing time. This partially cured matrix has to be strained at curing or post-curing temperature until the completion of curing is performed. After cooling down under load the residual (permanent) deformation of the matrix reveals an improved time and temperature stability, which is caused by chemical bonds built up during the final curing in strained state.

Applying Proper Treatment Conditions

A successful treatment of a laminate is only possible if suitable conditions are selected, specifically treatment temperature and load. The upper limit of the glass transition temperature range of the matrix may be used as an estimate of the minimum treatment temperature. Because of dependence of glass transition temperature on loading conditions [*3*], a proper test for determinations of this value has to be used. In-house researches indicate that a tension test on $\pm 45°$ laminates as described in Ref *4* allows the determination of glass transition temperature under different static loads and therefore it gives a close simulation of conditions applied during a thermomechanical treatment.

Information on the loadability of a laminate can be gained from its strength parameters at intended treatment temperature such as interlaminar shear strength, in-plane shear strength, and stress and strain at failure in the transverse tensile test. Suitable treatment conditions are further obtained by systematically changing these treatment conditions until the characteristics of the treated laminate have reached the optimum value desired.

Changes in Matrix Properties

Behavior During Treatment

The effectiveness of thermomechanical treatment of a composite depends on the stress-strain behavior of the matrix during treatment. Characteristics of neat resin that are important for the assessment of property improvements in the composite can be summarized as follows: (1) the strain to failure, (2) the stress increase up to maximum strain, (3) the stress reduction under constant strain (stress relaxation) up to the end of the maximum load duration as well as (4) the resulting modulus of elasticity. The realization of property improvements of polymer composites is mainly based on a low value of matrix modulus at the end of the loading period under treatment temperature. Figure 1 shows that at the treatment temperature, the modulus of completely cured neat resins (Treatment A) decreases with increasing strain. For the tested resin, this relationship can be approximated by a logarithmic or power function. In contrast to the completely cured resin (Treatment A), the modulus of elasticity of the partially cured neat resin (Treatment B) obviously increases with the strain. Since the modulus, and consequently, the stress values of the resin are very low during Treatment B, this increase of modulus is of only minor importance for the treatment of laminates.

Behavior After Treatment

After the above-mentioned treatment the permanent deformation, the coefficient of thermal expansion (CTE) and the changes in stress-strain behavior of the neat resin specimens were determined. The permanent deformation gives information about the effectiveness of the treatment while the other parameters measured give information about the state of the resin after this treatment.

Permanent Deformation—The measured results show that the permanent deformation is approximately proportional to the strains applied during the treatment at elevated temperatures (Fig. 2). For the completely cured neat resin (Treatment A) the ratio between the two values (permanent deformation/strains applied during the treatment) is approximately 0.8. The shortened curing time (partially cured neat resin, Treatment B) resulted in a ratio of nearly 1.0.

CTE Change—Proper samples were cut from the treated and untreated neat resin specimen to achieve these measurements. The results show that the treatment of neat resin specimen caused the CTE to decrease in the loading direction during treatment and to increase perpendicular to it (Fig. 3). After Treatment A, the change of the CTE was in the direction of treatment load approximately inversely proportional and perpendicular to the direction of

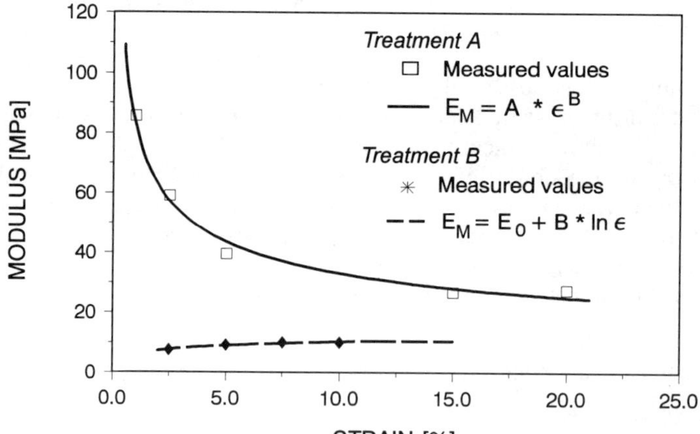

FIG. 1—*Modulus of neat resin NARMCO 5245C as a function of strain* ϵ *after a load duration of 5 min (Treatment A) and 4 h (Treatment B) at constant strain; approximation parameter for Treatment A:* A = *82.91 MPa/% and* B = *−0.399; approximation parameter for Treatment B:* E_0 = *5.98 MPa and* B = *1.9 MPa/%; temperature* = *220°C.*

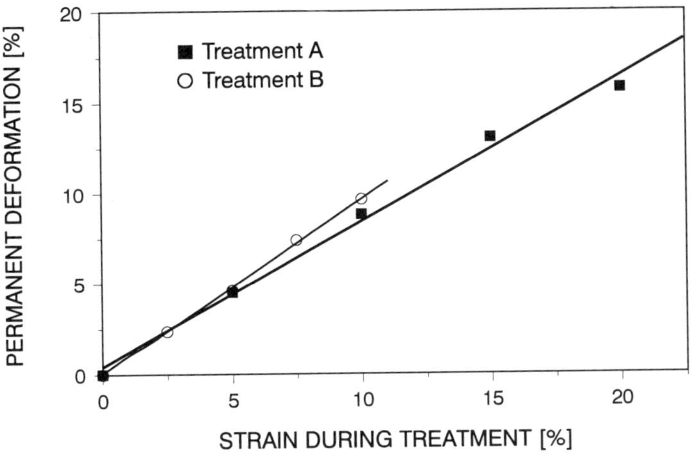

FIG. 2—*Permanent deformation of neat resin NARMCO 5245C after treatment at different strain.*

treatment load proportional to the strain applied during the treatment. On average, Treatment B led to smaller CTE changes than Treatment A. The changes in CTE can be interpreted as quantitative proof of the introduction of anisotropy in the resin by both treatments.

Modulus Changes—The tensile tests were mainly aimed to determine maximum changes in stress-strain behavior. For this reason and because of the CTE results concerning the anisotropy effect by the different treatment, these test were only achieved after Treatment A. After this treatment, the modulus tended to increase with increasing treatment strain, but did not do so continuously. A better correlation (linear relationship) is given between the modulus and the treatment stress that indicates that the treatment stress is decisive for the resulting changes of the modulus (Fig. 4). The maximum relative change of the modulus is approximately 10%. This demonstrates that the resin changes from an isotropic behavior in

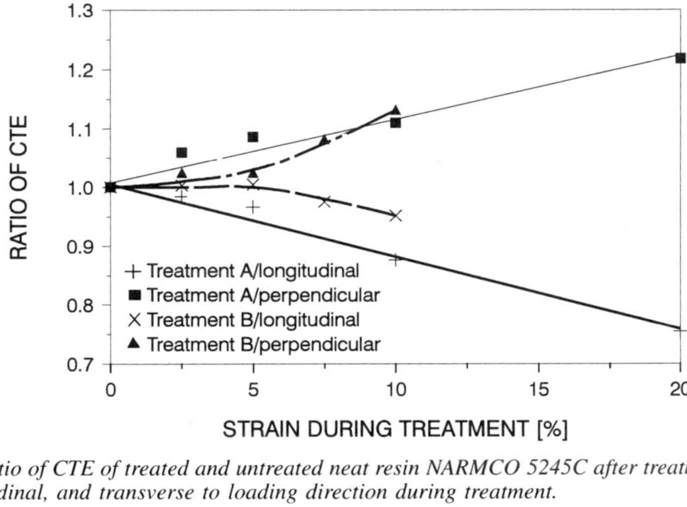

FIG. 3—*Ratio of CTE of treated and untreated neat resin NARMCO 5245C after treatment at different strain, longitudinal, and transverse to loading direction during treatment.*

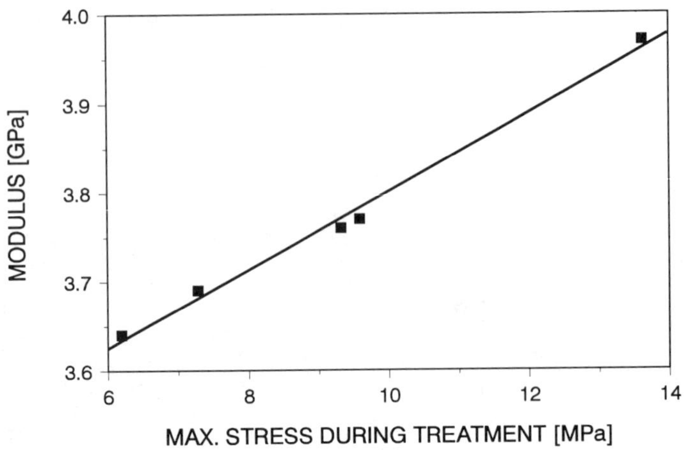

FIG. 4—*Secant modulus (between 0.1 and 1% strain) of neat resin NARMCO 5245C after treatment A at different stresses.*

the untreated state to an anisotropic behavior in the treated state. The strong tendency of the neat resin to stress relaxation counteracts such modulus increases. An intensified rearrangement of the molecular chains during this stress relaxation may be the cause.

Tensile Strength of Treated and Untreated Specimens—The individual values of tensile strength scatter widely (Fig. 5). This can be attributed to small material faults (for example, air bubbles) that could not be avoided with the selected manufacturing procedure. Because of the scatter it was difficult to determine any changes in strength caused by Treatment A. Under these conditions only a slight strength reduction along with an increasing treatment strain could be measured (Fig. 5).

Micromechanical Calculations

The local internal stresses transverse and longitudinal to the fiber direction in unidirectional fiber composite plies/laminates are different because of anisotropy of these kind of

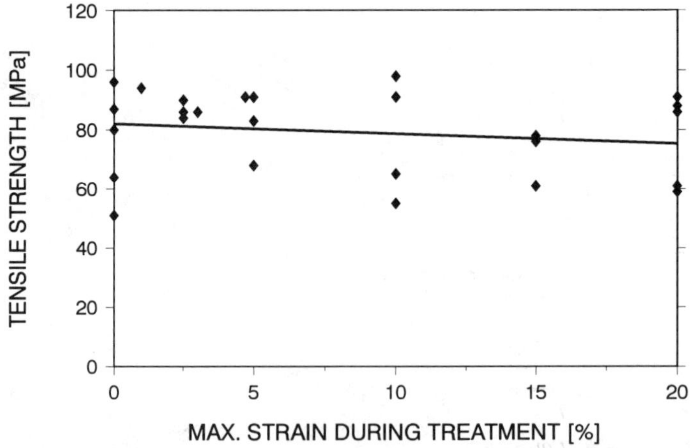

FIG. 5—*Single values of tensile strength of untreated and at different strain levels treated (Treatment A) neat resin NARMCO 5245C.*

materials. Due to the strengthening function of the fibers, the internal strains and stresses occurring in the direction of the fibers are not significant compared to the strength of the fiber composite. For this reason the deductions of this section are restricted to those local strains and stresses that occur perpendicular to the fiber direction. In this direction, due to the geometric conditions (circular cross-section of the fibers), considerable local strain/stress increases are produced. These increases depend upon the transverse microstructure of the fibers [*5*] and on the properties of the matrix itself. Most authors [*5–16*] assume a transverse isotropic behavior of carbon fibers. In addition, the values calculated by these authors depend upon the mechanical model used. Mainly there are two models: (a) the cylinder model [*5,8,16–19*] and (b) various geometric arrangements of fibers in a matrix completely filling all spaces between fibers [*6,20–29*]. The cylinder model is based on a hollow-cylinder matrix with a constant wall thickness that encloses the fibers. In contrast, the second model takes account of the various matrix accumulations around the complete circumference of a single fiber. Since these fiber/matrix relations approximate the real conditions, the second model was selected for theoretical considerations. In this model, the local stress and strain increases depend upon the "fiber distance/fiber diameter" ratio, which in turn, is determined by both the geometric arrangement (that is, triangular, hexagonal, square) of the fibers and the fiber content in the laminate. Laminates of continuous fiber-reinforced plastics with a realistic fiber content (approximately 60% by volume) usually have a mix of fiber arrangements. Since the square arrangement provides mean values of the above ratio as well as the most simple presentation of the relationship between the component and the laminate properties, most micromechanical analyses are based on this arrangement [*6,8,13,20–26,28*].

Therefore, this fiber arrangement was assumed for all the following cases considered. Similar to Refs *6* and *22*, the element of volume/area is divided into subelements (slices) with a very small thickness dy (Fig. 6). These subelements, which consist of different portions of fiber and matrix (with a rectangular shape) depending on angle ϕ, were used for calculation of local micromechanical strain and stress. They were considered as a series connection of materials with different properties. Because it was assumed, there is no con-

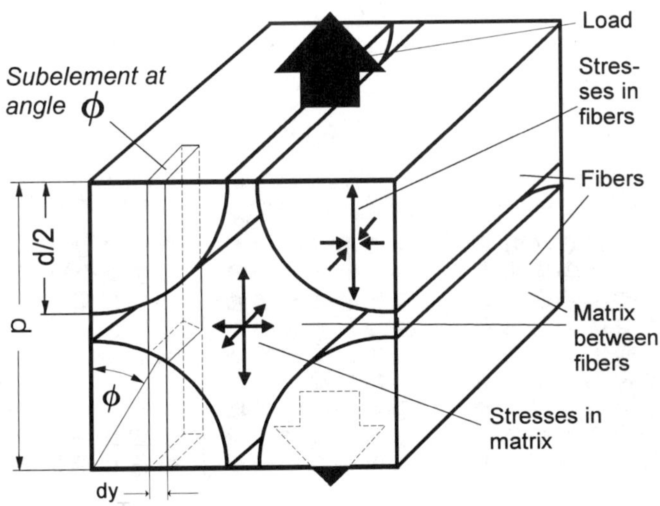

FIG. 6—*Volume/area unit of square arrangement of fibers with a sub-element at angle* ϕ *(or distance from fiber center), used for micromechanical calculations;* p = *distance of fibers (center-center),* d = *diameter of fiber, and* dy = *thickness of sub-element.*

nection between adjacent subelements, the shearing stresses and normal stresses are decoupled (shearing stresses are neglected). In addition and contrary to Refs *6* and *22*, the following facts were taken into consideration: (a) a three-dimensional local stress state due to different stiffness, CTE and Poisson's ratio of matrix and fiber (Fig. 6), (b) the anisotropic behavior of the carbon fiber, (c) the extreme cases of a statistical distribution of the fiber distances, and (d) the nonlinear stress-strain behavior of the matrix during treatment. As the extreme cases of statistical distribution of fiber distances were defined (Fig. 7):

- Case I — All fibers have a constant distance > fiber diameter.
- Case II — Some individual fibers are touching each other (fiber distance = fiber diameter) while the distance of the remaining fibers has an average constant value (> fiber diameter).
- Case III — The distance of individual fibers is several times larger than the prevailing fiber distance.

In order to evaluate the influence of thermomechanical treatment, calculation of micromechanical stress and strain has to be done for the three states of the laminate, namely, before treatment, during treatment and after treatment.

A complete derivation of micromechanical strain and stress respecting all conditions mentioned above is given in Ref *2*. In this paper, only some reasons for the occurrence of these parameters and the calculated results using material characteristics given in Table 1 are shown. The calculations were performed with a fiber content of 60% by volume. Because of mathematical reasons for the Case II (at $\phi = 0$ the length of matrix portion becomes 0 and the strain infinitive, Figs. 6 and 7) instead of a fiber distance/fiber diameter ratio of 1 (fibers in touch) a slightly higher value of 1.019 was assumed. For Case III, the results were determined using a value four times the mean fiber distance.

Local Stresses in Untreated Dry UD-Laminates

Production-Induced Local Stresses—Laminates made of high-performance composites are produced at relatively high temperatures (>175°C). However, the temperatures that these materials are subjected to in actual use, are usually far lower. The different expansion co-

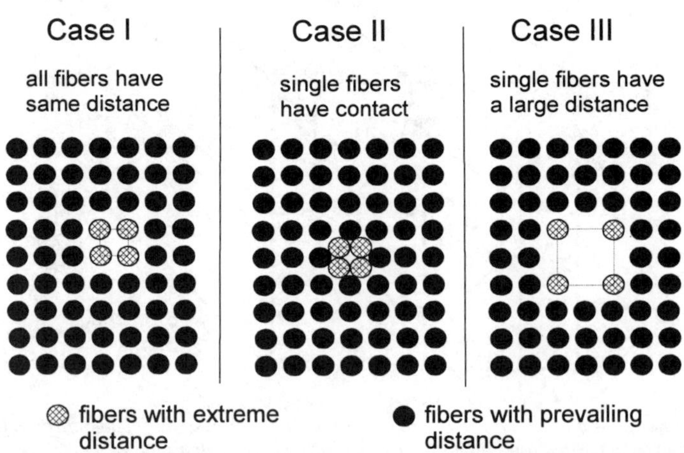

FIG. 7—*Definition of extreme cases of statistical distribution of fiber distances.*

TABLE 1—*Characteristics of fiber/matrix of studied laminate.*

Materials	Property	Temperature, °C	Value
NARMCO 5245C (Matrix)	E_M, MPa[a]	23	3625
	σ_{MB}, MPa[a]	23	76
	ϵ_{MB}, %[a]	23	2.3
	α_M, 1/K[b]	20–50	46.5 . . . 10^{-6}
	α_M, 1/K[b]	20–180	54.5 . . . 10^{-6}
	ν_M [a]	23	0.37
	ν_M [c]	210	0.47
T800 (Fiber)	E_{F1}, MPa[d]	. . .	294 000
	E_{F2}, MPa[e]	. . .	14 800
	α_{F1}, 1/K	. . .	−0.5 . . . 10^{-6}
	α_{F2}, 1/K[f]	. . .	8.6 . . . 10^{-6}
	ν_{F23} [g]	. . .	0.2
	ν_{F23} [g]	. . .	0.4
	d, mm[d]	. . .	5

[a] Values measured for this study at a test velocity of a 4 mm/min and at a strain of 1.25%; modulus of elasticity and Poisson ratio from the secant at a strain of 0.1 and 1.25%.
[g] Values taken from Ref *16*; assumption of a value equal to that of the fiber T300.
[b] Values measured for this study; heating rate = 4 K/min.
[c] Assumption of a nearly viscoelastic behavior because at this temperature the matrix is in the highly softened state (considerably above the onset of glass transition).
[d] Manufacturer's data.
[e] Value determined by calculation from the moduli of elasticity measured at 23°C on the laminate and on the pure resin.
[f] Value determined by calculation from the thermal expansion coefficients measured at temperatures between 20 and 50°C laminate and on the pure resin.

efficients of the matrix and the fiber (Table 1), including those in the longitudinal and transverse fiber directions (in the case of an anisotropic fiber), cause different length changes within the matrix and the fiber when the material is cooled down to the temperature of actual use. The solid connection between the matrix and the fiber, however, does not allow the individual components to change in length in accordance with their thermal properties. Therefore, the length changes within the matrix and the fibers of a fiber composite are mainly the same. These length change adaptations of the individual components give rise to internal stresses and strains within the composite.

For the three Cases (I to III) described, production-induced local stresses of the matrix and the fiber were calculated as a function of the distance from the fiber center (Fig. 8). The stress distribution in the fiber and the matrix calculated under the above-mentioned presumptions (the values are the same for both components) is shown in Fig. 9. The most significant characteristic of this distribution is the fact that in Cases I and II, compressive stresses are forming over the fiber center (distance fiber center/fiber radius = 0) while tensile stresses are forming in all cases within the region of the tangent at the fiber surface (distance fiber center/fiber radius = 1). According to the derivations in Ref *2* the integral of the stresses over the fiber distance assumes the value 0 for Case I. For Cases II and III, however, this is not true. In Case III, for example, only tensile stresses occur, while in Case II compressive stresses are predominant. This is caused by the higher (Case III) or the lower (Case II) integral thermal expansion of the element of area of these borderline cases due to higher respectively lower portion of matrix against that of the prevailing fiber distance (= Case I) (Fig. 7).

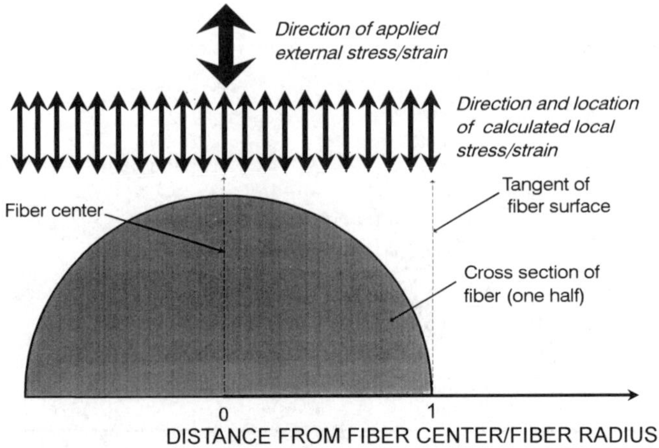

FIG. 8—*Schematic of position and direction of applied external load and calculated local stresses and strains, perpendicular to fiber direction of an unidirectional laminate.*

Load-Induced Local Stresses—If production-induced local stresses are neglected, perpendicular to the fiber direction, the unidirectional plies of fiber composite materials have a considerable lower strain at failure than in the direction of the fibers, or compared to the neat resin material [*6,20–30*]. This can be explained by the local strain magnification of the matrix (Fig. 10) which was first found by Kies [*20,30*] and later by others [*6,21–29*]. A local strain magnification means that within a laminate loaded perpendicular to the fiber direction, the local strain of the matrix is considerably higher than the average (macroscopic) strain of the laminate. This effect is caused by the large difference between the fiber and the matrix stiffness resulting in a predominant strain of the laminate by the extension of the matrix bridges between the individual fibers.

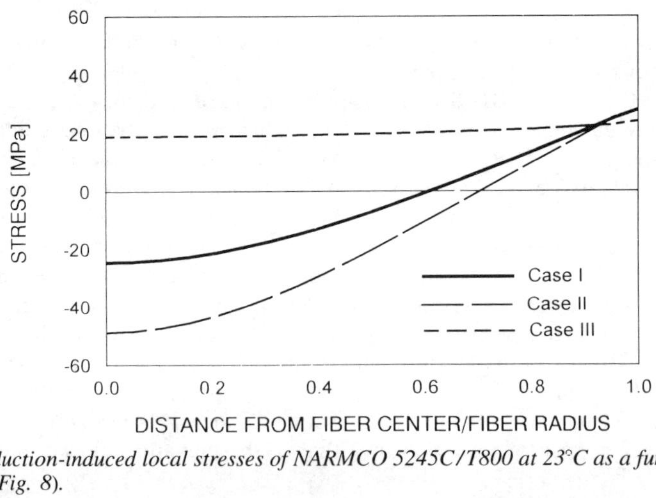

FIG. 9—*Production-induced local stresses of NARMCO 5245C/T800 at 23°C as a function of distance to fiber center (Fig. 8).*

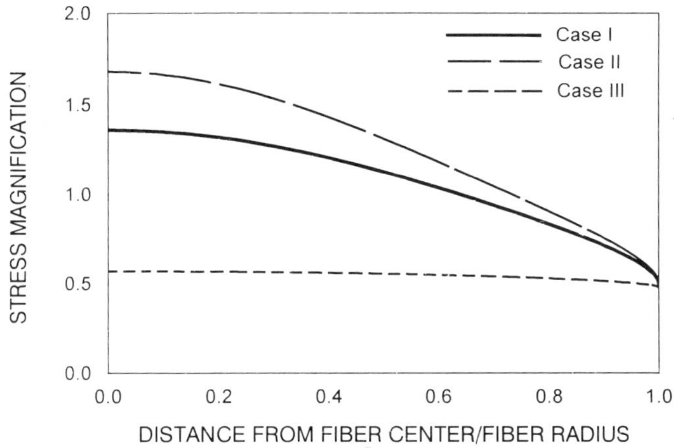

FIG. 10—*Load-induced stress magnification of NARMCO 5245C/T800 as a function of distance from fiber center (Fig. 8).*

The local stresses of the fiber and the matrix (the values are equal for both components) calculated under the above-mentioned suppositions are shown in Fig. 10. Similar to the production-induced stresses, these stress increases are highly dependent on the distance from the fiber center. In contrast to the production-induced stresses, however, they are higher above the fiber center than near the tangent at the fiber surface. In Case II, the largest stress increases are occurring above the fiber center. The second largest values at this point are found in Case I. The lowest values are obtained in Case III, where the absolute values of the local stresses are even smaller than those of the mean or external laminate stress (stress magnification = 1). In the region of the tangent at the fiber surface, the stresses are in all cases below the mean external laminate stress. In this region there are no differences between Cases I, II, and III.

Superposition of Production- and Load-Induced Stresses—The stress decisive for the failure of a unidirectional laminate subjected to a load perpendicular to the fiber direction is the sum of the production-induced stress and the load-induced stress, which is plotted in Fig. 11. It can be shown that in Cases I and II the thermal (production-induced) stresses reduce the load-induced stresses above the fiber center while they enhance them in the region of the fiber tangent. Therefore, in Case II with the external stresses used here (tensile strength of the laminate), the resulting stresses are higher within the region of the fiber surface tangent than they are above the fiber center. In Case I, the resulting stresses within the region of the fiber surface tangent are now nearly equal to those above the fiber center. In Case III, however, the largest stress increases are still above the fiber center. In contrast to the other two cases, the production-induced stresses here increase the resulting stresses across the total fiber radius. The maximum local stresses in Case III are, however, still clearly below those of the other cases. The largest local stresses occur for Case II in all cases.

Local Stresses in Treated UD-Laminates

Permanent Local Deformation of Matrix—Subjecting a laminate to thermomechanical treatments leads to permanent deformation especially in the matrix regions suffering the highest loads. As already mentioned, these highly loaded matrix regions appear in plies of composites reinforced with continuous fibers and extend perpendicular to the external load

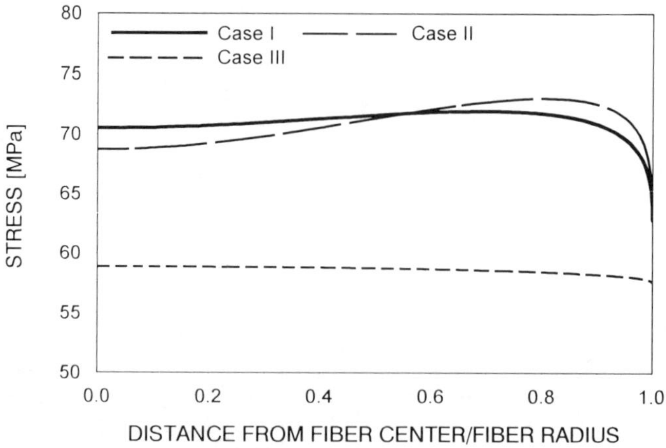

FIG. 11—*Superposition of production- and load-induced local stresses (Figs. 10 and 11) at 23°C (applied external tensile stress of laminate = 170 MPa).*

(during the treatment in this case). The usability of these permanent deformations for the strength increase of fiber composites mainly depends upon their average value and upon their local magnification between the individual fibers. These two quantities can be determined from local stresses produced between the fibers by the treatment. Since at the treatment temperature the modulus of the matrix largely depends on the prevailing stress (no linear relation between the stress and the strain, Fig. 1) the local stresses at treatment temperature and at room temperature have to be determined separately. This is done by calculating the local stresses and strains during the treatment. Figure 12 shows the calculated local strain versus distance from fiber center. Besides the expected dependence on this distance, the local strains in Case II are much higher than in Cases I and III. From these values the local permanent deformations are obtained. Figure 13 indicates that absolute values of local per-

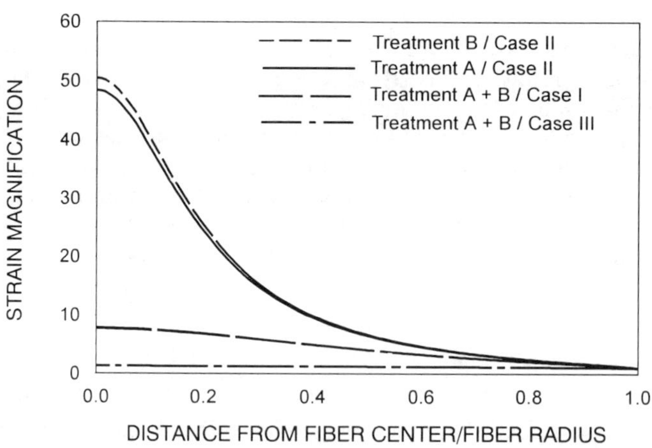

FIG. 12—*Strain magnification of matrix NARMCO 5245C as a function of distance from fiber center (Fig. 8) during treatment.*

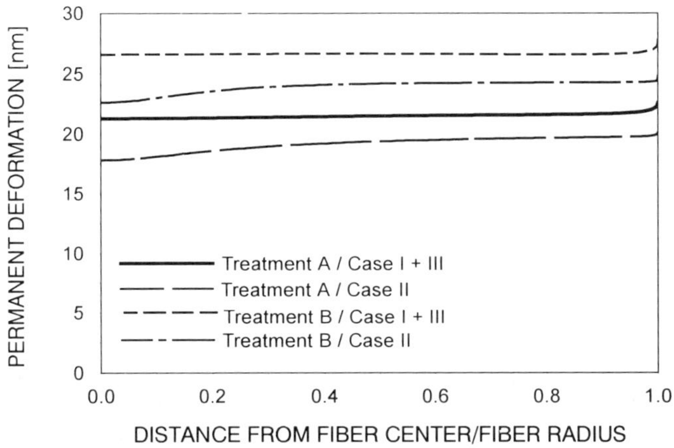

FIG. 13—*Local permanent deformation (absolute values) of matrix NARMCO 5245C as a function of distance from fiber center (Fig. 8) after treatment (calculated from strain magnification in Fig. 12, external transversal strain of laminate during treatment = 0.5%).*

manent deformation of the matrix material investigated here are approximately independent on distance from fiber center. This effect is caused by a contrary dependence of the relative matrix strain during treatment (~ relative permanent deformation, Fig. 12) and a dependence of the matrix portion length (sub-element, Fig. 6) on distance from fiber center. That infers thermomechanical treatment merely causes an increase of the fiber distance (Fig. 14), and combined with this effect a macroscopic permanent deformation of the laminate specifically the plies transverse to loading direction occurs.

Local Stresses after Treatment (Unloaded Laminate)—The calculation of local stresses after treatment requires that the properties of the matrix material after the treatment are also known. Although only a uniaxial load was applied during treatment, the determined treatment-induced changes in mechanical properties (strength and modulus) of the matrix were <10% and the changes in thermal expansion below the glass transition temperature were

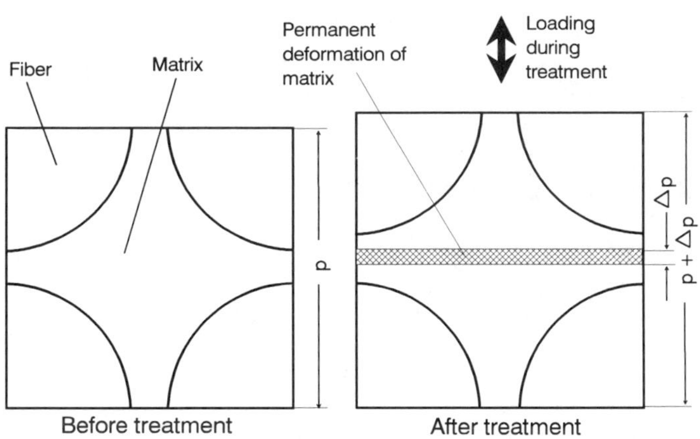

FIG. 14—*Influence of local permanent deformation of matrix (Fig. 13) on fiber distance.*

<20%. Since, during the treatment of a laminate, the matrix is subjected to a three-dimensional stress, the anisotropy of the matrix after treatment is lower than after the above mentioned uniaxial treatment of a neat resin. Furthermore, there should be a compensation due to the decrease in thermal expansion and an increase in modulus on local stresses. For these reasons, property changes of the matrix by treatment were neglected and the material characteristics of the untreated laminates were used as a basis.

The results obtained (Fig. 15) show that, even under the extreme conditions of Case II, the two thermomechanical treatments did not cause a significant change of the production-induced local stresses in the laminate (Fig. 9). The reason for this is that the absolute value of the permanent deformation of the matrix is nearly independent on the distance from the fiber center (Figs. 13 and 14).

Superposition of Local Stresses after Treatment and Load-Induced Stresses—The resulting local stresses built up in a treated laminate by loading are determined in the same way as those of the untreated laminate (Fig. 10). Figures 11 and 16 shows that the tension test (both treatment methods), as expected from local stresses after treatment (without external load), did not effect any significant change of the local stresses in the NARMCO 5245C/T800 laminate.

Influence of Fiber-/Matrix Properties on Treatment Results—The discussed results show that the matrix tested has suitable properties to perform a successful thermomechanical treatment. The question is now, what influence on local stresses after treatment can be expected, if other matrices and fibers with different specific properties are used. The specific properties, which mainly influence the results of the thermomechanical treatment, are the modulus of the matrix at treatment temperature, its dependence on the applied load and the transverse modulus of the fibers. For that reason calculation of local stresses transverse to the fiber direction applying different fiber/matrix moduli were performed. Figure 17 shows the local stresses at the fiber center versus the ratio of transverse fiber and matrix modulus after treatment in an unloaded and loaded unidirectional laminate. Additionally, no dependence of matrix modulus on load applied at treatment temperature was assumed, which closely agrees with the condition of Treatment B. It can be seen clearly that the influence of fiber

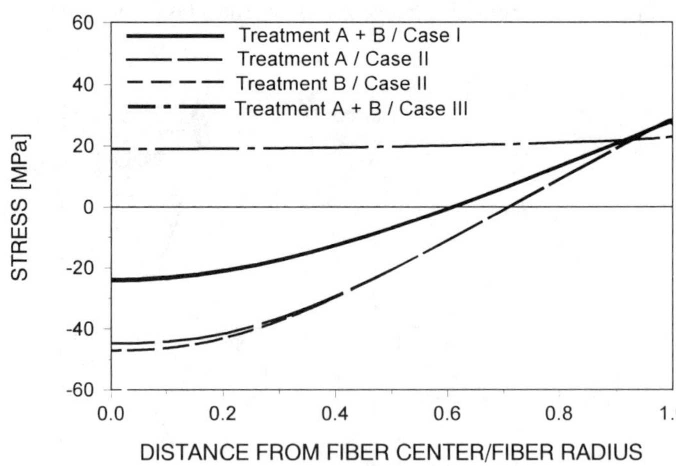

FIG. 15—*Local stresses after treatment (transversal tensile strain of laminate during treatment = 0.5%) as a function of distance from fiber center (Fig. 8) in an unloaded laminate NARMCO 5245C/T800 at 23°C.*

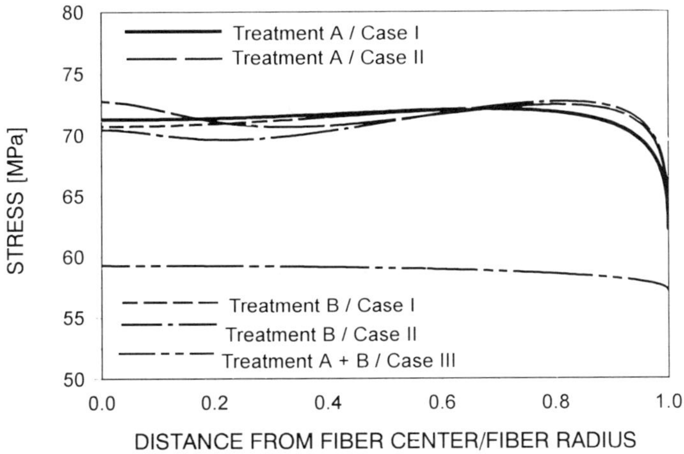

FIG. 16—*Superposition of local stresses after treatment (Fig. 15) and load-induced stresses at 23°C (applied external tensile stress of laminate = 70 MPa).*

matrix modulus depends on fiber arrangement (Case I or II). In Case I, there is no essential influence of fiber/matrix modulus ratio on local stresses, if extreme small values are not considered. In opposition to this, in Case II already at ratios smaller than 1000, a significant increase of local stresses compared to values for NARMCO 5245C/T800 (ratio = 1500) are given. Contrary to this behavior of a not completely cured matrix (before treatment), a completely cured matrix displays a load dependence of modulus at treatment temperature. In this case, for this research a simple ratio of fiber/matrix modulus can not be used. For demonstration of this matrix property influence, in the following calculations a load de-

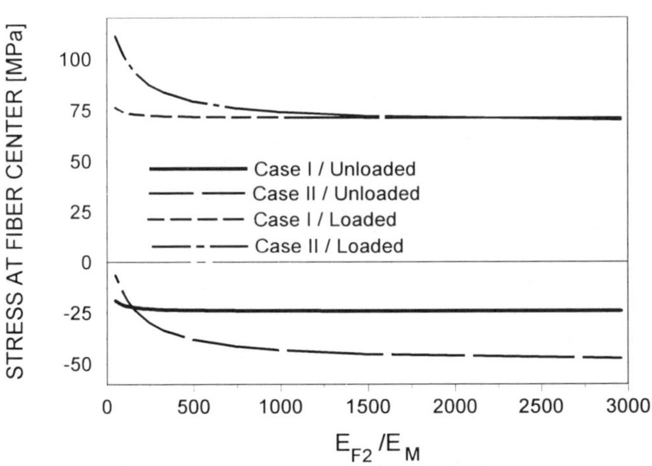

FIG. 17—*Local stress above fiber center (Fig. 8) as a function of the ratio transverse fiber modulus (E_F)/matrix modulus (E_M) for fiber arrangement Case I and Case II in a loaded (applied external tensile load = 70 MPa) and an unloaded laminate after treatment assuming a load independent matrix modulus during treatment.*

pendence of NARMCO 5245C was assumed and the fiber/matrix modulus ratio of NARMCO 5245C/T800 at varying load was multiplied just by a constant factor. Figure 18 shows the stresses at the fiber center calculated by applying this varying condition. Similar to the results in Fig. 17, there is no essential influence of fiber/matrix modulus ratio on local stresses in the fiber arrangement of Case I. In Case II, an increase of local stresses with decreasing fiber/matrix modulus ratio can also be seen. Compared to the matrix with a load-independent modulus, a load-dependent modulus leads to a smaller increase of local stresses with decreasing fiber/matrix modulus ratio. This effect is more pronounced for matrix materials with higher load dependence and less pronounced for materials with lower load dependence of modulus than the one of NARMCO 5245C.

Summary and Conclusions

Research carried out with the pure resin showed that a thermomechanical treatment of a laminate caused a permanent deformation of the matrix. The amount of this permanent deformation is almost proportional to the strain at the treatment temperature. Measurements of thermal expansion and the mechanical properties after the treatment revealed a distinct anisotropy of the treated neat resins. Although only a uniaxial load was applied during the treatment, the determined treatment-induced changes of the mechanical properties were <10% and the thermal expansion below the glass transition point was <20%. Since, during the treatment of a laminate, the matrix is subjected to a three-dimensional load, the anisotropy of the matrix is lower after the treatment. For this reason, such a small change in matrix properties can't be considered as a significant contribution to strength improvement of a laminate by thermomechanical treatment. Thus, the local strain during treatment and the subsequent permanent deformation of the matrix is essential for the local stresses occurring in the laminate after the treatment. An extended calculation of micromechanical stresses and strains showed, that the absolute value of the local permanent deformation of the matrix material is therefore almost constant across the entire fiber radius. This means the treatment

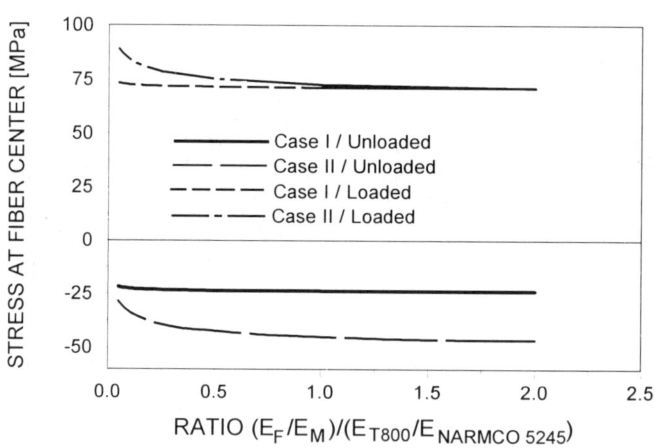

FIG. 18—*Local stress above fiber center (Fig. 8) as a function of ratio (transversal fiber modulus* (E_F)*/matrix modulus* $(E_M))/(E_{F(T800)}/E_{M(NARMCO\ 5245)})$ *for fiber arrangement Cases I and II in a loaded (applied external tensile load = 70 MPa) and an unloaded laminate after treatment assuming matrix with load dependent modulus during treatment like NARMCO 5245C.*

causes only a change of the fiber distances while the local stresses remained unchanged. In the case of treatment of multidirectional reinforced laminates, this increase of fiber distance causes a permanent deformation of the transverse plies, which in turn results in a decrease of the residual stresses in these plies; subsequently, changes of mechanical properties of the laminate are the consequence (that is, elongation at transverse cracking). In summary, it can be stated that the mechanisms of property improvement of polymer composites by thermomechanical treatment have been described here completely.

Acknowledgments

The authors thank the Director of the Institute for Material Research and Testing of German Armed Forces, Professor J. Guttenberger, for his support of this study.

Professor K. Friedrich, gratefully acknowledges the support of the Fonds der Chemischen Industrie, Frankfurt, for his personal research activities in 1996. Further thanks are given to the AGARD project G75 (Cooperation on Failure of Composites between Prof Friedrich and Prof Paipetis, University of Patras, Greece).

References

[1] Maier, G. and Friedrich, K., "Property Improvements of Polymer Composites After Loading at High Temperatures," *High Temperature and Environmental Effects on Polymeric Composites, ASTM STP 1174*, C. E. Harris and T. S. Gates, Eds., American Society for Testing and Materials, West Conshohocken, 1993, pp. 147–160.

[2] Maier, G., Deutsche Hochschulschriften 535, "Festigkeitssteigerung von ebenen Hochleistungsfaserverbund-Laminaten mit Polymerer Matrix durch Thermomechanische Behandlung, Verlag Hänsel-Hohenhausen, Egelsbach, Frankfurt, Washington, 1994, ISBN 3-89349-535-5.

[3] Maier, G., "Local Dependence of Glass Transition of Fibrous Composites," Kunststoffe, German Plastics, Vol. 80, No. 8, 1990, pp. 906–908.

[4] Maier, G., "Determination of Glass Transition in Fibre-Reinforced Materials by a Static Tensile Test," Kunststoffe German Plastics, Vol. 78, No. 4, 1988, pp. 330–333.

[5] Avery, W. B. and Herakovich, C. T., "Effect of Fiber Anisotropy on Thermal Stresses in Fibrous Composites," *Journal of Applied Mechanics*, Vol. 53, 1986, pp. 751–756.

[6] Spencer, A., "The Transverse Moduli of Fibre-Composite Material," *Composites Science and Technology*, Vol. 27, 1986, pp. 93–109.

[7] Schapery, R. A., "Thermal Expansion Coefficients of Composite Materials Based on Energy Principles," *Journal of Composite Materials*, Vol. 2, No. 3, 1968, pp. 388–404.

[8] Halpin, J. C. and Kardos, J. L., "The Halpin-Tsai Equations: A Review," *Polymer Engineering and Science*, Vol. 16, No. 5, 1976, pp. 344–352.

[9] Pagano, N. J. and Hahn, H. T., "Evaluation of Composite Curing Stresses," *Composite Materials: Testing and Design, ASTM STP 617,* American Society for Testing and Materials, West Conshohocken, PA, 1977, p. 317–329.

[10] Hashin, Z., "Analysis of Properties of Fiber Composites with Anisotropic Constituents," *Journal of Applied Mechanics*, Vol. 46, No. 9, 1979, pp. 543–550.

[11] Adams, D. F. and Crane, D. A., "Combined Loading Micromechanical Composite," *Composites*, Vol. 15, No. 3, 1984, pp. 181–192.

[12] Chamis, C. C., "Simplified Composite Micromechanics Equations for Hygral, Thermal, and Mechanical Properties," *SAMPE Quarterly,* No. 4, 1984, pp. 14–23.

[13] Collings, T. A. and Stone, D. E. W., "Hygrothermal Effects in CFRP Laminates: Strains Induced by Temperature and Moisture," *Composites*, Vol. 16, No. 4, Oct. 1985, pp. 307–316.

[14] Niederstadt, G., "Wärmeausdehnung/Quellung" in: Heißler, H., "Verstärkte Kunststoffe in der Luft- und Raumfahrt," Verlag W. Kohlhammer, Stuttgart Berlin Köln Mainz, 1986, pp. 172–179.

[15] Adams, D. F., "A Micromechanics Analysis of the Influence of the Interface on the Performance of Polymer-Matrix Composites," *Journal of Reinforcded Plastics and Composites*, Vol. 6, No. 1, 1987, pp. 66–88.

[16] Bowles, D. E. and Tompkins, S., S., "Prediction of Coefficients of Thermal Expansion for Unidirectional Composites," *Journal of Composite Materials*, Vol. 23, No. 4, 1989, pp. 370–388.
[17] Schneider, W., "Mikromechanische Betrachtung von Bruchkriterien Unidirektional Verstärkter Schichten aus Glasfaser/Kunststoff," Dissertation TH Darmstadt D17, 1974.
[18] Tsai, S. W., *Introduction To Composite Materials*, Technomic Publishing Co., 1980, pp. 379–433.
[19] Bowles, D. E. and Griffin, O. H., "Analysis of Thermal Stresses in Polymer Matrix Composites," *Proceedings*, 34th International SAMPE Symposium, 8–11 May, 1989, pp. 575–584.
[20] Kies, J. A., "Maximum Strains in the Resins of Fibre Glass Composites," U.S. Naval Research Lab. Report 5752, 1962.
[21] Puck, A., "Deformation Behavior and Failure Mechanisms of Unidirectional and Orthotropic Glass Fiber Composites," Kunststoffe (German Plastics), Vol. 55, No. 12, 1965, pp. 913–922.
[22] Puck, A., "Loading and Deformation of Glass Fiber Composites," Dissertation Technische Universität (TU), Berlin, 1966.
[23] Puck, A. and Schneider, W., "On Failure Mechanics and Failure Criteria of Filament-Wound Glass-Fibre/Resin Composites," *Plastics and Polymers*, Feb. 1969.
[24] Haberl, G., "Theoretical Determination of Properties of Fiber-Reinforced Materials Loaded Transverse to Fiber Direction," Dissertation Technische Universität (TU) Wien, 1972.
[25] Aveston, J. and Kelly, A., "Theory of Multiple Fracture of Fibrous Composites," *Journal of Materials Science*, Vol. 8, 1973, p. 352.
[26] Garrett, K. W. and Bailey, J. E., "Multiple Transverse Failure in 90° Cross Ply Laminates," *Journal of Materials Science*, Vol. 12, 1977, p. 157.
[27] Brintrup, H., Derek, H., Meffert, B., and Thebing, U., "Mechanical Behavior of Anisotropic Reinforced Resins," Berichte zum 9. Kunststofftechnisches Kolloquium des IKV (Institut für Kunststoffverarbeitung) in Aachen vom 8. bis 10.3.1978.
[28] Parvizi, A. and Bailey, J. E., "On Multiple Transverse Cracking in Glass Fibre Epoxy Cross-Ply Laminates," *Journal of Materials Science*, Vol. 13, 1978, p. 2131.
[29] Parvizi, A., Garret, K. W., and Baily, J. E., "Constraint Cracking in Glass Fibre Reinforced Epoxy Cross-Ply Laminates," *Journal of Materials Science*, Vol. 13, 1978, p. 195.
[30] Bader, M. G., Bailey, J. E., Curtis, P. T., and Parvizi, A., "The Mechanisms of Initiation and Development of Damage in Multi-Axial-Fibre-Reinforced Plastics Laminates," Vol. 3, ICM 3, Cambridge, England, Aug. 1979, pp. 227–239.

Scott W. Case,[1] *Richard B. Plunkett,*[1] *and Kenneth L. Reifsnider*[1]

Simulation Methods for Life and Remaining Strength Prediction of High-Temperature Polymeric Composites Subjected to Cyclic Loads

REFERENCE: Case, S. W., Plunkett, R. B., and Reifsnider, K. L., **"Simulation Methods for Life and Remaining Strength Prediction of High-Temperature Polymeric Composites Subjected to Cyclic Loads,"** *High Temperature and Environmental Effects on Polymeric Composites: 2nd Volume, ASTM STP 1302.* Thomas S. Gates and Abdul-Hamid Zureick, Eds., American Society for Testing and Materials, 1997, pp. 35–49.

ABSTRACT: The purpose of this paper is to establish and demonstrate a methodology for the prediction of the remaining strength and life of polymer-based composite laminates subjected to cyclic and sustained thermomechanical conditions. Conditions to be considered include elevated temperature (in air) and cyclic (fatigue) loading at various stress levels. The approach taken is to construct a simulation of the damage accumulation process and failure mode that controls the remaining strength and life for carbon fiber-reinforced K3B polymer matrix systems. The simulation is based on micromechanical and ply-level models of composite strength that combine the properties, geometry, and arrangement of the constituents to estimate the global strength as a function of those parameters. As applied conditions influence those parameters, those influences are combined to yield estimates of the remaining strength and life (when global strength is reduced to the level of the applied conditions). These predicted values for the remaining strength as a function of the number of fatigue cycles, temperature, and load level are in good agreement with available experimentally measured values.

KEY WORDS: polymer matrix composite, remaining strength, life prediction, durability, performance simulation, high temperature

The durability of composite materials, particularly the area of analytical modeling of damage and the resulting degradation in material properties, has received much attention in recent years [*1–5*]. Nevertheless, the prediction of the remaining strength and life in these inhomogeneous, anisotropic materials is inherently complicated; there are many different damage modes (such as delamination or matrix cracking) that may develop and interact before a final failure occurs. These damage modes, that occur before final failure, make it possible to design components from composite materials that are extremely damage tolerant. However, to use this damage tolerance to its fullest, we must have some method to predict the stiffness, strength, and life of these materials in differing environmental conditions. In recent works, Reifsnider et al. [*6–9*] have proposed a "micro-kinetic" approach to such predictions. In this paper, we consider the application of such an approach in the framework

[1] Research assistant professor, graduate research assistant, and Alexander Giacco professor, respectively, Engineering Science and Mechanics Department, Virginia Tech, Blacksburg, VA 24061-0219.

of the critical element model [*10*] to the performance simulation of polymeric composite materials subjected to fatigue loading.

Experimental Characterization

Test Specimens

To generate baseline inputs for the modeling process, a number of experiments were conducted. All specimens used in this work were supplied as precut specimens of IM7/K3B by NASA Langley Research Center. These specimens were of four different lay-up's ($[0]_{12}$, $[90]_{12}$, $[\pm45]_{3s}$, and $[0/\pm45]_{2s}$) and were designated for room-temperature testing and for testing at 350°F (177°C). The nominal dimensions of the specimens are given in Table 1. Each of the specimens was ultrasonically C-scanned to determine its integrity. All specimens were determined to be defect-free (within the limits of the technique) except for some of the $[0/\pm45]_{2s}$ specimens in which imperfections were noted in the ultrasonic C-scans and were also visible on the specimen surface. However, due to a shortage of material, these specimens were also tested and the imperfections noted. Subsequent to the C-scan, the specimens were placed in an oven and allowed to dry for 120 h at 120°C. This process was intended to serve two purposes: to remove any moisture that might have been absorbed during the C-scanning process and also to physically age the specimens so that the specimens were as close to thermodynamic equilibrium as possible.

Mechanical Testing

The unnotched $[0]_{12}$, $[90]_{12}$, and $[\pm45]_{3s}$ specimens were used to obtain baseline values for the quantities E_1, E_2, G_{12}, X_t, and Y_t at room temperature and 350°F (177°C). In addition, a value for a shear strength S was calculated, although this value is open to question due to the considerable amount of necking and scissoring that occurred during the test. The $[0]_{12}$ specimens were prepared by adhering end tabs, 1.5-in. (38.1-mm) long, 0.5-in. (12.7-mm) wide, and 0.1-in. (2.5-mm) thick, formed from high-pressure glass/epoxy cross-plied laminates to each specimen using a high toughness, high peel, and shear strength adhesive. A similar procedure was followed for the $[\pm45]_{3s}$ laminates and for the room temperature $[90]_{12}$ laminates. However, it was found that it was not necessary to place end tabs on the elevated temperature $[90]_{12}$ in order to achieve consistent gage-section failures. All of the elevated temperature tests were performed in a convection oven. Due to the nature of the oven, it was not possible to accurately control the heating rate, but all specimens were heated to the testing temperature within a period of 15 min. After the specimens reached the test temperature, a hold period of 10 min followed to ensure that the specimens were at a temperature equilibrium.

The $[0/\pm45]_{2s}$ specimens were prepared for quasi-static testing by placing type EP-08-045PG-120 strain gages along the hole axis as shown in Fig. 1. The gages were intended to

TABLE 1—*Nominal specimen dimensions.*

Lay-Up	Room Temperature, in. (mm)	350°F (177°C)
$[0]_{12}$	0.5 by 6 (12.7 by 152)	0.5 by 8 (12.7 by 152)
$[90]_{12}$	1 by 6 (25.4 by 152)	1 by 8 (25.4 by 203)
$[\pm45]_{3s}$	0.5 by 6 (25.4 by 152)	1 by 8 (25.4 by 203)
$[0/\pm45]_{2s}$	1.5 by 6 (38.1 by 152)	1.5 by 8 (38.1 by 203)
	0.25 (6.35) diameter center hole	0.25 (6.35) diameter center hole

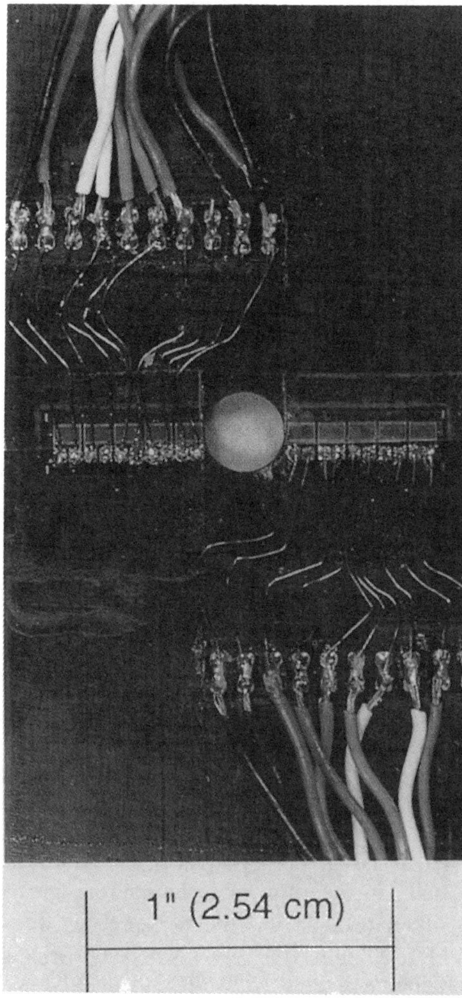

FIG. 1—*Illustration of strip strain gage location for strain distribution measurements.*

serve two purposes: to provide a validation of the use of Lekhnitskii's anisotropic plate solution [*11*] in conjunction with lamination theory for determining the strain distribution near the hole, and to provide a comparison for later measurements taken after fatigue. The results of these strength tests at room temperature and elevated temperature are given in Table 2. It should be noted that it was not possible to accurately measure the strain distribution at the elevated temperature because of degradation of the adhesive used on the gages and a lack of appropriate compensation for thermal apparent strains. Therefore, the measurements with these gages were conducted solely at room temperature.

After performing the quasi-static tests, the $[0/\pm 45]_{2s}$ specimens were fatigued at 10 Hz ($R = 0.1$) at 70 and 80% of their ultimate strengths. During this process, the maximum and minimum stroke signals were monitored using a computer-controlled data acquisition system to provide stiffness reduction measurements. These tests were then interrupted at various

TABLE 2—*Quasi-static strength test results.*

Property (Number of Replicates)	Room Temperature	350°F (177°C)
E_1 (5)	24.3 ± 0.3 msi (168 ± 2 GPa)	25.7 ± 1.6 msi (177 ± 11 GPa)
X_t (5)	392 ± 18 ksi (2700 ± 124 MPa)	325 ± 27 ksi (2240 ± 186 MPa)
E_2 (5)	1.38 ± 0.03 msi (9.51 ± 0.21 GPa)	1.27 ± 0.02 msi (8.76 ± 0.14 GPa)
Y_t (5)	8.63 ± 1.17 ksi (59.5 ± 8.1 MPa)	5.15 ± 0.42 ksi (35.5 ± 2.9 MPa)
G_{12} (5)	0.721 ± 0.043 msi (4.97 ± 0.30 GPa)	0.610 ± 0.040 msi (4.21 ± 0.28 GPa)
S (5)	26.4 ± 0.6 ksi (182 ± 4 MPa)	17.1 ± 0.5 ksi (118 ± 3 MPa)
$[0/\pm45]_{2s}$ notched tensile strength (2)	77.9 ± 3.4 ksi (537 ± 23 MPa)	76.6 ± 0.6 ksi (528 ± 4 MPa)

numbers of cycles for X-ray damage analysis and residual strength testing. The residual strength tests were conducted in a manner identical to that employed for the quasi-static strength tests (including the use of the EP-08-045PG-120 strain gages for strain distribution measurements). The results of these residual strength tests at room temperature are given in Table 3, and those at 350°F (177°C) are given in Table 4.

Penetrant Enhanced X-Ray Radiography

All of the fatigue specimens were X-ray radiographed in order to ascertain the damage development as the cyclic loading progressed. Sample radiographs for room temperature and elevated temperature specimens are shown in Fig. 2. As shown in the X-rays, similar damage development patterns occur in both the room temperature and the elevated temperature specimens. The X-rays, as well as visual observations during room temperature tests, suggest that the damage is principally in the form of delaminations growing longitudinally above the hole section as well as matrix cracks distributed uniformly throughout the laminate. The principal difference for the room temperature and the elevated temperature specimens appears to be that the rate of delamination growth is greater for the elevated temperature specimens than that for the room temperature specimens. Furthermore, the longitudinal splitting appears greater at the elevated temperature.

Critical Element Model

The essentials of the critical element model are shown schematically in Fig. 3. Such an approach is based upon the assumption that the damage associated with property degradation

TABLE 3—*Room temperature normalized residual strength results.*[a]

Cycles	70% Ultimate Strength	80% Ultimate Strength
10 000	0.95	0.93
260 000	1.09	1.21
1 500 000	1.35	1.21

[a] Normalized to 77.9 ksi (537 MPa unfatigued strength).

TABLE 4—*Elevated temperature normalized residual strength results.*[a]

Cycles	70% Ultimate Strength	80% Ultimate Strength
10 000	1.09	1.12
75 000	. . .	1.29
224 500	1.18	1.35

[a] Normalized to 76.6 ksi (528 MPa unfatigued strength).

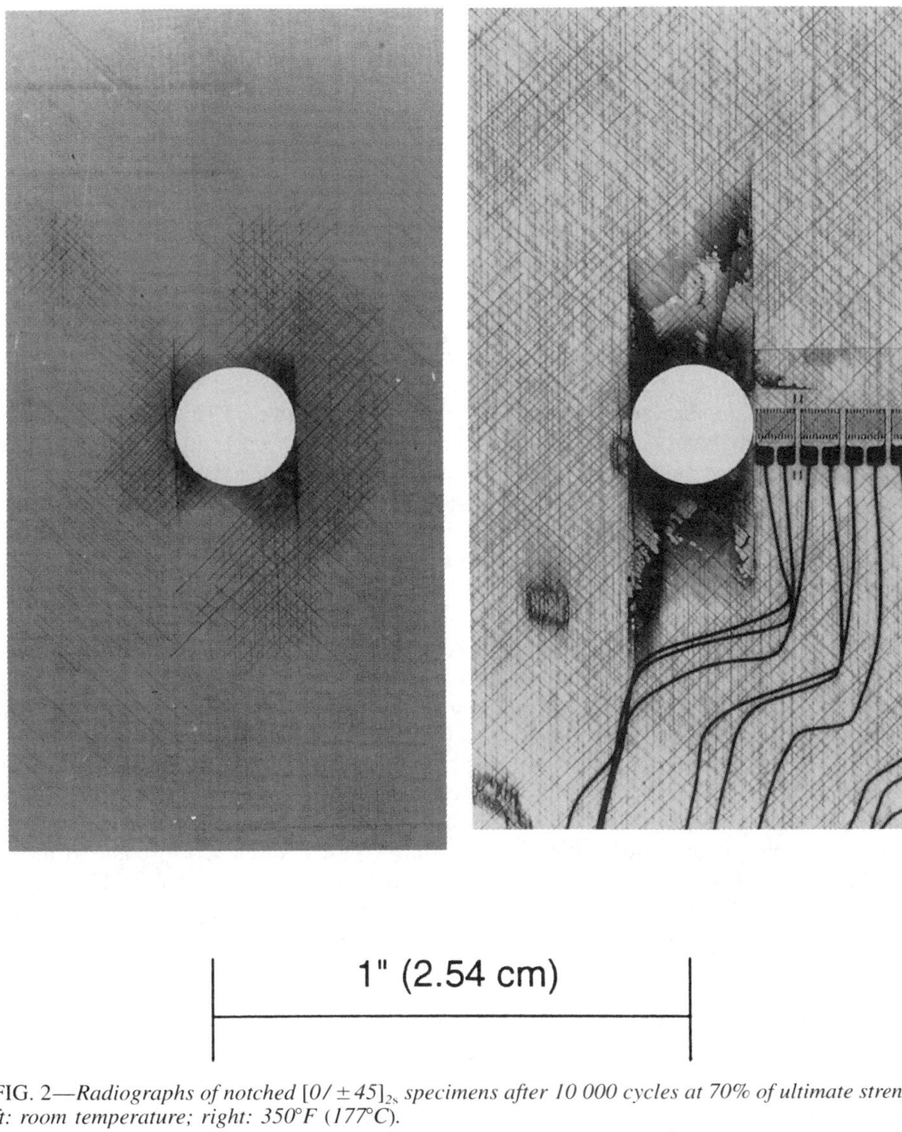

FIG. 2—*Radiographs of notched* $[0/\pm45]_{2s}$ *specimens after 10 000 cycles at 70% of ultimate strength. Left: room temperature; right: 350°F (177°C).*

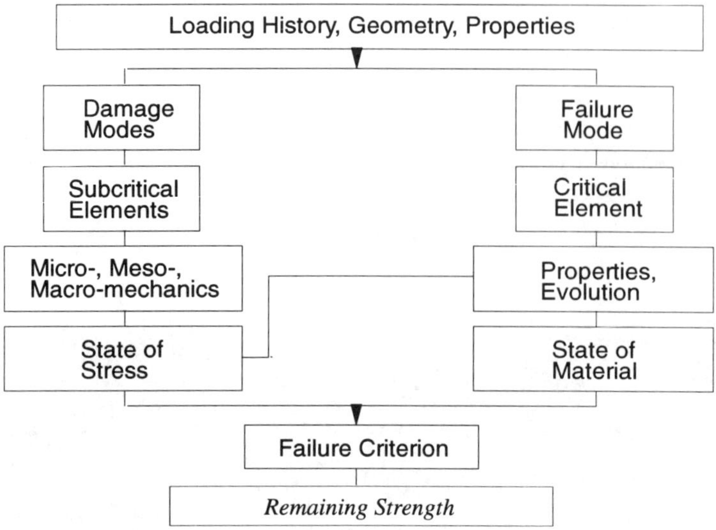

FIG. 3—*Schematic representation of the approach used to predict the remaining strength and life of composite components.*

is widely distributed within the composite laminate. In addition, it is assumed that a representative volume can be chosen such that the state of stress in that volume is typically of all other volumes in the laminate, and that the details of stress distribution and damage accumulation in that volume are sufficient to describe the final failure resulting from a specific failure mode. Thus, it is possible to select different representative volume elements for different failure modes. We proceed by further dividing the representative volume into "critical" and "subcritical" elements. The critical elements are selected in such a manner that their failure controls the failure of the representative volume, and therefore (by definition of the representative volume), of the laminated component. The remainder of the elements in the representative volume are regarded as sub-critical because their failure does not cause failure of the representative volume and, therefore, of the component. Their failure (due to such events as cracking or delamination) does, however, lead to greater stresses in the critical element that contribute to the eventual failure of the component. As an example of such a failure process, we may consider the case of tensile fatigue failure of a cross-ply laminate. During the fatigue process, matrix cracks develop in the 90° plies. However, these cracks do not cause failure of the laminate. They do increase the stress level in the 0° plies. But it is only when the 0° plies fail that the laminate fails. Thus, in this simple example, the 0° plies would correspond to the critical element and the 90° plies would correspond to the subcritical element. This example takes into account only ply-level knowledge. The inclusion of micromechanical models (such as that for tensile strength) is readily accomplished within the framework of the model [*5*].

Implementation of the Model

The mechanistic modeling begins with the determination of the stress state in the critical element. For unnotched laminates, classical lamination theory may be used to determine those stresses. In the case of quasi-static loading of notched laminates, the situation becomes somewhat more complicated. Any approach used should include the local geometry (such as hole size and laminate thickness) in order to obtain an accurate representation of the stress state in the representative volume element. For the present analysis, we use the method of Whitney and Nuismer [*12*] applied on a *ply-level* basis. This method is implemented as follows. The material properties (along with the lamination sequence and the applied global loads) are substituted into classical lamination theory to determine the mechanical properties of the effective anisotropic equivalent plate. The values, along with the hole size and applied load, are then used as inputs to the Lekhnitskii solution. The resoluting strain distribution is then used in conjunction with lamination theory to determine the stress distribution in the ply of interest. These stresses are then averaged over the distance a_o so that

$$\bar{\sigma}_{ij}(x) \equiv \frac{1}{a_o} \int_{R}^{R+a_0} \sigma_{ij}(x, y)dy \tag{1}$$

as illustrated in Fig. 4.

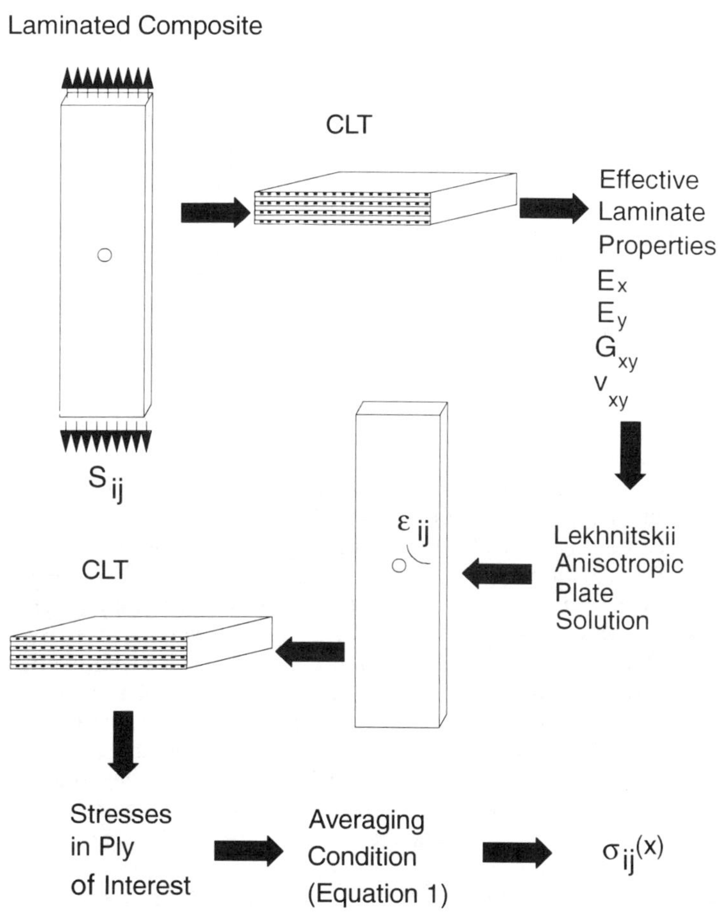

FIG. 4—*Schematic representation of the process used to determine stresses in the ply of interest using the strains predicted by the Lekhnitskii solution.*

All that remains is the selection of an appropriate failure criterion. For laminates containing 0° plies, we have found that a maximum stress criterion works well. Therefore, the averaging distance a_o is selected so that when the applied stress on the laminate is equal to the failure stress determined experimentally, we have

$$\bar{\sigma}_{11}^{0°}(0) \equiv \frac{1}{a_o} \int_R^{R+a_o} \sigma_{11}^{0°}(0, y)dy = X_t \tag{2}$$

where X_t is the unidirectional tensile strength in the fiber direction and the superscripts 0° denote that the calculations are to be performed in the zero degree ply. For the purposes of simplicity of discussion and notation, we have assumed that the maximum value of this stress occurs along the hole axis (this may easily be verified). Equation 2 may be viewed as the defining equation for a_0.

When the laminate is fatigue loaded, however, the situation is greatly complicated. As damage such as matrix cracking develops, the stiffness of the plies may change. In particular, it is known that relaxation of the stress concentration surrounding a notch in a composite laminate during fatigue often results in increased notched tensile strength. Although the exact nature of this relaxation is complex, it has been shown [*8*] that it is possible to represent the damage development in such a way that a typical relaxation zone in the region of the hole is created, so that the correct local stresses are determined. This is accomplished by representing the hole in the laminate by an "effective" hole having different physical dimensions. These physical dimensions are obtained in the following manner. The majority of the delamination surrounding the hole appeared to be in the longitudinal, rather than transverse direction, as is shown in Fig. 2. Therefore, only the hole axis parallel to the direction of loading was modified. The global stiffness reduction measured during the test was then used to calculate a stiffness reduction of the matrix-controlled stiffnesses (E_2 and G_{12}) based on the assumption that the value of E_1 did not change during the cyclic loading. For ease of analysis, these normalized values for E_2 and G_{12} were represented in the form of

$$\frac{G_{12}(n)}{G_{12}(0)} = \frac{E_2(n)}{E_2(0)} = C - (C - 1)(1 + \alpha n) \exp(-\beta n) \tag{3}$$

where n is the current cycle number and C, α, and β are fitting parameters. An example of such a fit is given in Fig. 5 based on average data taken at room temperature from four separate tests (two each at 70 and 80% of ultimate notched tensile strength). It should be noted that any values determined in this manner at elevated temperature contain the combined effects of stiffness reduction due to fatigue damage as well as that due to viscoelasticity. These values were then used as inputs to the Lekhnitskii solution, and the size of the major axis adjusted so that the error between the predicted strain values and the measured values was minimized in a least squares sense. Although this may seem to be a crude approach, the resulting strain distribution closely matches the experimentally measured one, as is demonstrated in Fig. 6. The elevated temperature specimens provided even more difficulties. As mentioned previously, it was not possible to accurately measure the strain distributions at the elevated temperatures. A compromise approach was therefore employed: the strain distributions were measured at room temperature, an effective hole size determined based on the room temperature data, and the assumption made that such a hole size could adequately represent the material behavior at the elevated temperature. Once these values for the effective hole size had been determined, they were also fit to the form of

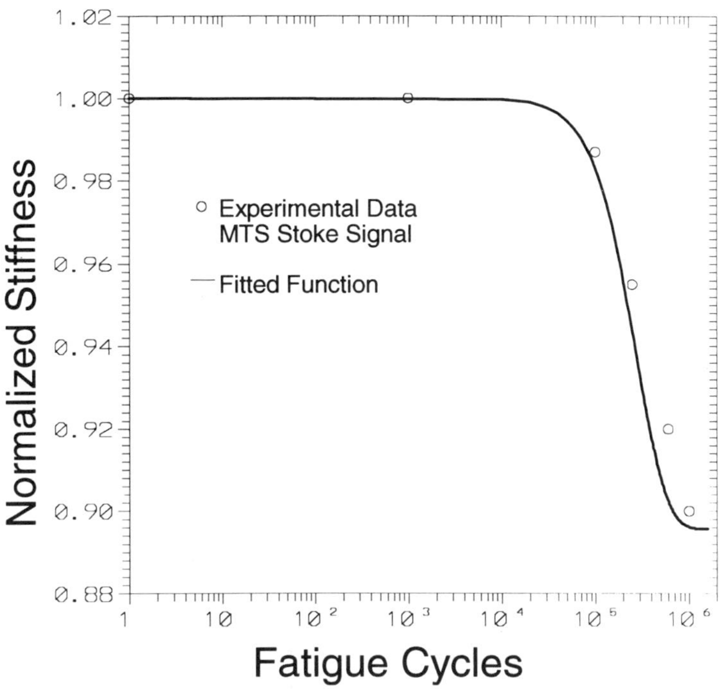

FIG. 5—*Comparison of experimentally measured stiffness reduction with that calculated using the stiffness reduction fits.*

$$\frac{b}{a} = A + B \exp[-D(Fa)^p n] \tag{4}$$

where b is the hole major axis, a is the hole minor axis, n is the current cycle number, Fa is the initial ratio of the applied stress level to the ultimate strength of the laminate, and A, B, D, and p are fitting parameters. The effective hole sizes, as well as the fitted values given by Eq 4 for room temperature and elevated temperature, are shown in Fig. 7.

Using Eqs 1, 3, and 4 in conjunction with lamination theory and the Lekhnitskii solution, it is then possible to calculate the stress state in the critical element at any point during the fatigue process. All that remains is to specify a rationale for combining all of the effects that occur during this process. To do so, we use the approach suggested by Reifsnider et al. [*6*]. In the interest of brevity, we will highlight here the important aspects of that approach. We begin by selecting the remaining strength of the critical element as a damage metric. We then assume that the remaining strength of the critical element can be determined as a function of loading level (all other features being held constant). For one load level, a given fraction of life corresponds to a certain reduction in the remaining strength. We claim that an "equivalent" fraction of life at another load level is one that gives an identical reduction in the remaining strength. Using this idea, it is then possible to consider sequenced loading effects as well as varying load levels. To do that, we could calculate the strength reduction at the first load level and then sum that with the strength reduction at the second load level. Such an approach is illustrated schematically in Fig. 8. For this particular loading case, we

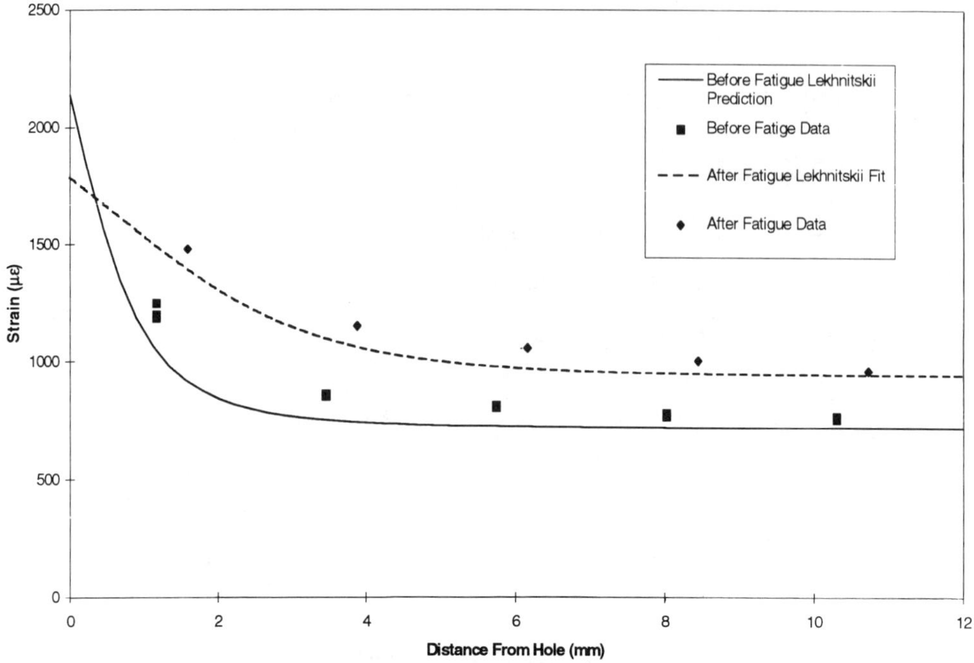

FIG. 6—*Comparison of measured strain distributions to those from the Lekhnitskii solution.*

begin by cycling for n_l cycles at a normalized applied stress level of S_a^l. This gives us a normalized remaining strength of S_r^l. If we then begin cycling at an applied stress of S_a^2, the total remaining number of cycles to failure is $N_2 - n_2$, where N_2 is the expected number of cycles to failure at an applied stress level of S_a^2 alone, and n_2 is the number of cycles at S_a^2 alone that would give us the same strength reduction as n_l cycles at S_a^l. In general, the strength reduction curves are nonlinear so that the resulting strength and life predictions are path as well as history dependent.

The next step is to postulate that the normalized remaining strength (the damage metric we are using) is an internal state variable [*6*]. For our purposes we will also postulate that our failure function *Fa* is given by

$$Fa = \frac{\overline{\sigma}_{11}{}^{0°}(0)}{X_t} \tag{5}$$

We then define our normalized residual strength of the critical element *Fr* in terms of *Fa* so that

$$Fr(n) = 1 - \int_0^{n/N(Fa)} (1 - Fa(n))j\left(\frac{n}{N(Fa)}\right)^{j-1} d\left(\frac{n}{N(Fa)}\right) \tag{6}$$

where *N*(*Fa*) is the number of cycles to failure at a normalized stress level of *Fa* alone. The quantity *j* is a material parameter. However, we have found that a value of 1.2 works well for most materials. The value of *N*(*Fa*) is represented as

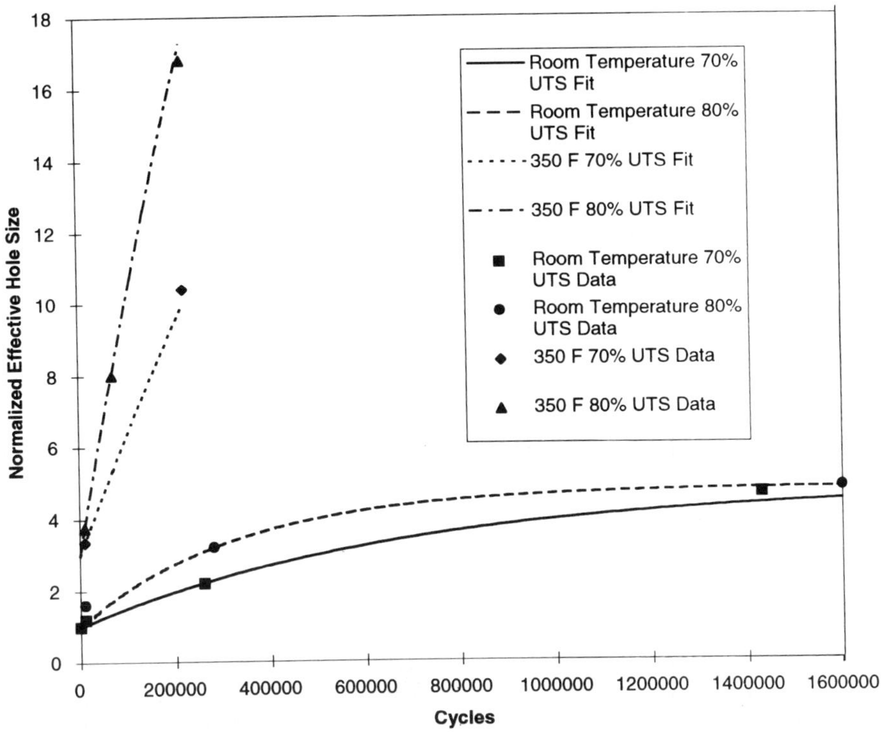

FIG. 7—*Effective hole sizes used for calculation of stress distributions surrounding the hole during fatigue.*

$$N = 10^{(Fa-A_n/B_n)1/P_n} \tag{7}$$

where A_n, B_n, and P_n are material parameters. Equation 7 may be viewed as the number of cycles to failure of the critical element under baseline conditions. Equation 6 is the basis for all the remaining strength calculations.

There is one final matter to be considered when calculating the remaining strength of composite laminates containing a hole. The above calculations have been used to calculate the remaining strength of the critical element. For the cases we will consider, this is a strictly decreasing quantity. However, we know from experience that the laminate level residual strength may increase due to relaxation of the stress concentration surrounding the hole. As this relaxation of stress concentration occurs, the stress state in the critical element is decreased leading to an apparent increase in remaining strength. However, the residual strength of the critical element (the 0° ply controlling failure) is decreasing due to an accumulation of fiber fractures during the fatigue loading. The laminate-level residual strength that results is a combination of these two competing effects. Any attempt to accurately model the behavior should include these effects.

Model Results

The values required to calculate the remaining strength of the critical element using Eq 6 can become quite complicated, even for simple cases such as an unnotched laminate. For

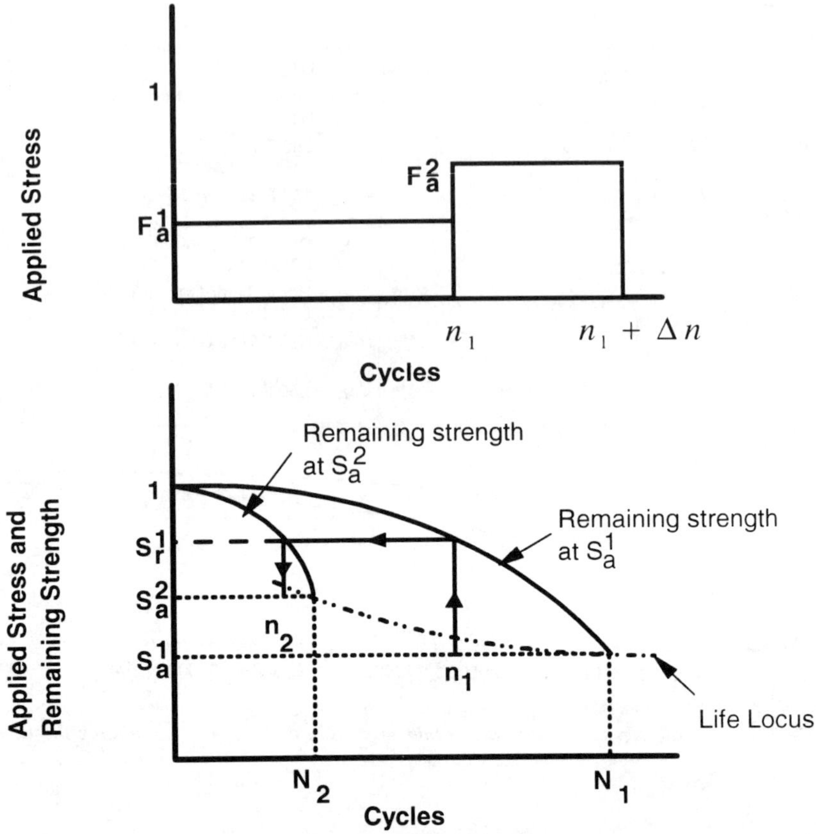

FIG. 8—*Illustration of the use of remaining strength as a damage metric and the concept of equivalent remaining strength.*

more complicated cases the problem does not admit closed-form solutions. As a result, our approach has been to construct a computer code, MRLife, which performs the calculations numerically. Using the inputs from the quasi-static tests, the effective hole size fit, Eq 4, as well as the stiffness reductions given by Eq 3, it is possible to simulate the laminate behavior for arbitrary loading conditions. As an example of its utility, we will compare the predicted remaining strength values with those measured experimentally for the IM7/K3B system.

First, we will consider the tests conducted at room temperature. As reported in Table 3, the laminates fatigue tested both at 70 and 80% of ultimate strength showed an initial decrease followed by an increase in residual strength. The predicted values based on the MRLife simulation are shown in Fig. 9, along with the experimentally determined values. It should be pointed out that while the calculated values are based upon a number of fits (such as those given by Eqs 3, 4, and 7), these are calculated, not fitted values. In fact, the predictions for the remaining strength at 10 000 cycles were made based on the other available data at greater numbers of cycles. These residual strengths are laminate-level predictions. To gain a greater insight into the nature of the predictions, it is helpful to look at the output from the MRLife calculations. Such an output for an applied stress level of 80% of ultimate strength is shown in Fig. 9. As the fatigue process begins, and fiber fractures accumulate in

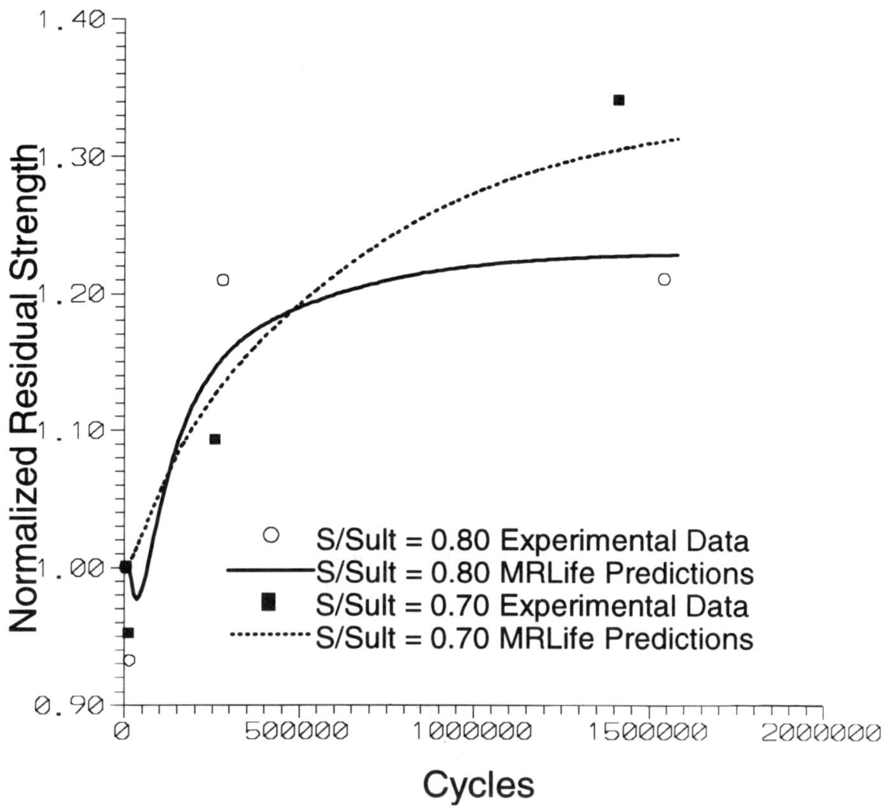

FIG. 9—*Comparison of predicted and experimentally measured remaining strength values at room temperature.*

the 0° plies, we see a steep decrease in the residual strength of the critical element. However, at the same time, the stress concentration is being decreased. (This decrease is represented by increase in the effective hole size.) The net result is that the predicted laminate level strength shows an initial decrease followed by an increase. Such a result is borne out experimentally.

At elevated temperature, we see a slightly different result in Fig. 10. In this case, the predicted strength at both 70 and 80% of the ultimate strength increases from the beginning of the fatigue cycling, without the decrease that was seen in the room-temperature tests. This is a result of the rapid rate that the material reduces the stress concentration. One interesting feature that is present in the predicted values as well as the experimental ones is that at room temperature the residual strength of the specimens fatigued at 80% is less than those fatigued at 70%, while at elevated temperature the opposite is true. The model is sensitive to these differences, suggesting that it is capturing many of the important physical features.

Conclusions

In this paper, a methodology has been presented for predicting the remaining strength and life of composite materials at room and elevated temperature. Using a number of phenom-

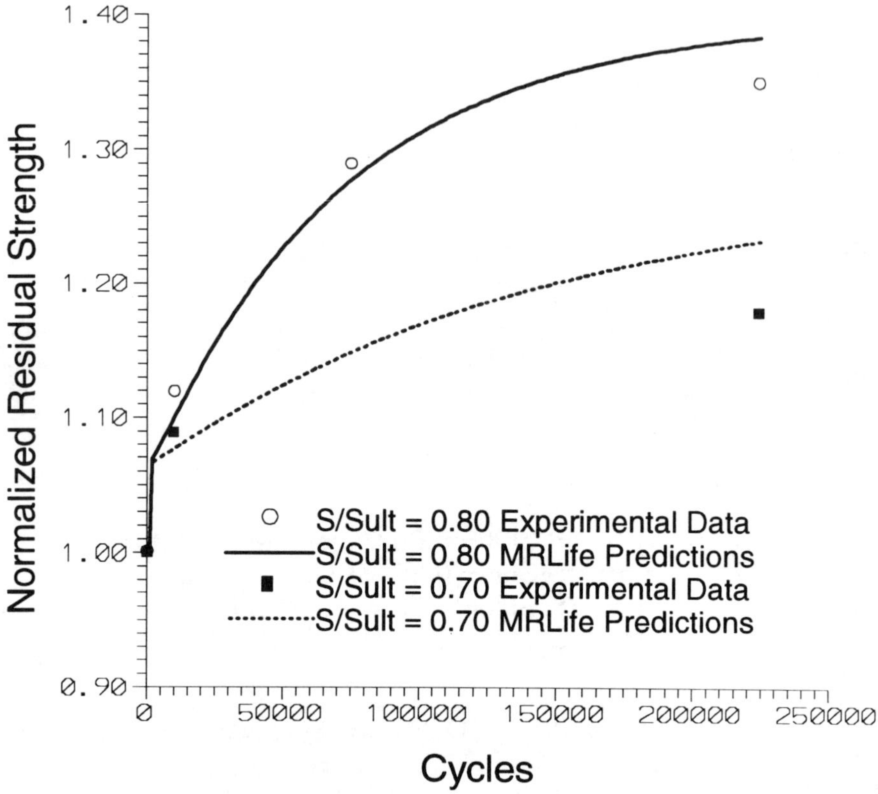

FIG. 10—*Comparison of predicted and experimentally measured remaining strength values at 350°F (177°C).*

enological representations for the stress state around a circular hole as well as stiffness reduction due to fatigue damage and viscoelasticity, it has been possible to construct numerical simulations of the remaining strength of the composite laminate. These simulations are conducted using mechanics-based models. The predicted values show good agreement with the experimentally determined values. In particular, the simulation has been able to predict details such as the difference in ordering of the residual strength data. (The specimens tested at 70% of ultimate strength at room temperature have a greater residual strength than those tested at 80% of ultimate strength; the opposite is true at elevated temperature.)

There are a number of ways that we are currently working on to improve the simulation. Most of the representations in the simulation are phenomenological. Improved simulations would require a more accurate depiction of the material behavior as a function of temperature. We are currently working to include accurate micromechanical models of strength at elevated temperature. Such models, as well as mechanics-based representations of the reduction of the stress concentration around the hole are badly needed. The approach used thus far is a framework that incorporates mathematical representations of physical material behavior. As these representations are improved, the model predictions will continue to improve.

Acknowledgments

The authors gratefully acknowledge the support of NASA Langley Research Center under Grant NAG1-343-37, the Air Force Office of Scientific Research under Grant FA9620-95-0217 and the National Science Foundation Science and Technology Center for High Performance Polymeric Adhesives and Composites under Grant DMR9120004.

References

[*1*] Talreja, R., "Damage Characterization," Chapter 3, *Fatigue of Composite Materials*, K. Reifsnider, Ed., Composite Materials Series, N. Pagano, Series Editor, Elsevier, 1991.

[*2*] Allen, D., Harris, C., and Nottorf, E., "A Thermomechanical Constitutive Theory for Elastic Composites and Distributed Damage," Parts 1 and 2, Texas A&M Report MM-5023-85-15, 1985.

[*3*] Joshi, S. and Frantziskonis, G., "Damage Evolution in Laminated Advanced Composites," *Composite Structures*, Vol. 17, 1991, pp. 127–139.

[*4*] Hwang, W. and Han, K., "Fatigue of Composite Materials — Damage Model and Life Prediction," *Composite Materials: Fatigue and Fracture, ASTM STP 1012*, American Society for Testing and Materials, West Conshohocken, PA, 1989, pp. 87–102.

[*5*] Subramanian, S., Reifsnider, K., and Stinchcomb, W., "A Cumulative Damage Model to Predict the Fatigue Life of Composite Laminates Including the Effect of a Fiber-Matrix Interphase," *International Journal of Fatigue*, Vol. 17, No. 5, 1995, pp. 343–351.

[*6*] Reifsnider, K. L., Case, S. W., and Xu, Y. L., "A Micro-Kinetic Approach to Durability Analysis: The Critical Element Method," Progress in Durability Analysis of Composite Systems, A. H. Cardon, H. Fukuda, and K. Reifsniders, Eds., A. A. Balkema (Rotterdam), 1996, pp. 123–144.

[*7*] Reifsnider, K., Iyengar, N., Case, S., and Xu, Y., "Damage Tolerance and Durability of Fibrous Material Systems: a Micro-Kinetic Approach," *Durability Analysis of Structural Composite Systems*, A. A. Balkema, Rotterdam, 1996, pp. 123–144.

[*8*] Reifsnider, K., Iyengar, N., Case, S., and Xu, Y., "Kinetic Methods for Durability and Damage Tolerance Design of Composite Components," Keynote Address, Conference on Composite Materials, Japan Society for Mechanical Engineers, 26 June 1995, Tokyo.

[*9*] Reifsnider, K., Lesko, J., and Case, S., "Kinetic Methods for Prediction of Damage Tolerance of High Temperature Polymer Composites," *Composites '95: Recent Advances in Japan and the United States*, I. Kimpara, H. Miyairi, and N. Takeda, Eds., Proceedings, Japan-U.S. CCM-VII, Kyoto, 1995, pp. 49–55.

[*10*] Reifsnider, K. and Stinchcomb, W., "A Critical Element Model of the Residual Strength and Life of Fatigue-Loaded Composite Coupons," *Composite Materials: Fatigue and Fracture, ASTM STP 907*, H. T. Hahn, Ed., American Society for Testing and Materials, West Conshohocken, PA, 1986, pp. 293–303.

[*11*] Lekhnitskii, S., *Anisotropic Plates*, S. W. Tsai and Cheron, translators, Gordon and Breach, 1968.

[*12*] Whitney, J. and Nuismer, R., "Stress Fracture Criteria for Laminated Composites Containing Stress Concentrations," *Journal of Composite Materials*, Vol. 8, July 1974, pp. 253–265.

Mark R. VanLandingham,[1] *Rushad F. Eduljee,*[2] *and John W. Gillespie, Jr.*[3]

The Effects of Moisture on the Material Properties and Behavior of Thermoplastic Polyimide Composites

REFERENCE: VanLandingham, M. R., Eduljee, R. F., and Gillespie, J. W., Jr., **"The Effects of Moisture on the Material Properties and Behavior of Thermoplastic Polyimide Composites,"** *High Temperature and Environmental Effects on Polymeric Composites: 2nd Volume, ASTM STP 1302*, Thomas S. Gates and Abdul-Hamid Zureick, Eds., American Society for Testing and Materials, 1997, pp. 50-63.

ABSTRACT: Thermoplastic polyimides are a relatively new class of polymers that exhibit high-temperature stability and are useful in composite applications. One such material is Avimid® K3B reinforced with Magnamite® IM7 graphite fibers. This composite system exhibits excellent strength and toughness, retains its strength and toughness after prolonged exposure to elevated temperatures, and resists microcracking at extremely low (liquid nitrogen) temperatures. Further characterization of IM7/K3B thermoplastic composites is focused on evaluating hygrothermal effects on the material properties and behavior. The environmental conditioning test matrix includes three temperatures (20, 40, and 80°C) and four relative humidity levels (75, 85, 97, and 100%). Observations and conclusions from these studies include the following: (1) the moisture diffusivity of IM7/K3B has a classic Arrhenius dependence on temperature; (2) the moisture saturation level depends on the relative humidity level to the power of 1.34 with a maximum value of 0.55% by weight; (3) the glass transition temperature T_g is lowered with moisture absorption but is recovered when the sample is redried; (4) the intralaminar fracture toughness G_{Ic} remains constant after extensive hygrothermal conditioning; (5) the diffusion kinetics are Fickian in general, except for a few non-Fickian anomalies that are related to development of transverse microcracks.

KEY WORDS: polymer matrix composites, polyimide matrix, graphite fibers, moisture, hygrothermal effects, glass transition temperature, fracture toughness, diffusion, transverse microcracking

[1] Graduate research assistant, Materials Science Program, Center for Composite Materials, University of Delaware, Newark, DE 19716.

[2] Assistant professor of Materials Science and research associate of the Center for Composite Materials, University of Delaware, Newark, DE 19716.

[3] Associate professor of Materials Science and associate director of the Center for Composite Materials, University of Delaware, Newark, DE 19716.

Nomenclature

D	Transverse diffusivity of the composite laminate (cm^2/s)
M_{sat}	Moisture saturation level, weight %
G_{Ic}	Intralaminar fracture toughness, J/m^2
a	Notch length of the fracture toughness test specimen, mm
δ	Notch width of the fracture toughness test specimen, mm
T_g	Glass transition temperature, °C
E''	Loss modulus, MPa
% RH	Percent relative humidity of the conditioning environment
wt%	Weight % or % by weight
v_f	Fiber volume fraction

Over the past 20 years, the effects of moisture on fiber-reinforced polymeric composite materials have been studied extensively [*1-9*]. In many of these studies, significant changes in the mechanical properties and various forms of moisture-induced damage have been reported [*4-8*]. For example, absorbed moisture has been shown to reduce the glass transition temperature T_g of the resin [*4,5*] reduce matrix-dominated properties of the composite such as transverse tensile strength and intralaminar shear strength [*4-6*], and cause swelling of the resin that induces residual stresses and causes the formation of microcracks [*5,7-10*]. These deleterious effects of absorbed moisture have been attributed to plasticization and degradation of the resin matrix and degradation of the fiber-matrix interface [*5-10*]. To date, most moisture studies have been concerned with thermosetting matrix composites (for example, graphite/epoxy) that absorb as much as 1.2 to 2% moisture by weight in 95 to 100% relative humidity environments (for fiber volume fractions v_f between 60 and 68%) [*1,2,5-7*]. Recently, thermoplastic (semicrystalline and amorphous) matrix composites have been developed that absorb very little moisture compared to their thermosetting matrix counterparts [*3,4*]. An example of such a system is a thermoplastic matrix composite composed of an amorphous polyimide matrix, Avimid®[4] K3B, reinforced with Magnamite®[5] IM7 graphite fibers (v_f = 62%) [*11*], that absorbs up to 0.55% moisture by weight during water immersion. In this study, extensive hygrothermal characterization of the IM7/K3B system is conducted to determine moisture-induced degradation, predict moisture contents in service environments, and develop accelerated environmental conditioning methods.

Experimental

Materials

The particular thermoplastic matrix composite used in this study employed an amorphous polyimide matrix, Avimid® K3B, reinforced with Magnamite® IM7 graphite fibers (v_f = 62%) [*11*]. Laminates were laid up in house and then processed by The DuPont Co. under optimized processing conditions [*12*]. Laminate configurations included the following: $[90]_{10}$, or unidirectional laminates; $[0_2/90]_s$, or thin cross-ply laminates; $[0_2/90_2/0_2]_s$, or thick cross-ply laminates; and $[45/0/-45/90]_s$, or quasi-isotropic laminates. After processing and prior to any testing or conditioning, ultrasonic c-scanning (transducer frequency = 15 MHz) was

[4] The DuPont Co., Wilmington, DE.
[5] Hercules, Inc., Wilmington, DE.

used to inspect laminate quality. No defects were detected in any of the panels. Also, baseline (dry) longitudinal and transverse tensile testing of unidirectional specimens was performed in accordance with Test Method for Tensile Properties of Fiber-Resin Composites (ASTM D 3039). The average strengths and moduli measured were within 5% of reported values [*11,13*].

Hygrothermal Conditioning

The hygrothermal conditioning test matrix consisted of three temperatures, 20, 40, and 80°C, and four relative humidity (RH) levels—75, 85, 97, and 100% RH. For 100% RH conditioning, samples were immersed directly in water contained in stainless steel tanks that were roughly 0.003 m^3 in volume. The remaining humidity conditions for 20 and 40°C environments were created within desiccators in accordance with Practices for Maintaining Constant Relative Humidity by Means of Aqueous Solutions (ASTM E 104). For the 40°C conditioning environments, the desiccators were then placed in the water baths used for the immersion studies, while the 20°C desiccators sat out in a controlled room-temperature environment. An environmental chamber was used to create 80°C humidity environments (except for 100% RH). The temperatures of the 40 and 80°C water baths were maintained by a temperature feedback controller, that employed J-type thermocouples (wrapped in a polyimide tape to limit galvanic corrosion) and flexible heating blankets as heating elements. Distilled, deionized water was used in all of the aqueous solutions and in the 100% RH baths. To date, conditioning times of the various specimens range from 1000 to 4400 h. (See the following subsections and the Results section for more specific information on conditioning times.)

During preliminary studies of moisture diffusion in the IM7/K3B system, the weights of specimens taken from each type of laminate reached equilibrium (dry) values after 2 to 3 days under vacuum at 120°C. Also, in an unpublished study,[6] IM7/K3B specimens were conditioned in 95% RH environments and then dried under vacuum at several elevated temperatures (>100°C). No significant degradation was observed after drying at 120°C. Therefore, all specimens were initially dried for 4 days under vacuum at 120°C prior to hygrothermal conditioning. Limited hygrothermal cycling studies were also performed which included 1200 to 1800 h of conditioning in the environments described previously, followed by desorption for 4 days under vacuum at 120°C, and finally reimmersion of the specimens into their original environments.

Weight Measurements

Square specimens with length dimensions of about 5 cm were cut with a diamond saw from each of the four types of laminates and used to measure moisture uptake during hygrothermal conditioning. At least three samples were conditioned in each constant-temperature, constant-humidity environment for 1200 to 4400 h. A few samples were used for limited hygrothermal cycling studies described in the previous subsection. Prior to conditioning, cross-ply and quasi-isotropic samples were polished on one edge to a 5-μm finish. This extra step allowed for the nondestructive detection of microcrack initiation by simply observing the polished edge of a sample under an optical microscope each time it was removed for weighing. Weight measurements were taken periodically during both absorption and desorption using a digital balance accurate to ±1 mg. The surfaces of samples immersed

[6] Personal communication with Dick Cornelia, The DuPont Co., Wilmington, DE, 1994.

in water baths were blotted dry with a cloth before they were weighed. Specimens were removed from their environments for weighing and microcrack inspection for no longer than 30 min (typically 15 to 20 min). Moisture loss during these removal periods was calculated to be insignificant.

Mode I Intralaminar Fracture Toughness

Intralaminar fracture toughness G_{Ic} was measured using a double edge notch (DEN) test [*14*]. Unidirectional panels with dimensions of 20.3 by 13 cm were end-tabbed and cut into four 20.3 by 2.5 cm samples using a slot grinder. Samples taken from conditioned panels were cut at least 1 cm from the panel's edges to eliminate any edge effects from the moisture absorption process. Notches were cut with a 0.5-mm diamond blade to approximately 6.9 mm in length on each side at the center of the sample. Notch lengths a and notch widths δ were measured to an accuracy of ± 0.01 mm using a microscope equipped with a micrometer-driven stage. In Ref *14*, notches were cut with a thinner 0.3-mm diamond blade. However, the thinner blade is less rigid and, due to out-of-plane vibration, yields equivalent notch widths to the thicker blade ($\delta \sim 0.7$ mm). Sample geometry is illustrated in Fig. 1.

Baseline (dry) panels were cut and notched and then dried for four days under vacuum at 120°C prior to testing. Samples taken from conditioned panels were cut and notched 2 to 3 days prior to testing. These samples were then placed back into their original environments until they were tested. Total exposure times ranged from 1200 to 3600 h (see Results section), and one panel (four specimens) was conditioned in each environment for each chosen exposure time. All specimens were pulled in tension until failure occurred using a mechanical load frame. All testing was performed at room temperature, and G_{Ic} values were calculated as specified in Ref *14*. After testing, the fracture surfaces of the G_{Ic} samples were observed under a scanning electron microscope (SEM) at magnifications up to 5000X at 30 kV.

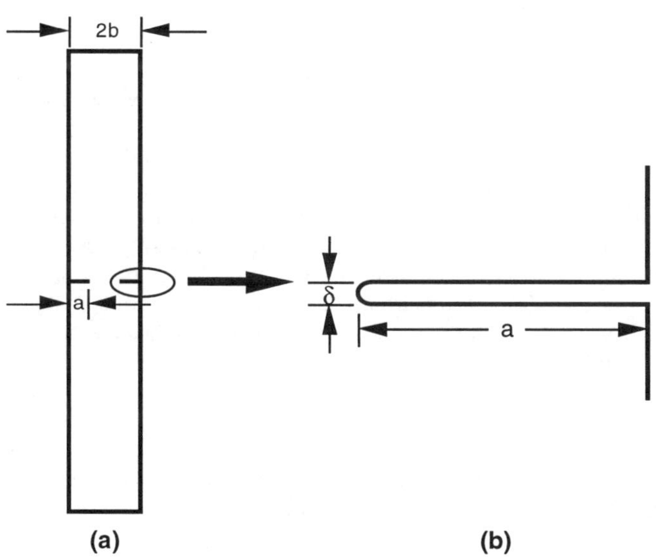

FIG. 1—*(a) Double edge notch (DEN) test geometry as described in Ref* 14 *and (b) illustration of actual notch geometry observed in this study.*

Glass Transition Temperature

Glass transition temperature T_g was measured by dynamic mechanical analysis (DMA). Samples were cut from conditioned and unconditioned unidirectional panels to roughly 6 by 1.25 cm using a diamond saw, and samples were taken at least 1 cm from the edge of a conditioned panel to avoid edge effects. Conditioned samples were tested after reaching moisture saturation levels in 20 and 80°C water baths and in the 20°C 75% RH environment, with the goal of producing bounds on the observed T_g reductions. DMA testing involved dynamic loading of samples in flexure (3-point bend) with a frequency of 1 Hz as the temperature was increased at a rate of 2°C/min. T_g values were measured with fibers oriented transversely and longitudinally to the neutral axis of the specimen. T_g is reported as the peak in the loss modulus, E''.

Damage Development

Ultrasonic c-scans were made of cross-ply panels immersed in 20, 40, and 80°C water for at least 1300 h. The scans included normal incidence and 30° polar backscattering, which enables determinations of transverse cracks [*13*]. After observing the onset of microcracking in the edge-polished samples under an optical microscope, moisture weight gain specimens were removed periodically from their conditioning environments, sectioned with a diamond saw, mounted in epoxy, ground and polished to a 5-μm finish. These polished cross-sectional surfaces of conditioned (and also unconditioned) samples were then observed using an optical microscope (magnifications up to 1000X). Damage was characterized as the number of transverse microcracks per ply per centimeter of cross-sectional material. Thus, a microcrack was counted only if it extended through the thickness of at least one ply.

Results

Characterization of Moisture Diffusion

Absorption and desorption of moisture for the IM7/K3B composite system can be characterized, in general, by Fick's second law [*2*]. In Figs. 2 and 3, average moisture uptake data are shown for 20°C conditioning of unidirectional specimens and 40°C conditioning of thick cross-ply specimens, respectively. The fitted curves were generated using an approximate solution to Fick's second law [*1*]. Generally, the standard deviation for each data point and the scatter of the data points about the curves are within the experimental accuracy of the balance used to weigh the samples, for example, ±1 mg corresponds to ±0.017 and ±0.015 wt%, respectively, for unidirectional and thick cross-ply specimens near saturation levels. However, a significant increase in the saturation levels of thick cross-ply specimens conditioned in 97 and 100% RH environments was observed after 4400 h (see Fig. 3). To date, moisture diffusion into unidirectional laminates has not deviated from Fickian behavior in any of the environments. (Conditioning in 40°C 75% RH and in 80°C 85 and 95% RH has not yet been completed.) Deviations from Fickian behavior, consisting of increases in saturation level after long times, have been observed during elevated-temperature conditioning of some cross-ply and quasi-isotropic specimens (see Discussion section).

Moisture saturation levels M_{sat} for the four laminate types increased with increasing humidity level, as shown in Fig. 4. This dependence was fitted with a power law function of the form [*1,2*]

$$M_{sat} = c(\%\ \mathrm{RH})^n \tag{1}$$

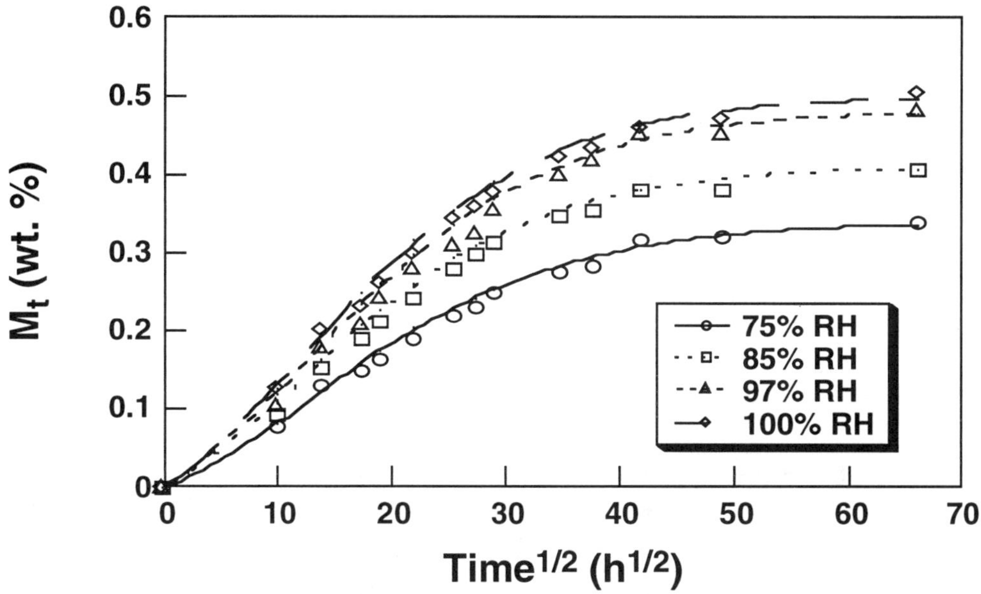

FIG. 2—*Moisture absorption curves for unidirectional IM7/K3B specimens conditioned in 20°C environments. Each data point is an average of 4 to 7 specimens. The curves shown approximate Fickian Behavior.*

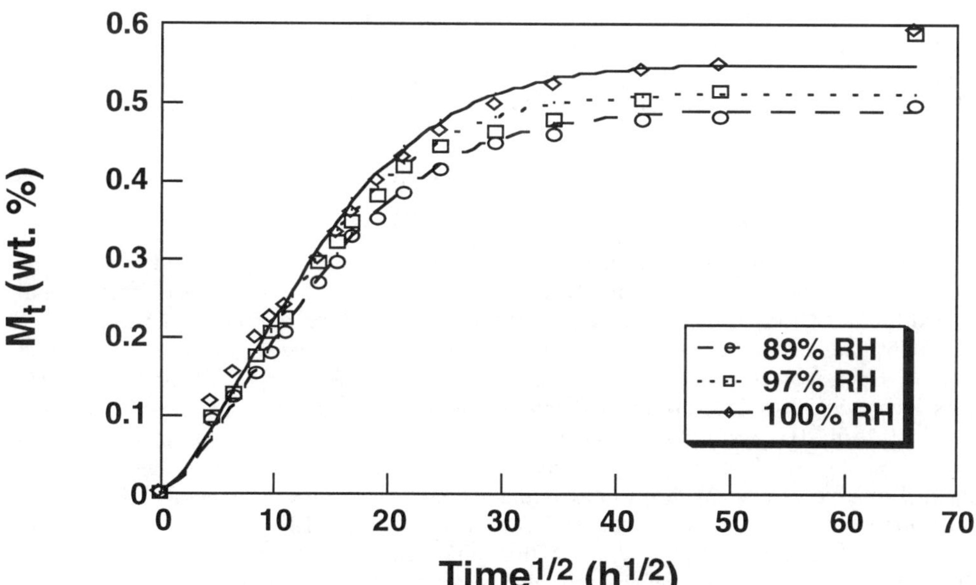

FIG. 3—*Moisture absorption curves for thick cross-ply IM7/K3B specimens conditioned in 40°C environments. Each data point is an average of 9 specimens. The curves shown approximate Fickian behavior.*

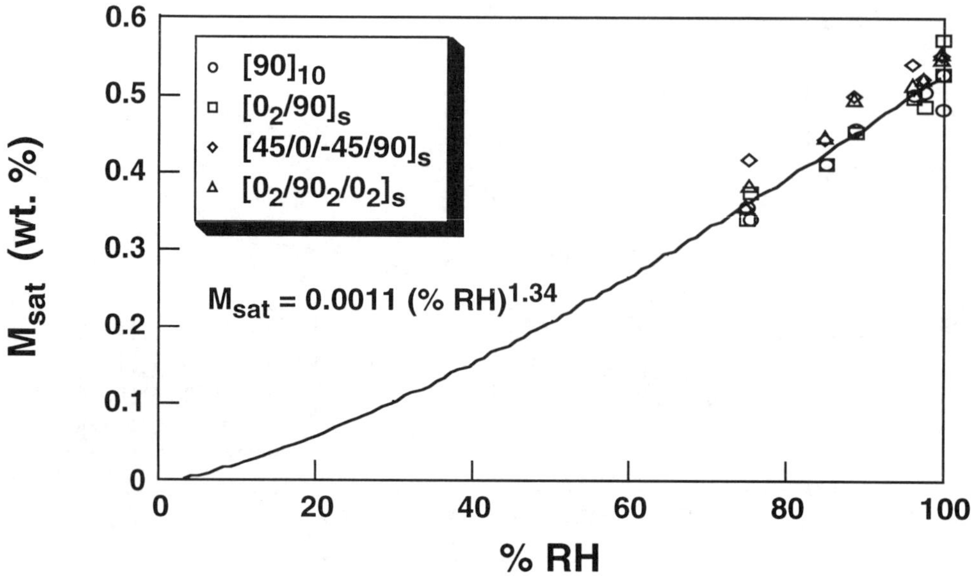

FIG. 4—*Moisture saturation levels plotted as a function of relative humidity for IM7/K3B laminates. Each data point is an average of 3 to 9 specimens.*

where $c = 0.0011$ and $n = 1.34$ are empirical parameters. No dependence of the saturation level on temperature has been observed over the investigated temperature range. Diffusivities were calculated for absorption in 20, 40, and 80°C environments and for desorption at 120°C using the solution for Fick's second law at short times [*1,2*]. A plot of diffusivity D as a function of inverse temperature is shown in Fig. 5. The diffusivity can be characterized by classical Arrhenius behavior and is given as a function of temperature, T (K), by

$$D = 0.01 \exp [-4750/T] \tag{2}$$

Deviations from this diffusion behavior occurred during the desorption and re-absorption of moisture in cross-ply specimens. Desorption diffusivities of cross-ply specimens conditioned for 2400 h in 20°C environments and for 1200 to 1800 h in 40°C environments were 20 to 50% higher than those of unidirectional specimens that had received the same conditioning. Upon reabsorption of cross-ply specimens conditioned at 40°C, absorption diffusivities increased 30 to 65% above their initial values. Thin cross-ply specimens conditioned for 2400 h in 20°C environments showed similar increases in absorption diffusivities, but the diffusivities of thick cross-ply specimens that had received the same conditioning remained relatively constant. Moisture cycling seems to also have a moderate effect on the diffusion behavior of unidirectional laminates conditioned for 1800 at 40°C, as diffusivities increased between 15 and 25% above initial values. In several cases, slight increases (5 to 10%) in moisture saturation levels were also observed for both unidirectional and cross-ply specimens during re-absorption. A comparison of diffusivities and saturation levels between the first and second absorption cycles is shown in Table 1.

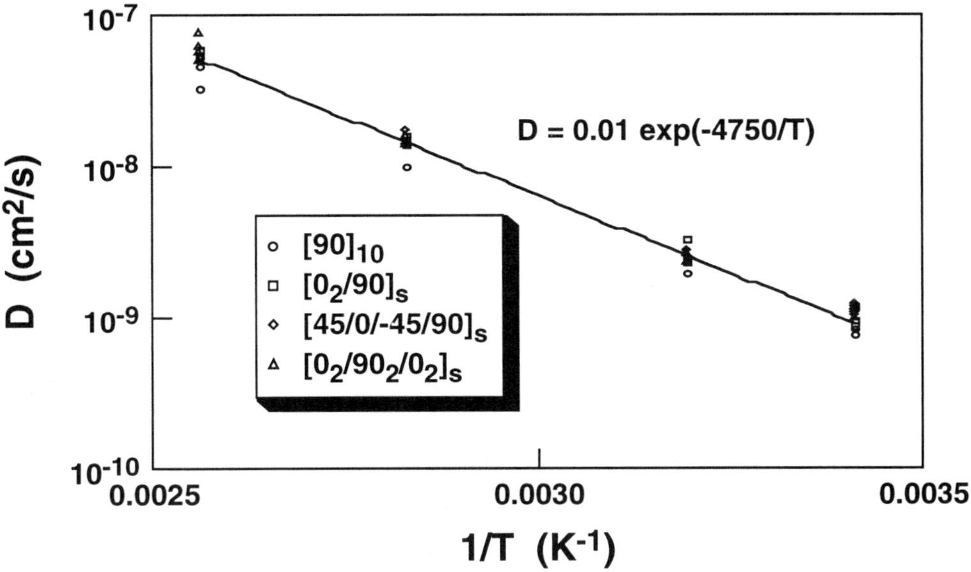

FIG. 5—*Arrhenius plot of diffusivity* D (cm^2/s) *as a function of inverse temperature for IM7/K3B laminates. Each data point is an average of 3 to 9 specimens.*

Mode I Intralaminar Fracture Toughness

The results of fracture toughness testing of samples conditioned at 20 and 40°C are shown in Figs. 6 and 7, respectively. Unidirectional specimens conditioned in room-temperature, high-humidity environments for up to 3600 h have exhibited no measurable toughness loss. Similarly, fracture toughness values have remained constant for specimens conditioned in 40°C environments for up to 3000 h. Also, no toughness loss has been observed for specimens conditioned in 80°C 75% RH for up to 1500 h. Specimens conditioned to saturation (300 h) in 80°C 100% RH did exhibit a moderate loss in toughness of 30 to 40%. However, the degradation mechanism (unknown at this time) responsible for this toughness loss might be active only during elevated-temperature water immersion and may not be representative of an "accelerated" conditioning effect.

Glass Transition Temperature

To date, only limited glass transition temperature T_g testing has been performed, the results of which are shown in Table 2. Standard deviations of the given values are of the order of 0.5°C. T_g values for dry specimens are similar to a T_g value of 255°C reported by Cornelia [*4*] for the same composite system. T_g values for moisture-saturated specimens were as much as 11% lower than the baseline value for a sample conditioned in the 80°C 100% RH environment. However, this effect was reversible, as T_g returned to its baseline value when the specimen was redried. These results indicate that moisture acts as a plasticizer to the polymer matrix. Similar results have been obtained by Cornelia [*4*] for several polyimide composite systems. More extensive characterization of T_g as a function of the conditioning time, temperature, and humidity level is currently under evaluation (see Discussion section).

TABLE 1—*Comparison of diffusivities* D *and moisture saturation levels* M_{sat} *for IM7/K3B specimens selected for limited cycling studies.*

Laminate/Conditioning Environment	Conditioning Time, h	D, cm²/s, Cycle 1	M_{sat}, wt %, Cycle 1	D, cm²/s, Cycle 2	M_{sat}, wt %, Cycle 2
$[90]_{10}$ 40°C, 89% RH	1200	2.05×10^{-9}	0.430	2.36×10^{-9}	0.460
$[90]_{10}$ 40°C, 89% RH	1800	1.81×10^{-9}	0.445	2.25×10^{-9}	0.465
$[90]_{10}$ 40°C, 97% RH	1200	2.90×10^{-9}	0.470	2.80×10^{-9}	0.490
$[90]_{10}$ 40°C, 97% RH	1800	2.50×10^{-9}	0.465	3.20×10^{-9}	0.510
$[90]_{10}$ 40°C, 100% RH	1800	2.40×10^{-9}	0.510	2.80×10^{-9}	0.540
$[0_2/90]s$ 40°C, 89% RH	1800	2.80×10^{-9}	0.470	2.82×10^{-9}	0.470
$[0_2/90]s$ 40°C, 97% RH	1800	2.60×10^{-9}	0.500	3.50×10^{-9}	0.540
$[0_2/90]s$ 40°C, 100% RH	1800	3.20×10^{-9}	0.510	4.06×10^{-9}	0.540
$[0_2/90_2/0_2]s$ 40°C, 89% RH	1200	1.96×10^{-9}	0.450	2.60×10^{-9}	0.450
$[0_2/90_2/0_2]s$ 40°C, 89% RH	1800	2.80×10^{-9}	0.470	2.82×10^{-9}	0.470
$[0_s/90_2/0_2]s$ 40°C, 97% RH	1200	2.59×10^{-9}	0.480	3.89×10^{-9}	0.530
$[0_2/90_2/0_2]s$ 40°C, 97% RH	1800	2.60×10^{-9}	0.500	3.50×10^{-9}	0.540
$[0_2/90_2/0_2]s$ 40°C, 100% RH	1800	3.20×10^{-9}	0.510	4.06×10^{-9}	0.540
$[90]_{10}$ 20°C, 75% RH	2400	0.85×10^{-9}	0.345	0.86×10^{-9}	...
$[90]_{10}$ 20°C, 85% RH	2400	0.94×10^{-9}	0.420	1.15×10^{-9}	...
$[90]_{10}$ 20°C, 97% RH	2400	0.86×10^{-9}	0.495	0.91×10^{-9}	...
$[0_2/90]s$ 20°C, 75% RH	2400	1.31×10^{-9}	0.380	1.63×10^{-9}	0.410
$[0_2/90]s$ 20°C, 85% RH	2400	0.83×10^{-9}	0.425	1.25×10^{-9}	0.440
$[0_2/90]s$ 20°C, 97% RH	2400	0.97×10^{-9}	0.470	1.08×10^{-9}	0.500
$[0_2/90_2/0_2]s$ 20°C, 75% RH	2400	1.17×10^{-9}	0.400	1.15×10^{-9}	...
$[0_2/90_2/0_2]s$ 20°C, 85% RH	2400	1.24×10^{-9}	0.470	1.18×10^{-9}	...
$[0_2/90_2/0_2]s$ 20°C, 97% RH	2400	0.95×10^{-9}	0.520	1.02×10^{-9}	...

Damage Development

Preliminary moisture studies included conditioning of cross-ply laminates in 80°C 100% RH for 1300 to 1500 h. During conditioning, moisture absorption was characterized by Fickian behavior up to saturation followed by a second stage uptake to a new saturation level, as reported in a previous section. After conditioning, both thin and thick cross-ply

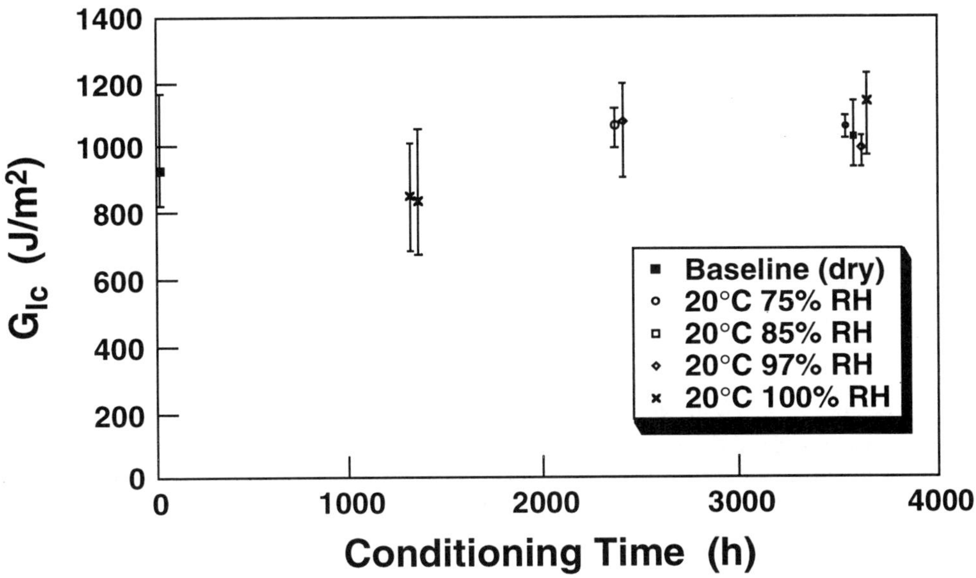

FIG. 6—*Intralaminar fracture toughness* G_{Ic} *(J/m²) of unidirectional IM7/K3B specimens conditioned in 20°C environments for up to 3600 h. Each data point is an average of 4 specimens. No toughness loss was observed.*

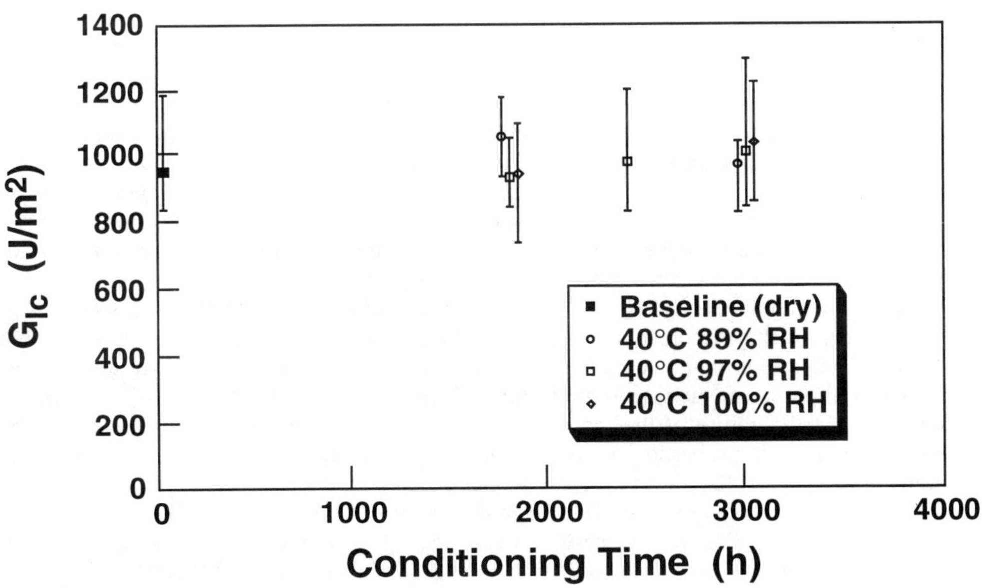

FIG. 7—*Intralaminar fracture toughness* G_{Ic} *(J/m²) of unidirectional IM7/K3B specimens conditioned in 40°C environments. Each data point is an average of 4 specimens. No toughness loss was observed.*

TABLE 2—T_g *measured after conditioning unidirectional IM7/K3B samples in various environments to saturation, and then again after redrying the samples.*

Conditioning	T_g Longitudinal Specimen, °C	T_g, Transverse Specimen, °C
Baseline, dry	262°	252°
20°C, 75% RH	245°	247°
→ Redried	260°	251°
20°C, 100% RH	236°	246°
→ Redried	260°	252°
80°C, 100% RH	233°	245°
→ Redried	262°	252°

laminates were sectioned, mounted in epoxy, and polished to reveal microcrack densities of 6 cracks/ply/cm or more. Subsequent conditioning studies have included periodic checks for microcracks. Except for a few anomalous specimens (see Discussion section), no significant microcracking (<1 crack/ply/cm) has developed in any laminates conditioned for up to 4400 h in 20 and 40°C environments, or in samples conditioned for up to 1500 h in the 80°C 75% RH environment.

Discussion

Moisture diffusion in IM7/K3B laminates is generally Fickian in nature. In a few cross-ply and quasi-isotropic specimens conditioned in 40°C 97% RH and 80°C 100% RH environments, some deviations from Fickian behavior were observed. After reaching saturation, these specimens would begin to take up more moisture and reach a new saturation level. This anomalous behavior was correlated to the microcrack density. For example, the second uptake of moisture for a thin cross-ply specimen conditioned in the 40°C 97% RH environment for 1200 h corresponded to a microcrack density of 3 to 4 cracks/ply/cm. However, these high crack densities and the non-Fickian diffusion was only observed for a few specimens, and was not representative of the overall sample population. Microcrack densities of only 1 crack/ply/cm or less have been observed in specimens that have not shown second uptakes in moisture. Thus, a large number of microcracks must form before any significant increases in moisture absorption can be measured. However, even low crack densities allow for increases in diffusivities during desorption and re-absorption. This observation is consistent with other moisture studies that have shown increasing deviations from Fickian behavior in graphite/epoxy composites during moisture cycling [*8*].

Generally, conditioning of polymer composites in high humidity environments has a significant effect on the macroscopic properties of the composite system, particularly at high temperatures due to a reduction in the glass transition temperature T_g [*5,6*]. For amorphous, uncrosslinked polymers such as Avimid® K3B, T_g represents a characteristic softening of the resin, reducing the ability of the matrix to transfer load to the fibers and thus reducing the elevated-temperature properties of the composite [*5,15*]. Our results indicated that T_g for unidirectional IM7/K3B specimens was reduced due to moisture saturation but recovered when the composite was redried. The magnitude of this reduction in T_g, however, was dependent on the fiber direction of the DMA specimen. In fact, the T_g was lowered only 2 to 3% for transverse specimens compared to a 6 to 11% reduction for longitudinal specimens. Also, the baseline T_g values differ by 10°C between the two specimen types. Further, Wedgewood [*11*] has reported a dry resin T_g of 237°C using dynamic scanning calorimetry, and a moisture-saturated T_g value for unidirectional composite of around 200°C based on a drop in flexure strength at elevated temperatures. These differences in reported T_g values might be due to the interaction between the mechanics of flexure loading and specimen anisotropy and can lead to discrepancies in the magnitude of the effect of moisture. Because the glass transition occurs over a range of temperature and can depend on test parameters such as heating rate, the choice of T_g could be another source of differences between reported values. Further characterization of the effects of moisture on T_g will include testing isolated polymer and unidirectional composite specimens using several different loading configurations.

Intralaminar fracture toughness G_{Ic} remained unchanged after four to five months of both room temperature and 40°C conditioning. Microscopic observation of conditioned unidirectional specimens revealed no microcracks or fiber-matrix debonding, which is consistent with no observed toughness loss. A moderate reduction in G_{Ic} was found for specimens conditioned for 300 h in 80°C 100% RH. The fracture surfaces of these specimens (observed using SEM) were characterized by a reduced amount of plastic flow of resin over the fibers, as compared to specimens that showed no toughness loss. Thus, some degradation of the fiber-matrix interphase region might have occurred, possibly resulting in the observed toughness loss. However, this degradation occurred after a limited amount of conditioning in an elevated-temperature water bath that may not be representative of a possible service environment. Other studies involving the conditioning of carbon- and glass-reinforced epoxies in environments with similar temperatures and humidity ranges as this study have observed anomalous behavior during elevated-temperature water immersion that was not observed in other environments [*2,16*]. Therefore, the toughness loss might not be an accelerated effect of the elevated-temperature, high-humidity conditioning.

Some moisture-induced degradation has been observed in the form of transverse microcracking in cross-ply and quasi-isotropic laminates. Microcracking in thin cross-ply specimens (6-ply) and quasi-isotropic specimens (8-ply) usually begins in the surface plies on the rough "bag" side of the specimen. A density of 2 to 3 cracks/cm develops in the bag-side plies before microcracking is seen in the surface plies on the smooth "tool" side. (Here, the side of the laminate laid down on the autoclave tool and the opposite side which faces up during autoclave processing are referred to as the tool side and bag side, respectively.) This effect could be due to resin bleeding from the bag side during processing, which could reduce the local toughness and cause locally high residual stresses. The thick cross-ply specimens (12-ply) do not show the same microcracking tendencies, indicating that the microcracking behavior may depend somewhat on thickness. In general, microcracks appear in the surface plies before any interior microcracking is observed. This behavior does not seem to be caused by moisture gradients, however, as microcracking occurs after saturation levels have been reached.

The formation of transverse microcracks in the absence of an externally applied load is puzzling, especially considering no toughness loss has been measured in panels conditioned for much longer periods of time than is required for microcrack initiation. Using the effective-flaw model developed by Wang [*13*] and assuming no swelling of the polyimide matrix as reported in Ref *4*, a fracture toughness of between 10 and 100 J/m^2 can be estimated for a thin cross-ply laminate that microcracks without applied load. This value is one to two orders of magnitude below the measured G_{Ic} values. This finding is consistent with Tsotsis and Weitsman [*17*], who reported a difference of three orders of magnitude between measured interlaminar G_{Ic} values and predicted G_{Ic} values for interfacial debonding and microcracking in moisture-saturated graphite/epoxy laminates. Other reasons for the formation of moisture-induced transverse microcracks include the constraint imposed by large fiber volume fractions on resin swelling [*18*], the ability of moisture to break chemical bonds and alter the surface-free energy at the fiber-matrix interface [*8*], and the attraction of moisture to the fiber-matrix interphase region due to stoichiometric or molecular weight gradient [*19,20*].

However, the latter two possibilities would result in a loss in fracture toughness during hygrothermal conditioning, which is not observed. The former explanation would result in local residual stresses that could be significant enough to cause microcrack initiation. Further study of possible effects is needed, including studies of moisture-induced swelling and of the effect of moisture on the fiber-matrix interface.

Conclusions

Hygrothermal conditioning of IM7/K3B thermoplastic composite laminates in several high-humidity environments has been performed. The moisture diffusion process was shown to be characterized by Fickian behavior, except in a few cases where moisture-induced degradation was significant. The moisture saturation level was shown to have a power law dependence on relative humidity but was not significantly affected by temperature. Also, a classical Arrhenius dependence of diffusivity on inverse temperature was observed. Moisture absorption induced the development of microcracks in the absence of an applied stress in quasi-isotropic and cross-ply laminates. Generally, microcracks were observed only in the surface plies and the microcrack densities were small. Observations from limited moisture cycling studies indicated that even small microcrack densities might be important during desorption and reabsorption of moisture. In a few cases involving cross-ply specimens, microcrack densities reached significant levels and caused non-Fickian, second-stage moisture absorption. However, these few cases may have been affected by other factors, such as resin bleed during processing. Further, these cases occurred in elevated-temperature (40 and 80°C) 97% and 100% relative humidity environments that might introduce degradation mechanisms that are not active in more realistic (lower humidity) environments.

Intralaminar fracture toughness values remained unchanged after 3000 to 3600 h of conditioning, and only a small reduction in glass transition temperature was observed in moisture-saturated specimens, which was recovered upon redrying the specimens. Also, moisture-induced degradation has not been observed in unidirectional laminates. Therefore, the macroscopic properties and behavior of unidirectional laminates seem relatively unaffected by steady-state hygrothermal conditioning.

Future Work

Continuation of this research includes further conditioning of composite laminates in various environments and further characterization of T_g, G_{Ic}, and microcrack resistance as a function of conditioning time, temperature, and humidity level. Also, the effects of combined moisture and temperature cycling need to be addressed, as well as the effect of applied stress on the moisture diffusion process. Microscopic studies, including fragmentation testing of hygrothermally conditioned single fiber test specimens, will also be performed. Finally, efforts will be made to develop test methodology that can evaluate the long-term durability of polymeric composite systems.

References

[1] Shen, C.-H. and Springer, G. S., "Moisture Absorption and Desorption of Composite Materials," *Journal of Composite Materials*, Vol. 10, 1976, pp. 2-20.

[2] Shirrell, C. D., "Diffusion of Water Vapor in Graphite/Epoxy Composites," *Advanced Composite Materials — Environmental Effects, ASTM STP 658*, American Society for Testing and Materials, West Conshohocken, PA, 1978, pp. 21–42.

[3] Wang, Q. and Springer, G. S., "Moisture Absorption and Fracture Toughness of PEEK Polymer and Graphite Fiber Reinforced PEEK," *Journal of Composite Materials*, Vol. 23, 1989, pp. 434–447.
[4] Cornelia, D., "Hygrothermal Performance of Polyimides," *Proceedings of the 39th International SAMPE Symposium*, 1994, pp. 917–929.
[5] Browning, C. E., Husman, G. E., and Whitney, J. M., "Moisture Effects in Epoxy Matrix Composites," *Composite Materials: Testing and Design, ASTM STP 617*, American Society for Testing and Materials, West Conshohocken, PA, 1977, pp. 481–496.
[6] Springer, G. S., "Environmental Effects on Epoxy Matrix Composites," *Composite Materials: Testing and Design, ASTM STP 674*, American Society for Testing and Materials, West Conshohocken, PA, 1979, pp. 291–312.
[7] Whitney, J. M. and Browning, C. E., "Some Anomalies Associated with Moisture Diffusion in Epoxy Matrix Composites," *Advanced Composite Materials — Environmental Effects, ASTM STP 658*, American Society for Testing and Materials, West Conshohocken, PA, 1978, pp. 43–61.
[8] Jackson, S. P. and Weitsman, Y., "Moisture Effects and Moisture Induced Damage in Composites," *Proceedings of the Fifth International Conference on Composite Materials*, 1985, pp. 1435–1452.
[9] Pipes, R. B., Vinson J. R., and Chou, T.-W., "On The Hygrothermal Response of Laminated Composite Systems," *Journal of Composite Materials*, Vol. 10, 1976, pp. 129–148.
[10] Tsotsis, T. K. and Weitsman, Y., "Energy Release Rates for Cracks Caused By Moisture Absorption in Graphite/Epoxy Composites," *Journal of Composite Materials*, Vol. 24, 1990, pp. 483–496.
[11] Wedgewood, A. R., "Melt Processible Polyimides for High Performance Composite Applications," *Proceedings of the 24th International SAMPE Technical Conference*, 1992, pp. 385–398.
[12] Wedgewood, A. R., "Autoclave Processing of Condensation Polyimide Composites Based on Prepregs of Avimid® K," *Proceedings of the 19th International SAMPE Technical Conference*, 1987, pp. 420–434.
[13] Huang, X. and Gillespie, J. W., Jr., "Fracture Test and Prediction of High Speed Civil Transport Composite Materials in Service Temperatures," *Proceedings of the Eighth Technical Conference of the American Society for Composites*, 1993, pp. 573–582.
[14] Coxon, B., Walker, T., Ilcewicz, L., and Seferis, J. C., "Intralaminar Fracture Toughness Characterization for Matrix Cracking in Composites," *Proceedings of the Society for Experimental Mechanics Spring Conference*, 1987, pp. 144–150.
[15] Kunz, S. C., "Thermomechanical Characterization of Graphite/Polyimide Composites," *Composites for Extreme Environments, ASTM STP 768*, American Society for Testing and Materials, West Conshohocken, PA, 1982, pp. 33–53.
[16] Dewimille, B. and Bunsell, A. R., "Accelerated Ageing of a Glass Fibre-Reinforced Epoxy Resin in Water," *Composites*, Vol. 14, 1983, pp. 35–40.
[17] Tsotsis, T. K. and Weitsman, Y., "Energy Release Rates for Cracks Caused By Moisture Absorption in Graphite/Epoxy Composites," *Journal of Composite Materials*, Vol. 24, 1990, pp. 483–496.
[18] Farrar, N. R. and Ashbee, K. H. G., "Self-Stress-Enhanced Water Migration in Composites," *Resins for Aerospace* (ACS Symposium Series), Vol. 132, 1980, pp. 435–447.
[19] Palmese, G. R. and McCullough, R. L., "Kinetic and Thermodynamic Considerations Regarding Interphase Formation in Thermosetting Composite Systems," *Journal of Adhesion*, Vol. 44, 1994, pp. 29–49.
[20] Pangelinan, A. B., McCullough, R. L., and Kelley, M. J., "Analytic Model for Surface Induced Molecular Weight Segregation in Thermoplastic Composites," *Journal of Polymer Science, Part B: Polymer Physics*, Vol. 32, 1994, pp. 2383–2394.

Barbara A. Pratt[1] *and Walter L. Bradley*[1]

Effect of Moisture on the Interfacial Shear Strength: A Study Using the Single Fiber Fragmentation Test

REFERENCE: Pratt, B. A. and Bradley, W. L., **"Effect of Moisture on the Interfacial Shear Strength: A Study Using the Single Fiber Fragmentation Test,"** *High Temperature and Environmental Effects on Polymeric Composites: 2nd Volume, ASTM STP 1302,* Thomas S. Gates and Abdul-Hamid Zureick, Eds., American Society for Testing and Materials, 1997, pp. 64–72.

ABSTRACT: The effect of absorbed moisture on the interfacial strength of a carbon fiber/epoxy matrix composite has been studied using the single fiber fragmentation test. The absorption of 1.40% moisture produced a reduction less than 20% in the interfacial shear strength. This reduction in interfacial strength may be associated with a reduction in the residual compressive stresses at the fiber/matrix interface associated with the cooldown from the curing temperature, a weakening of the chemical adhesion, or a softening of the matrix giving failure by interphase yielding. These results suggest that long-term durability of carbon fiber/epoxy matrix composites in seawater may not be much affected by absorbed seawater.

KEY WORDS: polymeric composites, seawater, interfacial strength, single-fiber fragmentation test

Geologists claim that there is more oil at 2000 m of ocean depth off the continental shelf of the United States than in all of the middle east. However, drilling and producing oil at such ocean depths cannot be done economically using current technology. A free-floating production platform or a tension leg production platform would have to have sufficient buoyancy to support 45 000 000 kG of steel tubulars, or risers, through which the oil would be produced. The replacement of these steel risers with composite materials risers would dramatically reduce the weight to be supported by the platform. Production platforms that have been constructed for use at 500 to 1000 m could be used in these new deep-water applications, since the specific gravity of a carbon fiber/epoxy matrix composite is less than 1.8, compared to steel at 7.9. The economic feasibility of such a substitution has already been demonstrated by Tamarelle and Sparks [*1*]. The possible use of polymeric composite materials for off-shore structures has been recently reviewed by Gibson [*2*].

Polymeric matrix composites are known to have excellent fatigue resistance, do not corrode, and can be tailored to give optimal combinations of stiffness, strength, and thermal expansion by proper ply sequencing. The two technical obstacles to the use of polymeric composite materials in drilling and production risers are joining and long-term durability. While polymeric composite materials do not corrode like metals, they do absorb moisture,

[1] Research assistant and professor of mechanical engineering, respectively, Texas A&M University, Department of Mechanical Engineering, College Station, TX 77843-3123.

generally 1 to 3% of the weight of the matrix. This absorbed moisture can potentially soften the matrix, reduce the strength of the glass fibers (carbon fibers appear to be unaffected by moisture), and/or degrade the strength of the fiber/matrix interface. Since production risers will be immersed continuously for 10 to 20 years, they will certainly reach saturation levels of moisture absorption eventually. Thus, the effect of absorbed moisture needs to be determined.

The successful use of polymeric composites for long-term service in contact with water, including pressurized water distribution systems, frigates' superstructures, mine sweeping ships, and submarines for sonar domes and nose cones is well known [*3,4*]. However, these successful long-term applications of polymeric composites in direct and continuous contact with moisture share in common design stresses that are 20% or less of the tensile strength of the polymeric composite. Whether satisfactory long-term service can be obtained at more substantial fractions of the tensile strength remains to be demonstrated. Some studies on graphite/epoxy composites have indicated minimal reductions in transverse tensile strength, indicating little adverse effect of moisture on either the matrix or the interface [*5,6*], while other studies on similar systems have shown more dramatic reductions in transverse tensile strength [*7,8*]. Both the fiber strength and the interfacial strength in glass/epoxy composites have been found to be degraded when soaked for extended times at higher temperatures, while more modest reductions in fiber and interfacial strength have been observed at ambient temperature [*9,10*].

Most of the studies done to date have used macroscopic mechanical properties such as transverse tensile strength or compression strength to evaluate the effect of moisture on the composite material. The purpose of this study has been to more directly study the effect of moisture on the fiber/matrix interfacial strength using the single fiber fragmentation test developed by Drzal [*11–13*].

Experimental and Analytical Procedures

A system to do single fiber fragmentation testing has been constructed following the general approach developed by Drzal [*11*], but with some modifications. The system that has been constructed for this work includes the following features: (1) continuous displacement applied through an electric motor and a gear box, allowing displacement-rates from 2.83×10^{-6} to 0.0226×10^{-6} m/s; (2) direct measurement of load, displacement, and strain as a function of time; and (3) real-time determination of the number of breaks in the fiber using acoustic emission sensors. The load measurements were made using a small load-cell placed in the loading line that had a capacity of 450 N and a resolution of 1 N. Displacement measurements were made using an LVDT while strain measurements were made using a miniature clip gage. The simultaneous measurement of displacement and strain allowed the effective gage length to be determined. Subsequently, displacement measurements alone were used in some tests.

The reliability of the acoustic emission sensor to accurately determine the number of fiber breaks was established initially using resins that were transparent. However, the system is now routinely used for opaque specimens as well. Direct observations were also used to confirm that fiber breaks seems to be confined to the uniform gage section with no breaks found in the transition zone between the reduced cross-section and the larger ends.

The epoxy used in this study is a proprietary system based on diglycidyl ether bisphenal-A (DGEBA) with a cure temperature of 422 K. The carbon fibers used were Hercules AS4.

Test specimens with the geometry seen in Fig. 1 were prepared by carefully placing a graphite fiber along the length of a silicon mold with spues cut in each end to allow the graphite fiber to be positioned at mid-thickness in the mold. The silicon mold was in turn

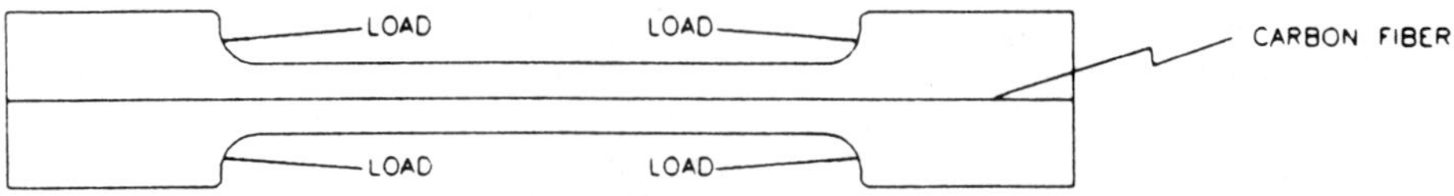

FIG. 1—*Single fiber fragmentation test specimen geometry.*

placed in a shallow aluminum holder approximately at the same height of the carbon fiber as it left the sprue in the silicon mold. The carbon fibers were gently stretched to remove all sag and then lightly taped to the aluminum shoulder prior to casting the epoxy. The epoxy and cross-linking agent were heated to 339 K until both were liquid, mixed, and vacuum degassed before casting in the silicon molds with single graphite fibers in place. After casting the epoxy was cured at 366 K for 2 h and 422 K for 4 h.

The cast specimens were subsequently divided into three groups for conditioning prior to testing. One group was dried at 338 K in a vacuum furnace for 24 h to remove all moisture that was present at the time of fabrication. Periodic weighing was used to determine when the residual moisture had been removed. A second group of cast test specimens were immersed at ambient temperature in simulated seawater prepared using *Instant Ocean* mix from a pet store. Periodic weighing was again used to monitor the seawater absorption, and this group of specimens were soaked until saturation was approached, which took 110 days. The third group of specimens were tested without any conditioning. The moisture contents of the three groups of specimens prepared for testing were 0, 0.65, and 1.40%, respectively.

The stress-strain behavior of the cured epoxy with one graphite fiber as calculated from the measured load-time and strain time data is seen in Fig. 2. Typical experimental results for number of breaks versus strain are presented in Fig. 3.

The calculation of the interfacial strength from the measured number of breaks (cumulative) versus applied mechanical strain was made using the Kelly-Tyson approach [*14*].

$$\tau = \frac{1}{2}\frac{d}{l_c}\sigma_f \tag{1}$$

where d is the fiber diameter and l_c is the critical fiber length and is calculated to be $4l/3$,

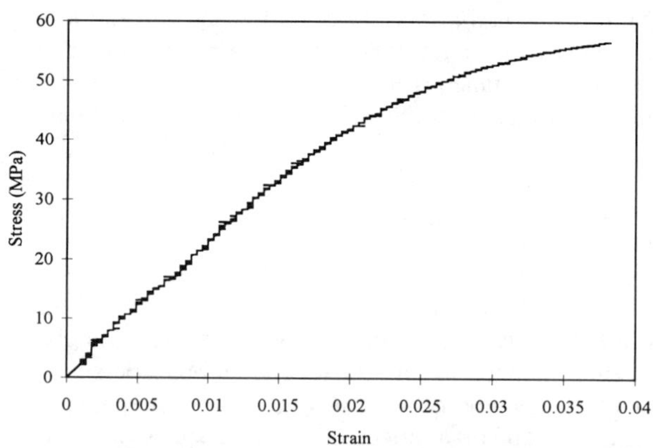

FIG. 2—*Typical stress versus strain behavior observed during the SFFT.*

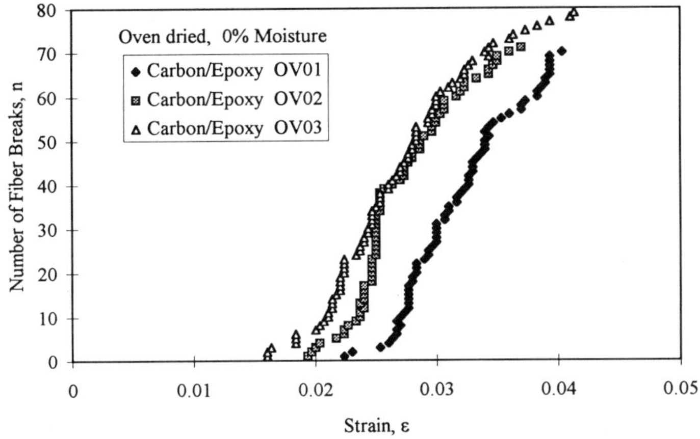

FIG. 3—*Number of fiber breaks as a function of strain for carbon/epoxy at a 0% moisture condition.*

with l being the mean fiber fragmentation length as saturation is reached in the test [*15*]. The mean fiber fragmentation length l is calculated as the specimen gage length/$\{N + 1\}$ where N is the number of fiber breaks. In these tests, saturation was not always reached (that is, the number of breaks versus strain curves did not reach a plateau before the test specimen failed) but for each of the tests it was possible to estimate by extrapolation the saturation value with minimum ambiguity.

The fiber strength σ_f was calculated using the approach of Wagner et al. [*16*]. They assume that the fiber strength distribution as a function of fiber length can be inferred from the average fragment length as a function of fiber stress using Weibull statistics. If one assumes that the fragmentation test behavior of the fiber is similar to multiple tensile tests with independent samples, which is at least approximately true far below saturation, and also assumes a Weibull weakest link model, the relationship between the mean fragmentation length and the mean fiber tensile strength is given by

$$\bar{\sigma}_f = \alpha\left(\frac{l_0}{l_c}\right)^{(1/\beta)} \Gamma\left(1 + \frac{1}{\beta}\right) \tag{2}$$

where α and β are Weibull scale and shape parameters, respectively, Γ is the Gamma function, l_0 is the original fiber length, and l_c is the critical fiber length. In this calculation, Γ $(1 + 1/\beta)$ is approximately one. A log-log plot of σ_f versus l_0/l_c can be used to determine α and β, using only the linear portion of the plot. The deviation from linearity occurs as the saturation limit is approached and end effects on the shorter fibers violate the assumptions implicit in Eq 2.

To calculate the fiber strength, the number of breaks versus strain data as illustrated by Fig. 3 must be first converted into fiber stress versus number of breaks. The conversion can be made by assuming that the fiber stress is given by the fiber modulus E_f times the fiber strain, which is assumed to be equal to the matrix strain, or

$$\sigma_f = E_f\,(\epsilon_m + \epsilon_t + \epsilon_s) \tag{3}$$

where ϵ_m, ϵ_t, and ϵ_s are the mechanical strains, the thermal cool-down strains, and the swelling strains due to moisture absorption, respectively. This result is an approximation of a more complex result obtained by Cox [1] and should be applicable when the fragment lengths are long compared to the critical fragment length, which they will be early in the test before saturation is reached. Treating the total strain as the sum of the mechanical strain, the thermal cool-down strain and the moisture-induced swelling strain is based on measurements by Huang and Young [18]. They found the mechanical and thermal strains to be linearly additive, with the thermal strain in the fiber simply calculated as the difference in the thermal coefficients of expansion for the fiber and the matrix times the temperature drop from the cure temperature to the test temperature.

The number of breaks as a function of mechanical strain from Fig. 3 is converted into average fragment length as a function of fiber stress, calculated using Eq 3. The average fragment length l is used to determine the critical fiber length l_c. The normalized length is plotted as a function of fiber stress as shown in the log-log plot of Fig. 4. The linear portion of these curves in Fig. 4 are used to determine the Weibull scale and shape parameters. The fiber strength at the mean fragment length at saturation is then calculated from Eq 2. It is in turn used in Eq 1 to calculate the interfacial shear strength.

Results and Discussion

The results for the specimens tested with three different moisture levels (0.0, 0.65, and 1.40%) are presented in Figs. 3, 5, and 6. These results have then been converted to normalized fragment length versus fiber stress using Eq 3 and plotted on a log-log plot, as seen in Figs. 4, 7, and 8. The failure in the dry specimens is seen to start at a higher fiber strain than the wet specimens, which presumably have a less negative value of axial thermal strain. The Weibull values, calculated fiber tensile strength at saturation and the mean fragmentation length to diameter ratio are summarized in Table 1 for each of the three moisture contents investigated in this study. The values of interfacial strength calculated from these results using Eq 1 are presented in Fig. 9. It is clearly seen that the effect of absorbed moisture on the interfacial strength in this carbon fiber/epoxy matrix system is less than 20%.

These results are consistent with earlier results in which the transverse tensile strength has been used to monitor degradation in interfacial strength, carbon fiber/thermal setting resins

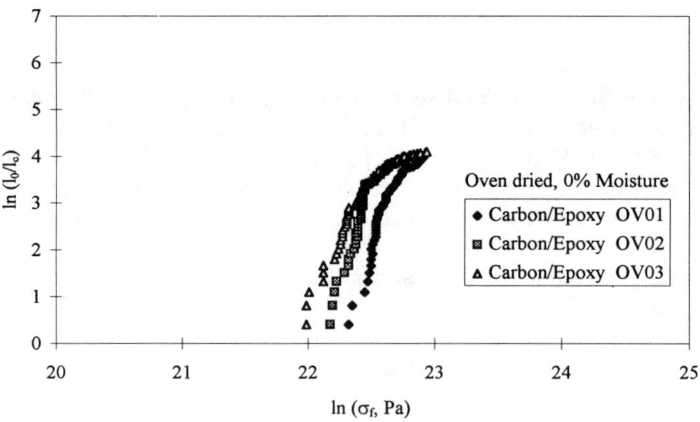

FIG. 4—*Natural log of the normalized length as a function of the natural log of the fiber stress for 0% moisture carbon/epoxy specimens.*

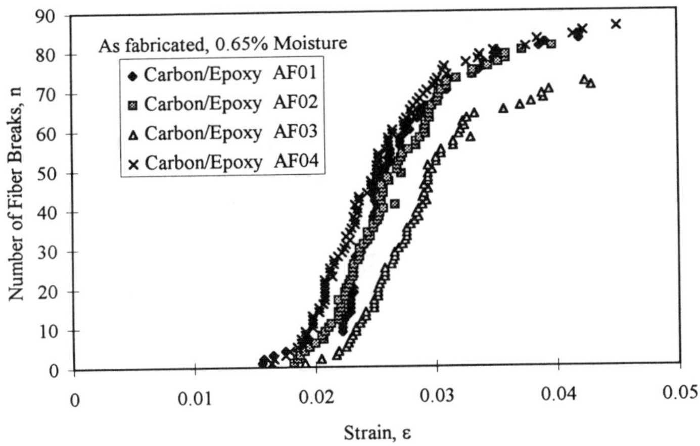

FIG. 5—*Number of fiber breaks as a function of strain for carbon/epoxy at a 0.65% moisture condition.*

with absorbed moisture showing a degradation of at most 17%. These results were confirmed using microbonding tests, which gave similar results [*5,6*].

One would expect that moisture effects could have an adverse effect on interfacial strength in two different ways. First, the coefficient of thermal expression (CTE) in epoxy resins is found to be much greater than in graphite fibers. This difference in CTE in epoxy compared to carbon fibers results in a residual compressive stress at the fiber matrix interface after cooldown. Interfacial strength is a combination of the effects of chemical adhesion and mechanical interlocking, with the latter being aided by residual compressive stresses at the interface. Absorbed moisture with its associated swelling will reduce the magnitude of these residual stresses, potentially reducing the interfacial strength as well. In our epoxy, the cure temperature is 422 K, and the thermal coefficient of expansion is 1.802×10^{-5}/C for the

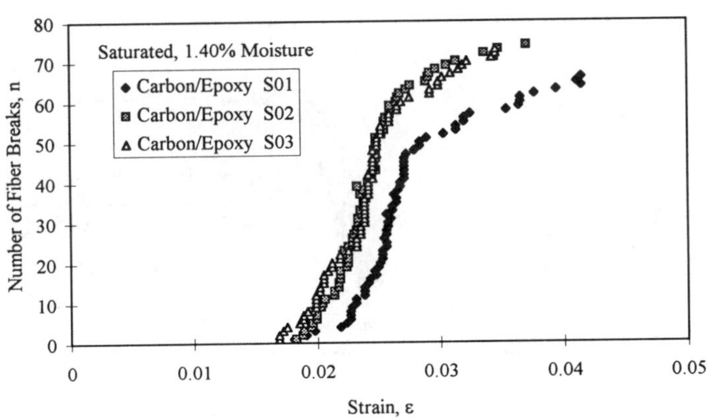

FIG. 6—*Number of fiber breaks as a function of strain for carbon/epoxy at a 1.40% moisture condition.*

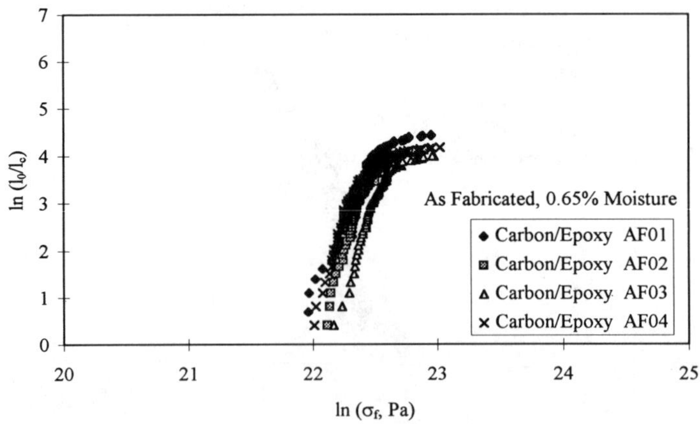

FIG. 7—*Natural log of the normalized length as a function of the natural log of the fiber stress for 0.65% moisture carbon/epoxy specimens.*

carbon fiber and 7.56×10^{-5}/C for the epoxy. One can simply calculate the radial residual stresses at the interface as a function of moisture content (and inter-fiber spacing) using classical elasticity [*19*]. The residual stresses at the fiber/matrix interfaces in the radial direction are found to be compressive with a magnitude of less than 20 MPa for the specimens with 1.40 wt% moisture with higher magnitudes of compressive stress for the two specimen conditions with less absorbed water and no water. Thus, the residual stresses at all moisture contents used in this study are compressive. It appears that either the mechanical interlocking is either a small fraction of the total interfacial shear strength or the mechanical

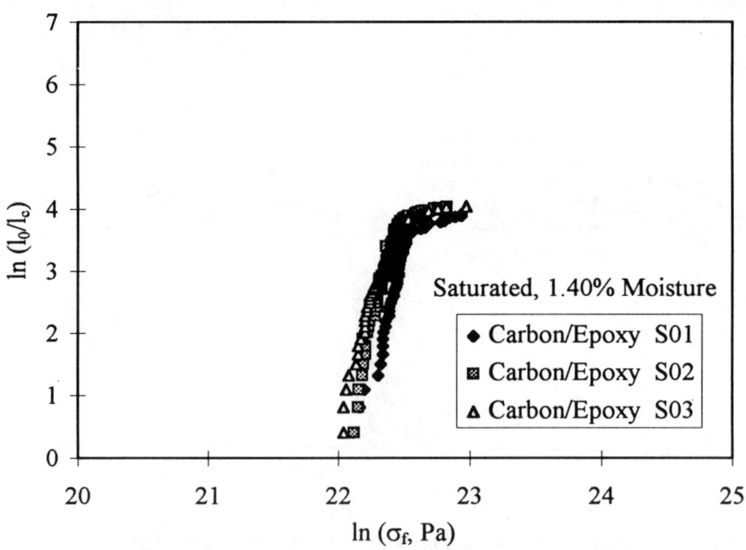

FIG. 8—*Natural log of the normalized length as a function of the natural log of the fiber stress for 1.40% moisture carbon/epoxy specimens.*

TABLE 1—*Calculated values from 0, 0.65, and 1.40% moisture test data.*

	Weibull Parameters			
Moisture, %	α	β	σ_f, MPa	$1/d$
0	3.59	7.14	6541	58.7
0.65	3.29	6.86	6039	53.8
1.40	3.55	8.45	5763	63.1

interlocking is not a sensitive function of the residual thermal stresses in the radial direction until the compressive stresses are reduced to a level that begins to reduce the effectiveness of the mechanical interlocking on interfacial strength. In our tests, this only occurred at 1.40 wt % moisture. The much greater reduction in interfacial strength noted by some other investigators may be the result of either a lower curing temperature and/or more moisture absorption giving a net tensile residual thermal stress in the radial direction at the fiber/matrix interface.

Finally, the chemical adhesion of the matrix to the fiber may not be much affected by the absorbed moisture, as the changes in residual stresses might be sufficient to account for the observed small degradation in the interfacial strength. However, there is no model at present to quantitatively relate residual stresses in the radial direction to the interfacial shear strength through its effect on mechanical interlocking. Furthermore, if failure of the interface is due to interphase yielding, the compressive residual stress actually lowers the interphase shear stress required to give yielding using a von Mises yield criteria.

Conclusion

The effect of absorbed moisture on the interfacial strength of a carbon fiber/epoxy matrix composite has been studied using the single fiber fragmentation test. The absorption of 1.40% moisture produced a reduction of less than 20% in the interfacial shear strength. This reduction in interfacial strength may be associated with a reduction in the residual compressive stresses at the fiber/matrix interface associated with the cool-down from the curing temper-

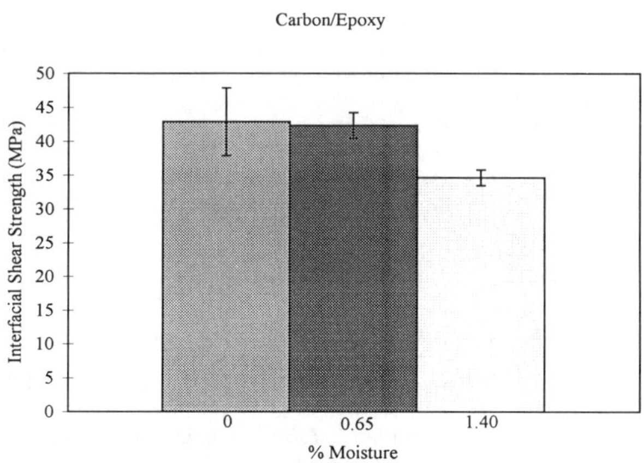

FIG. 9—*Interfacial shear strength as a function of moisture content for carbon/epoxy.*

ature. If, however, failure is due to yielding of the interphase, then the effect of moisture in softening the resin is more significant than the reduction in driving force for yielding that would accompany relaxation of the residual stresses.

Acknowledgments

The financial support of the National Science Foundation through the Offshore Technology Research Center at Texas A&M University is gratefully acknowledged. The assistance of Larry Drzal in the design of parts of the experimental apparatus and for several helpful discussions regarding the analysis of the results is also acknowledged.

References

[*1*] Tamarelle, P. J. C. and Sparks, C. P., *Proceedings of Nineteenth Annual Offshore Technology Conference*, Houston, TX, May 1987 (Richardson, TX: Offshore Technology Conference), p. 255.
[*2*] Gibson, A. G., "Composite Structures in Offshore Applications," *Composite Materials in Marine Structures*, Chpt. 3, 1993, R. A. Shenoi, Ed., Cambridge Ocean Technology Series.
[*3*] Le Lan, J. Y., Parneix, P., and Gueguen, P. L., "Composite Material Superstructures," *Nautical Construction with Composite Materials*, P. Davies and L. Lemoine, Eds., IFREMER, France, 1992, pp. 399–411.
[*4*] Lemiere, Y., "The Evolution of Composite Materials in Submarine Construction," *Nautical Construction with Composite Materials*, P. Davies and L. Lemoine, Eds., IFREMER, France, 1992, pp. 441–449.
[*5*] Bradley, W. L. and Grant, T. S., "The Effect of the Moisture Absorption on the Interfacial Strength of Polymeric Matrix Composites," *Journal of Materials Science*, Vol. 30, 1995, pp. 5537–5542.
[*6*] Grant, T. S. and Bradley, W. L., "In-Situ Observations in SEM of Degradation of Graphite/Epoxy Composite Materials Due to Seawater Immersion," *Journal of Composite Materials*, 1995, Vol. 29, No. 7, pp. 852–867.
[*7*] Pomies, F. and Carlsson, L. A., *Journal of Composite Materials*, 1994, p. 28.
[*8*] Lee, M. E. and Peppas, N. A., "Water Transport in Graphite/Epoxy Composites," *Journal of Applied Polymer Science*, 1993, Vol. 47, pp. 1349–1459.
[*9*] Schutte, C., McDonough, W., Shioya, M., McAuliffe, M., and Greenwood, M., "The Use of a Single-Fiber Fragmentation Test to Study Environmental Durability of Interfaces/Interphases Between DGEBA/mPDA Epoxy and a Glass Fiber. The Effect of Moisture," 1975, in press in *Composites*.
[*10*] Grami, M. E., Moilanen, M. J., and Rosenow, M. W. K., "Environmental Degradation of Glass Fibers in Polyurethane Matrix Composites," 1995, *Proceedings of ICCM-10*, Whistler, B. C., Canada, pp. 207–214.
[*11*] Drzal, L. T., Rich, M. J., and Lloyd, P. E., "Adhesion of Graphite Fibers to Epoxy Matrices: I. The Role of Fiber Surface Treatment," *Journal of Adhesion*, Vol. 16, 1982, pp. 1–30.
[*12*] Ho, H. and Drzal, L. T., "Comparison of Single Fiber Fragmentation and Microindentation Tests for Composite Interfacial Shear Strength Measurement," 1995, *Proceedings of ICCM-10*, Whistler, B. C., Canada, pp. 203–210.
[*13*] Herrera-Franco, P. J. and Drzal, P. J., "Comparison of Methods for the Measurement of Fiber/Matrix Adhesion in Composites," *Composites*, 1992, Vol. 23, No. 1, pp. 1–27.
[*14*] Kelly, A. and Tyson, W. R., *Journal Mech. Phys. Solids*, Vol. 13, 1965, p. 329.
[*15*] Ohsawa, T., Nakayama, A., Miwa, M., and Hasegawa, A. J., *Applied Polymer Science*, Vol. 22, 1978, p. 3203.
[*16*] Yavin, B., Gallis, H. E., Scherf, J., Eitan, A., and Wagner, H. D., "Continuous Monitoring of the Fragmentation Phenomenon in Single Fiber Composite Materials," *Polymer Composites*, Vol. 12, No. 6, 1991, pp. 436–446.
[*17*] Cox, H. L., *British Journal of Applied Physics*, Vol. 3, 1952, p. 72.
[*18*] Huang, Y. and Young, R. J., "Interfacial Behaviour in High Temperature Cured Carbon Fiber/Epoxy Resin Model Composite," *Composites*, Vol. 26, No. 8, 1995, pp. 541–550.
[*19*] Wood, C. and Bradley, W., "A New Technique to Study the Interfacial Strength and Transverse Cracking Scenario in Composite Materials," *Fiber, Matrix and Interfacial Properties*, American Society for Testing and Materials, West Conshohocken, PA, 1994.

D. Choqueuse,[1] *P. Davies,*[1] *F. Mazéas,*[1] *and R. Baizeau*[1]

Aging of Composites in Water: Comparison of Five Materials in Terms of Absorption Kinetics and Evolution of Mechanical Properties

REFERENCE: Choqueuse, D., Davies, P., Mazéas, F., and Baizeau, R., **"Aging of Composites in Water: Comparison of Five Materials in Terms of Absorption Kinetics and Evolution of Mechanical Properties,"** *High Temperature and Environmental Effects on Polymeric Composites: 2nd Volume, ASTM STP 1302*, Thomas S. Gates and Abdul-Hamid Zureick, Eds., American Society for Testing and Materials, 1997, pp. 73–96.

ABSTRACT: This paper presents results for wet aging of five glass fiber-reinforced polymer matrix composites. Specimens were immersed in distilled water for 2 years at four temperatures (5, 20, 40, and 60°C), at both atmospheric pressure and 10 MPa hydrostatic pressure. Absorption kinetics and the evolution of properties were followed through periodic T_g measurements, modal analysis, and destructive mechanical testing. There is no direct correlation between the quantity of water absorbed and the loss of mechanical properties; this emphasizes that absorption curves are of limited use for material selection. A hydrostatic pressure of 10 MPa accelerates moisture absorption in the epoxy composites. The establishment of a large data base of results has enabled accelerated test methods to be critically evaluated and the role of temperature and pressure as accelerating parameters is discussed.

KEY WORDS: composite material, aging, hydrostatic pressure, absorption, degradation, water, accelerated test

Organic matrix composites are being increasingly used in naval (including underwater) applications, where high specific properties and good corrosion resistance are essential. As a consequence the influence of the marine environment, characterized by the presence of water and water pressure, on composite behavior is a key parameter for the success of such materials. However, long-term immersion data are still lacking, and the acceleration procedures necessary to obtain such data are not well established.

The aim of this paper is to provide extensive data concerning five composite materials that have been studied within the European MAST (Marine Science and Technology) program. The materials investigated include four thermoset resins (polyester, vinyl ester, epoxy and prepreg epoxy) reinforced with E-glass fibers, and a carbon fiber/PEEK composite. Panels of these materials have been immersed for two years in distilled water at various temperatures and pressure levels (up to 60°C and 10 MPa).

Fiber-reinforced plastics (FRP) show good chemical stability in the presence of seawater and the use of these materials in marine applications (surface ships, offshore and in sub-

[1] IFREMER, Marine Materials Laboratory, Centre de Brest, BP 70, 29280 Plouzané, France.

marines) is increasing. These applications will now be briefly described, and previous work on absorption, degradation, and accelerated testing in water will then be summarized.

Marine Applications

FRP has been used in the manufacture of pleasure boats for over 40 years. Military usage of composites took off with the minesweeper programs in the United Kingdom and France in the 1960's [*1*] and significant tonnages of FRP are now used for frigate superstructures [*2*]. The off-shore industry is also currently examining a large number of applications and these have been described recently [*3*]. A third application of composite materials is in underwater structures. Experience exists with torpedoes and sonar domes and considerable amounts of secondary structures on military submarines are presently produced from composites [*4*]. Large military programs are currently studying the possibilities of using composites for primary hull structures. In parallel with this military effort, there is much potential for civil underwater use of composites, in applications ranging from deep sea containers for electronic apparatus [*5*] and protective outer casings of submersibles (for example, the Nautile, Cyana) to the hulls of underwater vehicles. The latter application has been extensively studied recently in programs such as the AUSS (Advanced Unmanned Search System) vehicle in the United States [*6*] and the AUV (Autonomous Underwater Vehicle) program in Europe [*7,8*]. Much work has also been performed in Russia. One of the outstanding questions for these applications is whether it is feasible to produce large, thick structures in carbon fiber-reinforced composites. It is not yet clear whether carbon fibers offer significant advantages over glass fibers when large-scale parts must be fabricated so work is continuing on both types of reinforcement. Another materials aspect to be resolved is the suitability of thermoplastic matrix resins to these applications and this avenue is also currently being explored.

It is clear from this brief overview that the marine use of composites is a very active area of study at present and underwater applications are attracting particular research interest. However, in spite of their good chemical stability composite materials immersed in water are subject to absorption phenomena that are controlled by the laws of diffusion. A thorough understanding of water-composite interactions is necessary if reliable long-term performance is to be assured and these will be discussed below.

Absorption Phenomena in Composites

Absorption phenomena have been studied in detail by many authors [*9-14*], and several models have been proposed. Weitsman has reviewed this area recently and concluded that the number of papers related to composite-fluid interactions probably runs into the thousands [*12*], so an exhaustive review is not possible here. Absorption is generally modeled by Fick's laws of diffusion, where only one phase is considered using the notion of free water, but some authors also propose the use of a Langmuir model where two phases are considered, free water and fixed water [*14*]. For these models, the degradation or change of the material due to the presence of water inside the material is not taken into account. In general close fits to experimental data can be obtained for neat resins using Fickian diffusion, but modeling is more difficult for the composite material. In the composite the role of the fiber-matrix interface is critical and no model considers this satisfactorily.

The effect of hydrostatic pressure on diffusion has been studied by several authors [*15-18*]. In the resin two competing mechanisms are invoked. First, the pressure can cause a reduction of free volume that will reduce the space available for moisture; secondly, the imposed pressure can drive more water into the material and hence enhance the diffusion and increase moisture levels. Tucker et al. found no effect of pressures of 10 MPa on absorption in polyester and vinyl ester resins [*15*]. Published results for absorption in composites generally show small effects for pressures up to 70 MPa. Avena and Bunsell noted a small decrease in absorption with increase in pressure up to 40 MPa in glass-reinforced epoxy [*16*], while Pollard et al. [*17*] and Grant [*18*] found virtually no influence.

Degradation in Water

With time, due to the presence of water inside the composite, the physico-chemical and mechanical properties can be affected by attack of resin and degradation of the interface between matrix and fibers. The effects of moisture on properties have been studied by several authors [*19-25*]. However, few long-term data are available on this degradation, and the existing studies have mainly concentrated on aerospace composites. These differ significantly from the materials considered for most marine applications both by their fabrication techniques and their constituents, so the available data are not directly transferable. Nevertheless, some data on long-term behavior of marine composites are available, including results obtained by the French navy for aging in sea air over 9-, 13-, 17-, and 21-year periods [*22*].

Accelerated Testing

In order to predict long-term behavior, "accelerated aging" of the material is frequently performed [*26-28*]. The accelerating factor can be the temperature or the applied stress. In terms of water diffusion and chemical degradation, the accelerating factor generally retained is the temperature. The degradation processes are assumed to be thermally activated following an Arrhenius law as

$$s = A \exp(-B/T)$$

with s = rate of the phenomena, A and B are constants, and T is the absolute temperature.

Acceleration of tests is justifiable if the accelerating parameter does not cause an artificial change in the aging mechanism, but in order to verify this assumption, tests on a reasonably long-time scale are necessary. The data in Ref 22 indicated that 1000 h of accelerated aging at 70°C caused a similar loss in flexual properties to 15 years in natural conditions. This is one of very few examples for which the effect of accelerated aging has been checked.

Materials

Five composite materials have been studied, and their details are shown in Table 1. The first three materials are representative of commonly-used marine materials in terms of resin and manufacturing route, the fourth is prepreg based with the same fibers, and the fifth material is a carbon fiber-reinforced thermoplastic resin composite. The latter has no current application in this field but it shows good potential particularly for submarine application.

For Materials 1 to 4, the reinforcement is E Glass woven with 90% of fibers in the 0° direction (longitudinal or L direction) and 10% in the 90° direction (transverse or T direction). These materials were manufactured in Brest. Material 5 is PEEK (polyetheretherketone) reinforced with carbon fibers and was produced as 0°/90° panels from unidirectional prepreg by Westland Helicopters, England. Sample thickness is nominally 3 mm, but the epoxy prepreg composite is rather thinner (2.3 mm). The materials obtained by contact molding were postcured at 60°C for 24 h in order to stabilize them before aging.

TABLE 1—*Materials tested.* (L *and* T *refer to longitudinal and transverse directions, respectively.*)

Material	1	2	3	4	5
Nature of resins and source	isophthalic polyester CDF Nosodyne G703	Vinyl ester Dow Derakane 411-45	Epoxy Sicomin SR 1500/SD25 05	Epoxy ES 67	PEEK ICI
Nature of fibers	E glass Hexcel 1543 TF	E glass Hexcel 1543 TF	E glass Hexcel 1543 TF	E glass Hexcel 1543 TF	Carbon XAS prepreg 0°/90°
Fiber content, by weight	59%	61%	52%	60%, nominal	50%, nominal
Thickness, mm	2.9	2.8	3.3	2.3	3.3
L modulus, GPa	29.2	27.0	20.4	31.9	54.3
T modulus, GPa	10.3	10.3	8.8	11.6	
L strength, MPa	652	766	671	908	818
T strength, MPa	114	134	132	166	
Fabrication technique	contact molding	contact molding	contact molding	oven vacuum, prepreg	autoclave, prepreg

Experimental Methods

Test Environment and Specimen Geometry

The samples were immersed in tanks at two pressures, atmospheric pressure and 10 MPa hydrostatic pressure, and at four different temperatures: 5, 20, 40, and 60°C. Special pressure vessels were developed, placed in the temperature tanks, allowing pressures up to 60 MPa to be reached at temperatures up to 60°C. Pressure vessels, made of stainless steel, are of cylindrical shape, 500-mm in length and of 100-mm internal diameter. The pressure is maintained by means of a *Nova Swiss* pump and controlled by *Bourdon* manometers to within ±1%. The temperature is controlled using Polystat regulators, with ±1°C accuracy. In the current study pressures up to 10 MPa were applied (corresponding to an immersion at about 1000-m depth). The size of the samples were 5 by 20 cm. For tests at ambient temperatures, an additional pressure vessel was used to allow experiments on large samples (20 by 30 cm) to be performed. The temperature tanks and pressure vessels were filled with distilled water. Natural sea water is available at the test site but distilled water was preferred as it simplifies experimental procedures and has been shown to be more aggressive than sea water. The temperature of 5°C was chosen since it is representative of the general deep sea water temperature, while sea surface temperature is closer to 20°C. The tests at 40 and 60°C are considered to be accelerated tests. For each material, small (5 by 20 cm) and large (20 by 30 cm) plate specimens were cut using a diamond cutting wheel. For Materials 1 to 4, two specimen orientations were examined, namely longitudinal *L* and transverse *T*, as these materials do not have the same percentage of fibers in the two directions: 90% of fibers in the 0° *L* direction against 10% in the 90° *T* direction. In the following, specimen orientation will be defined according to whether they are cut with their length parallel *L* or perpendicular *T* to the 90% fiber direction. The specimens tested are presented in Table 2.

Weight Gain

At regular intervals the samples were removed from the temperature tanks and pressure vessels, and the mass was determined at atmospheric pressure for each sample by weighing

TABLE 2—*Details of the specimens tested. Number of plates for water uptake and modal analysis, for each material. (In brackets, number of specimens cut from the plates for flexural and ILSS destructive tests, for each material.* L *and* T *refer to longitudinal and transverse directions, respectively. For carbon/PEEK, only one direction was investigated.)*

Pressure	0.1 MPa				10 MPa			
Temperature	5°C	20°C	40°C	60°C	5°C	20°C	40°C	60°C
Large plates	1	1	1	1		1		
Small plates								
L direction	2 (5)	2 (5)	2 (5)	2 (5)	2 (5)	2 (5)	2 (5)	2 (5)
T direction	1 (5)	1 (5)	1 (5)	1 (5)	1 (5)	1 (5)	1 (5)	1 (5)

on a Sartorius balance before replacing in water. After 24 months of immersion, the samples were dried in a vacuum oven at 60°C until obtaining a constant weight and then cut into specimens and tested. Cast epoxy resin samples (the contact molding composition) 5 by 5 cm by 4-mm thick were also immersed, at 20 and 50°C.

Physicochemical Properties

The evolution of the glass transition temperature T_g may be used as an indicator of changes induced by aging (evolution of the free volume, plasticization). In this study, two methods for measuring the T_g have been used. First, the ASTM standard involving DSC (differential scanning calorimetry) was employed, D 3418. The second technique used a commercial dynamic mechanical analyzer (called "Kinemat") on which the T_g is determined on the torsion stiffness versus temperature curve. DSC measurements require very small samples (typically 10 mg) while the torsion specimens are larger, 5- by 1-cm rectangular specimens are used.

Mechanical Properties

Both nondestructive and destructive tests have been used to estimate the degradation of the mechanical properties. The former is based on the modal analysis of the bending vibration of the sample. The sample is suspended freely and excited by tapping. The natural frequencies are measured by up to 5 accelerometers bonded to the plate and the model parameters are identified by using a specific method [*29*]. With the help of preliminary estimations of properties a modeshape is associated with the measured frequencies. According to the small deflections classical lamination theory, the natural frequencies, in Hertz, of a free-free rectangular orthotopic plate are given by [*30*]

$$f_{ij} = \frac{\pi}{2\sqrt{\gamma}}\sqrt{\frac{G_1^4 D_{11}}{a^4} + \frac{G_2^4 D_{22}}{b^4} + \frac{2H_1H_2D_k}{a^2b^2} + \frac{4D_{66}(J_1J_2 - H_1H_2)}{a^2b^2}}$$

with f_{ij} frequencies of modeshape i, j, γ the density, a, b the length and width of the beam D_{11}, D_{22}, D_{66}, D_k flexural rigidities and G, H, J constants depending on mode type. The rigidities can then be identified by means of specific software using an optimization method [*31,32*].

All the small samples were tested using modal analysis. The frequencies of the pure bending modes are directly related to the rigidities D_{11} and D_{22} and the first three frequencies of the pure bending mode were used for the determination of D_{11} and D_{22}. In order to determine the evolution of the D_{66} rigidity, the first torsional modes of the L and T samples were used. For some samples, the determination of the frequencies has also been performed on the large plates after 9 and 24 months of immersion.

In addition to these nondestructive measurements the following destructive tests were performed before immersion and at the end of the test period, on 5 specimens for each condition: three point flexure test (EN 63) and short beam shear test (ISO 4585).

Results

Weight Gain

Figure 1 shows the weight gain curves plotted for each temperature and pressure condition. The results presented are the data corresponding to the mean of percentage weight gains for the three small panels at each condition. The curves have not been fitted and no points have been excluded. This will be discussed further below.

Physicochemical Properties

Measurements of evolution of glass transition temperature from DSC and Kinemat are presented in Tables 3 and 4. There are differences between results from the two methods, both in terms of absolute values and in the evolutions for each material. This is not unexpected as the two methods do not follow the same phenomena. The DSC value of T_g is determined by a slope change in enthalpy, measured on specimens weighing a few milligrams. It indicates the chemical state of the resin but is not sensitive to mechanical degradation such as cracking or interface debonding. The DMA value of T_g is determined from the mechanical small-strain response of larger samples and will reflect mechanical changes, but the way in which T_g is defined is by a change in mechanical response as temperature is raised, with respect to the room-temperature response. The loss of material and degradation, particularly of the polyester during aging, will affect the room-temperature response so that relative changes are very difficult to interpret.

Mechanical Properties

Figure 2 shows mean property evolution with aging for each material as a function of aging temperature both for atmospheric and 10 MPa pressure aging conditions. These plots contain a very large amount of data (each point corresponds to the average of 5 test results) and their interpretation and implications will be discussed.

Discussion

Kinetics of Absorption

The kinetics of absorption will be discussed in terms of scatter, the suitability of modeling, and the influence of resin type.

The absorption curves (Fig. 1) are drawn using the conventional $\sqrt{t}/h$ x axis, where t denotes the time of immersion (in hours), and h the sample thickness (in mm). The y axis indicates the mass of water absorbed in the material as a percentage of the initial sample mass. The absorption values presented are means for the three small plates. The variation in weight gain between the plates is very low and an example is presented (Fig. 3). No significant difference was noted between the weight gains of the plates cut in the direction L and

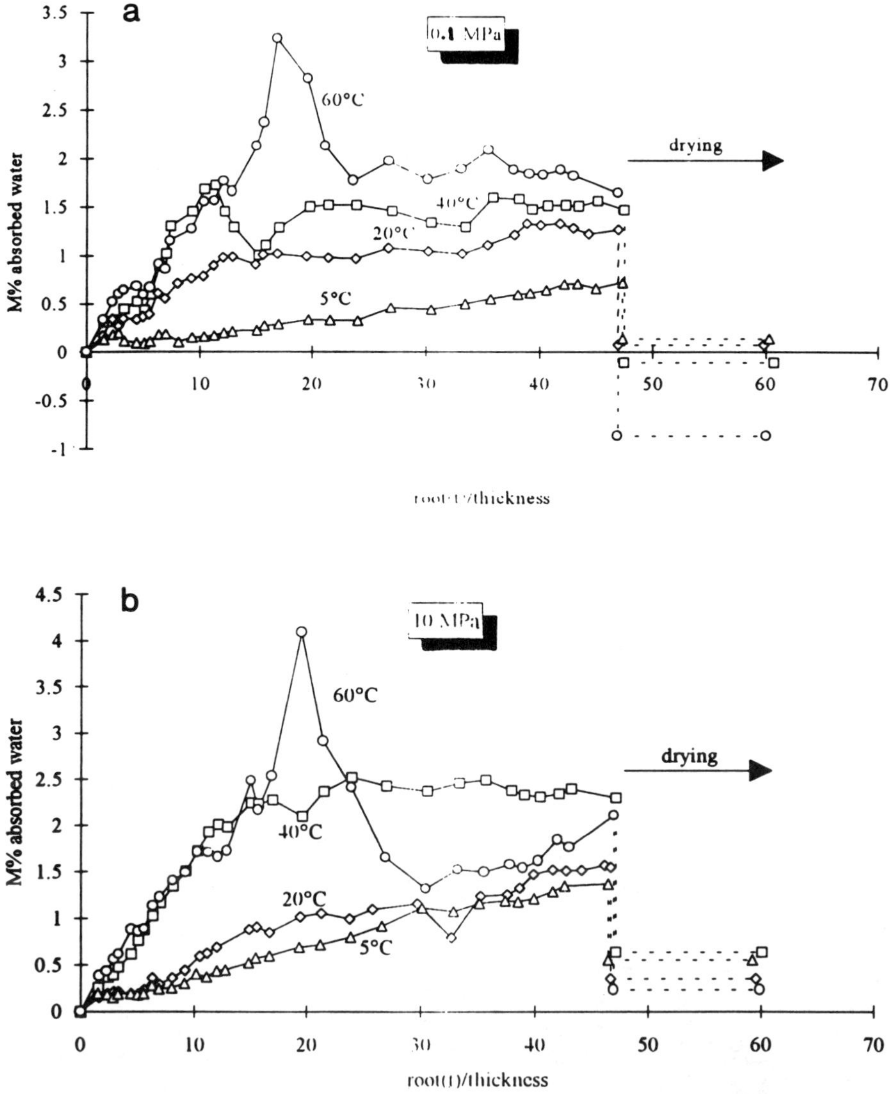

FIG. 1*a–b*—*Isophthalic polyester resin composite. Absorption curve (% by $h^{1/2} \cdot mm^{-1}$) after 2 years' immersion at 0.1 and 10 MPa.*

T, nor between the small and large plates. The modeling of the absorption curves is not easy for these materials. The description of the absorption phenomena by a Fickian model, which requires the moisture level to reach a maximum value asymptotically, is not appropriate as a saturation level is not generally reached even after two-years immersion (Fig. 1). The maximum moisture content cannot be determined, so the determination of the mass diffusion

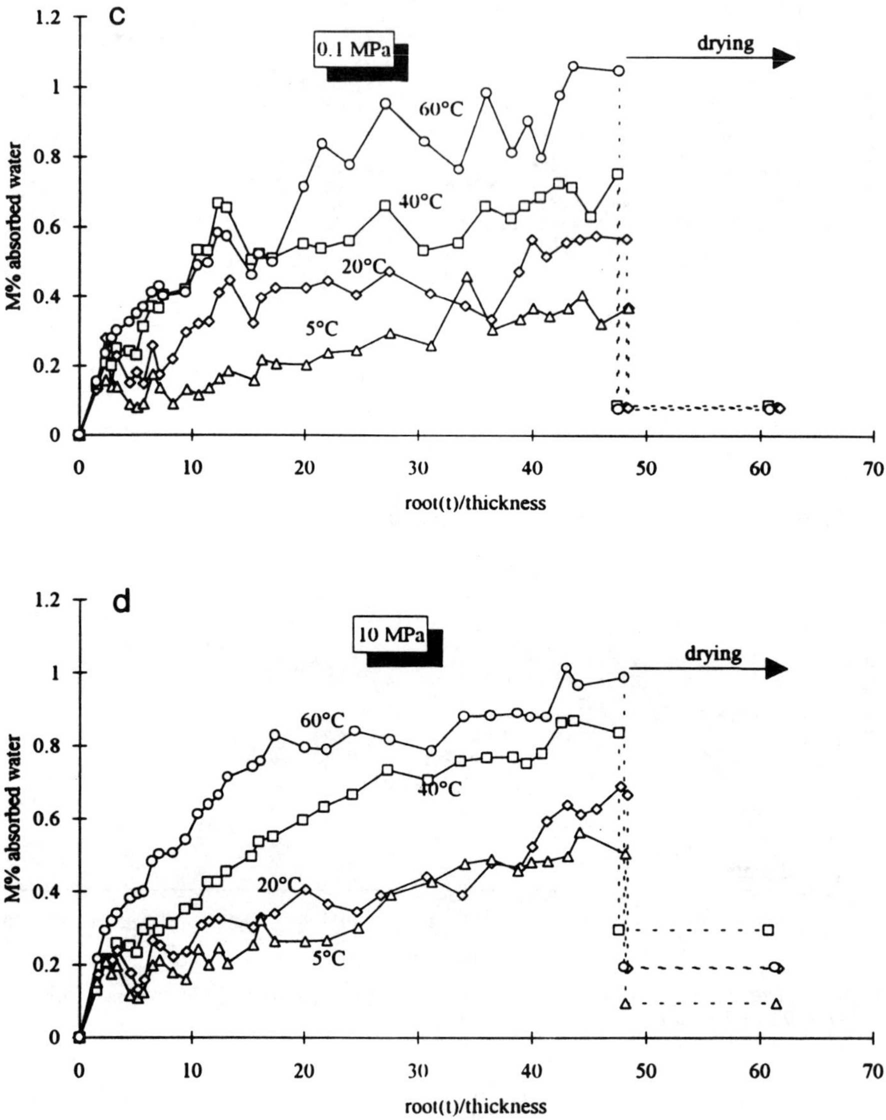

FIG. 1*c–d—Vinyl ester resin composite. Absorption curve (% by* $h^{1/2} \cdot mm^{-1}$*) after 2 years' immersion at 0.1 and 10 MPa.*

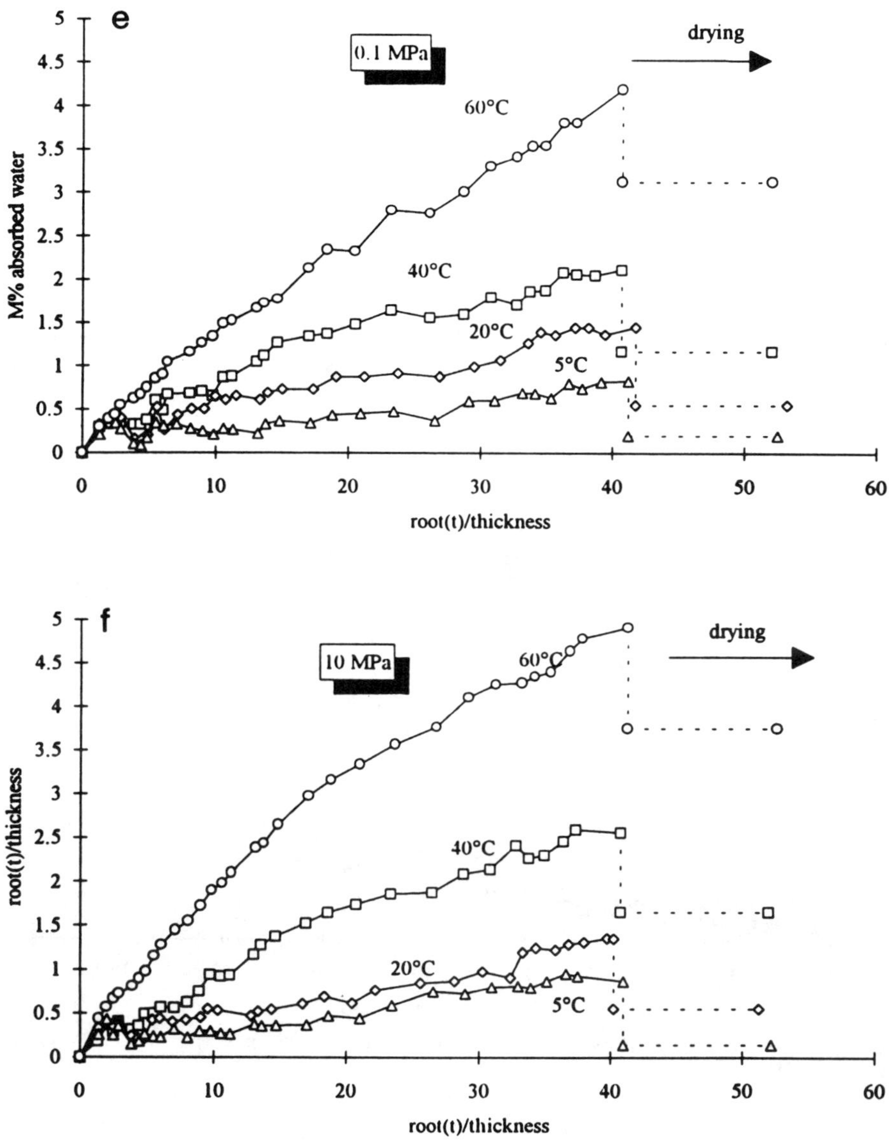

FIG. 1*e–f—Epoxy resin composite. Absorption curve (% by* $h^{1/2} \cdot mm^{-1}$) *after 2 years' immersion at 0.1 and 10 MPa.*

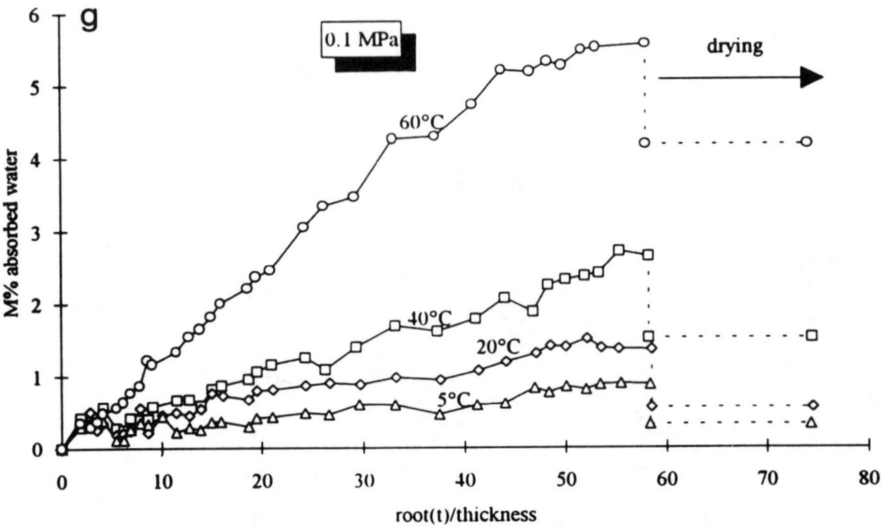

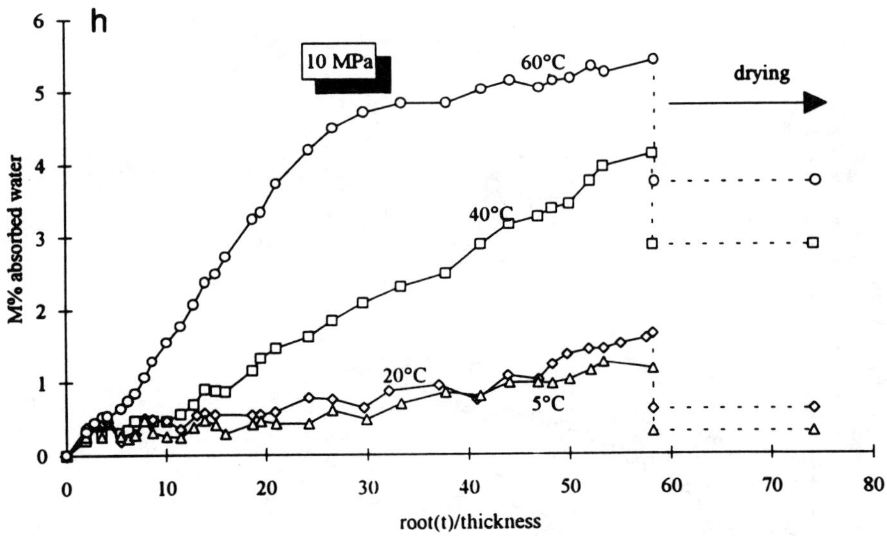

FIG. 1*g–h—Epoxy prepreg composite. Absorption curve (% by* $h^{1/2} \cdot mm^{-1}$*) after 2 years' immersion at 0.1 and 10 MPa.*

coefficient *D* is not possible. Attempts to fit with a Fickian model are shown in Fig. 4. This works quite well for the epoxy but not for the polyester. No model could be validated for all the samples. An initial diffusion rate parameter has nevertheless been determined for each material, as the original slope of the water uptake curve, in order to allow comparisons to be made (Table 5).

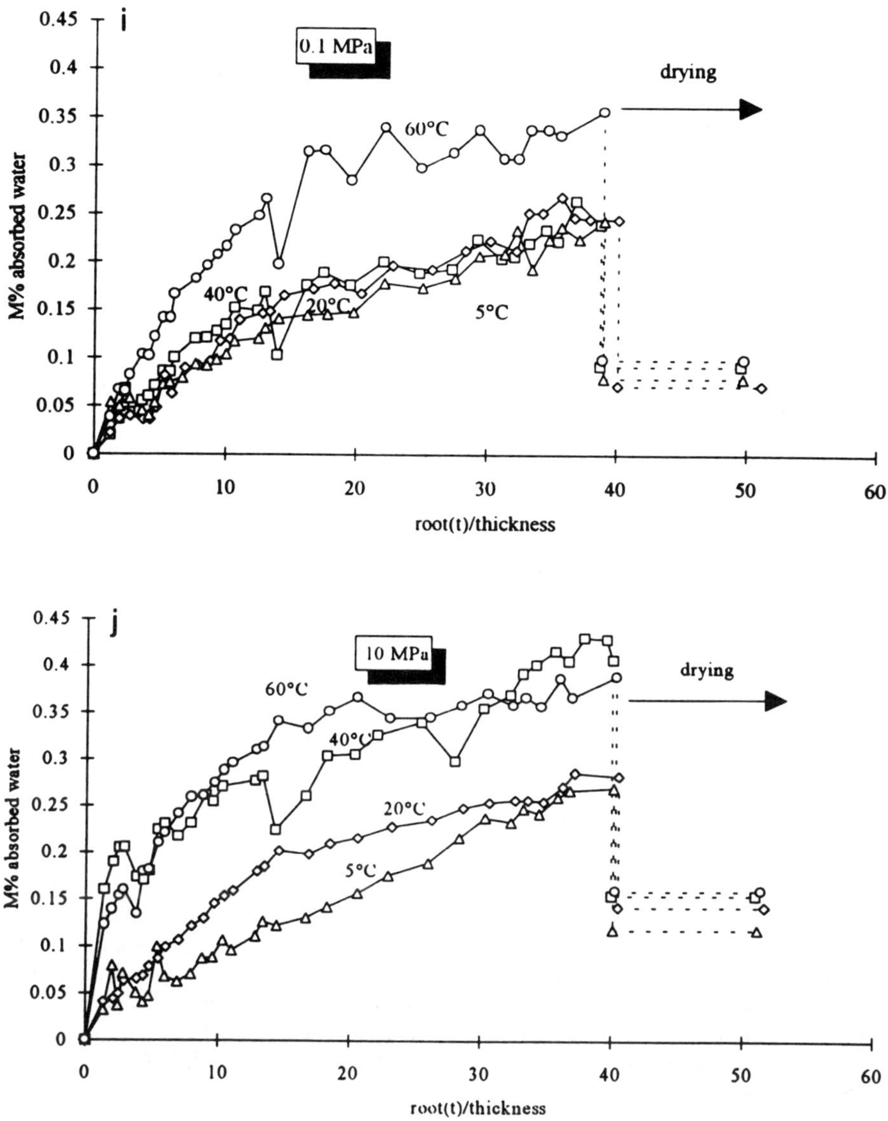

FIG. 1*i–j—PEEK matrix composite. Absorption curve (% by $h^{1/2} \cdot mm^{-1}$) after 2 years' immersion at 0.1 and 10 MPa.*

With respect to resin type, the epoxy materials (epoxy and prepreg) absorbed more water than the others (up to 5% after 2 years at 60°C) and after drying, a residual weight gain was noted. The T_g values for the epoxies measured by both methods are lowered after immersion (Tables 3 and 4) suggesting that plasticization of the resin has occurred and water is now fixed in the polymer. For the vinyl ester material, the weight gains are much smaller and no

TABLE 3—*Evolution of* T_g *of composites determined by DSC.*

	Polyester 110°C		Vinyl Ester 114°C		Epoxy 82°C		Prepreg 123°C	
T_g Initial	0.1 MPa	10 MPa	0.1 MPa	10 MPa	0.1 MPa	10 MPa	0.1 MPa	10 MPa
5°C	+1°C	+0°C	+3°C	+3°C	−0°C	+1°C	+0°C	+2°C
20°C	+0°C	+1°C	+3°C	+3°C	−1°C	−0°C	−5°C	−5°C
40°C	−1°C	−5°C	+0°C	+2°C	−0°C	−1°C	−10°C	−10°C
60°C	−5°C	−7°C	−1°C	+0°C	−0°C	−2°C	−9°C	−10°C

significant variation in T_g is observed. A particular behavior is observed for the isophthalic polyester composite, for which a decrease of weight is noted for the sample immersed at 60°C. Anomalous data for T_g are also measured for this material, with DSC indicating a drop in T_g but DMA showing large increases. It is clear that the T_g is test-dependent and great care is required in using its evolution to follow aging. Finally, the carbon/PEEK composite absorbs very little moisture, as has been found previously [*33*].

From a parallel study, absorption data for the neat epoxy resin corresponding to Composite 3 have been obtained and these are reported in Fig. 5 with data for the composite for comparison. If we assume the mass at saturation in the resin M_{sr} to be 3%, then the predicted mass at saturation in the composite M_{sc} is M_{sr} multiplied by the volume fraction of resin in the composite and by the ratio of density of resin over density of composite. This gives a value of M_{sc} of 1.3% in this case, but it is clear in Fig. 5 that the composite at 60°C absorbs much more water than this value (>4%). This suggests that the fiber-matrix interface is a preferential path for migration of the water though its behavior is not well defined. This indicates that it is impossible to extrapolate the absorption behavior of the composite from the behavior of the neat resin, even if the composite appears to follow Fickian diffusion.

Evolution of Properties

Figure 2 presents the evolution of properties. Except for the thermoplastic material, for which there is no significant evolution of the properties, a degradation of the properties by up to 50% was noted after 2 years of immersion at 60°C. The tests have been performed on samples after drying so the degradation can be considered irreversible [*34*]. This degradation is anisotropic and affects both rigidities and strengths. The most affected parameters are the D_{22} and D_{66} rigidities, and the flexural strengths. It must be emphasized that degradation is not homogenous through the thickness of the material as the external surface is subjected to water longer than the center of the specimens, so flexural tests are very detrimental.

TABLE 4—*Evolution of* Tg *of composites measured by torsion DMA.*

	Polyester 102°C		Vinyl Ester 111°C		Epoxy 80°C		Prepreg 104°C	
T_g Initial	0.1 MPa	10 MPa	0.1 MPa	10 MPa	0.1 MPa	10 MPa	0.1 MPa	10 MPa
5°C	−3°C	−1°C	+1°C	+2°C	−3°C	+7°C	+1°C	+1°C
20°C	+2°C	−1°C	=	−2°C	−8°C	−3°C	=	−7°C
40°C	+5°C	+2°C	−1°C	+1°C	−11°C	−6°C	−10°C	−12°C
60°C	+16°C	+10°C	+4°C	+4°C	−9°C	−2°C	−17°C	−14°C

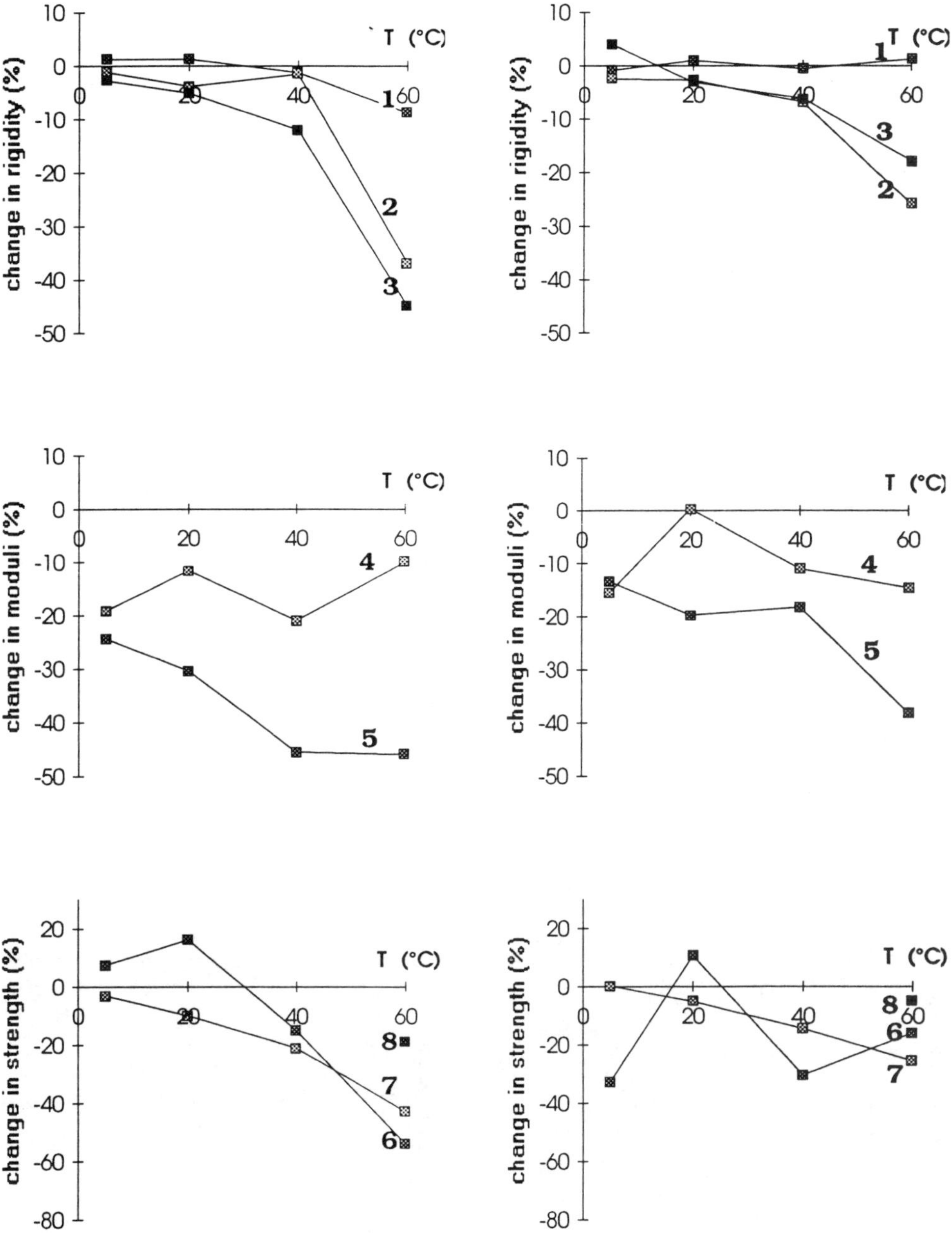

FIG. 2*a*—*Isophthalic polyester resin composite. Loss of mechanical properties after 2 years' immersion at 0.1 MPa* (left) *and 10 MPa* (right). *1*—D_{11}, *2*—D_{22}; *3*—D_{66}; *4*—E_{fl}; *5*—E_{ft}; *6*—σ_{fl}; *7*—σ_{ft}; *8*—τ_{13}.

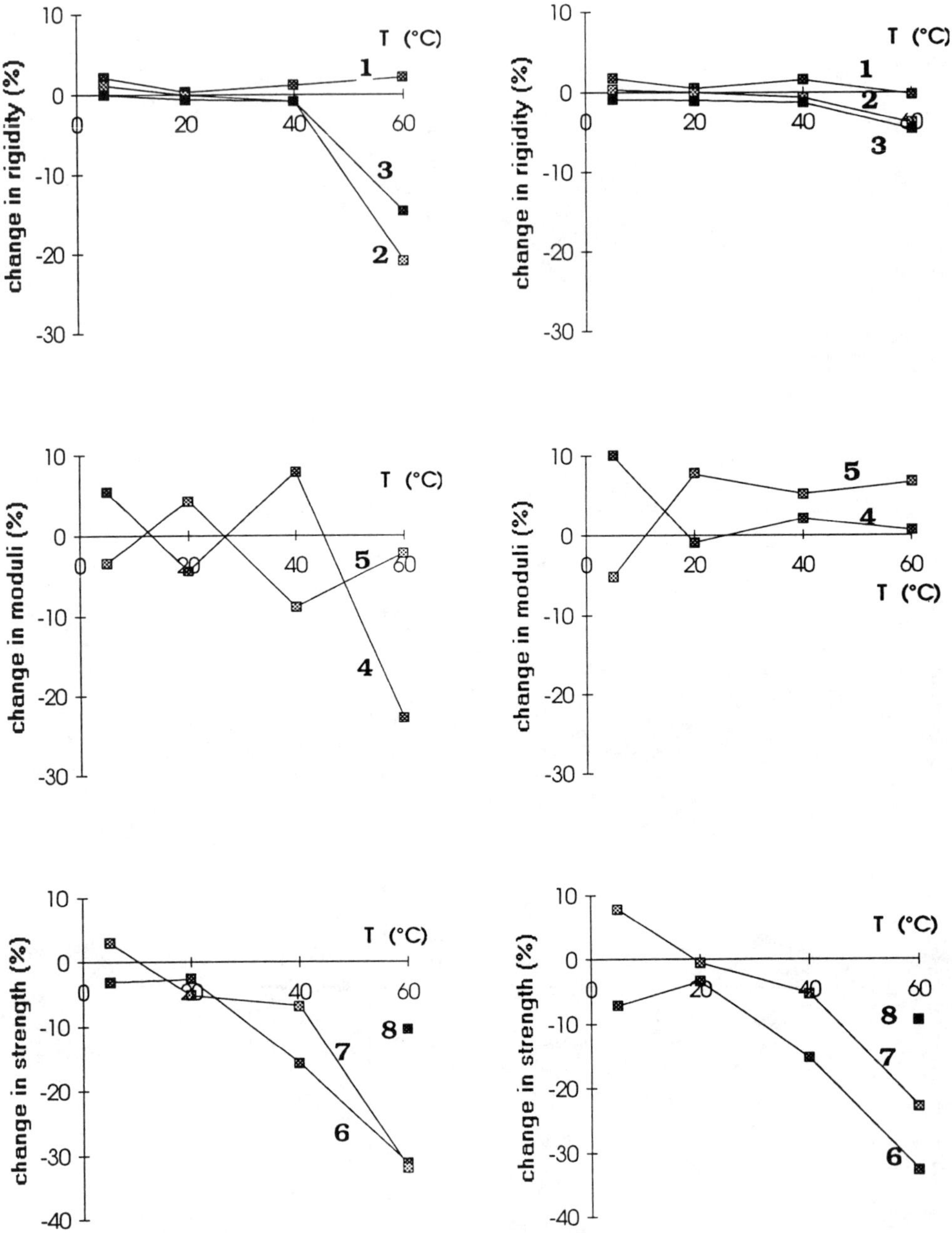

FIG. 2*b*—*Vinyl ester resin composite. Loss of mechanical properties after 2 years' immersion at 0.1 MPa* (left) *and 10 MPa* (right). *1*—D_{11}; *2*—D_{22}; *3*—D_{66}; *4*—E_{fl}; *5*—E_{ft}; *6*—σ_{fl}; *7*—σ_{ft}; *8*—τ_{13}.

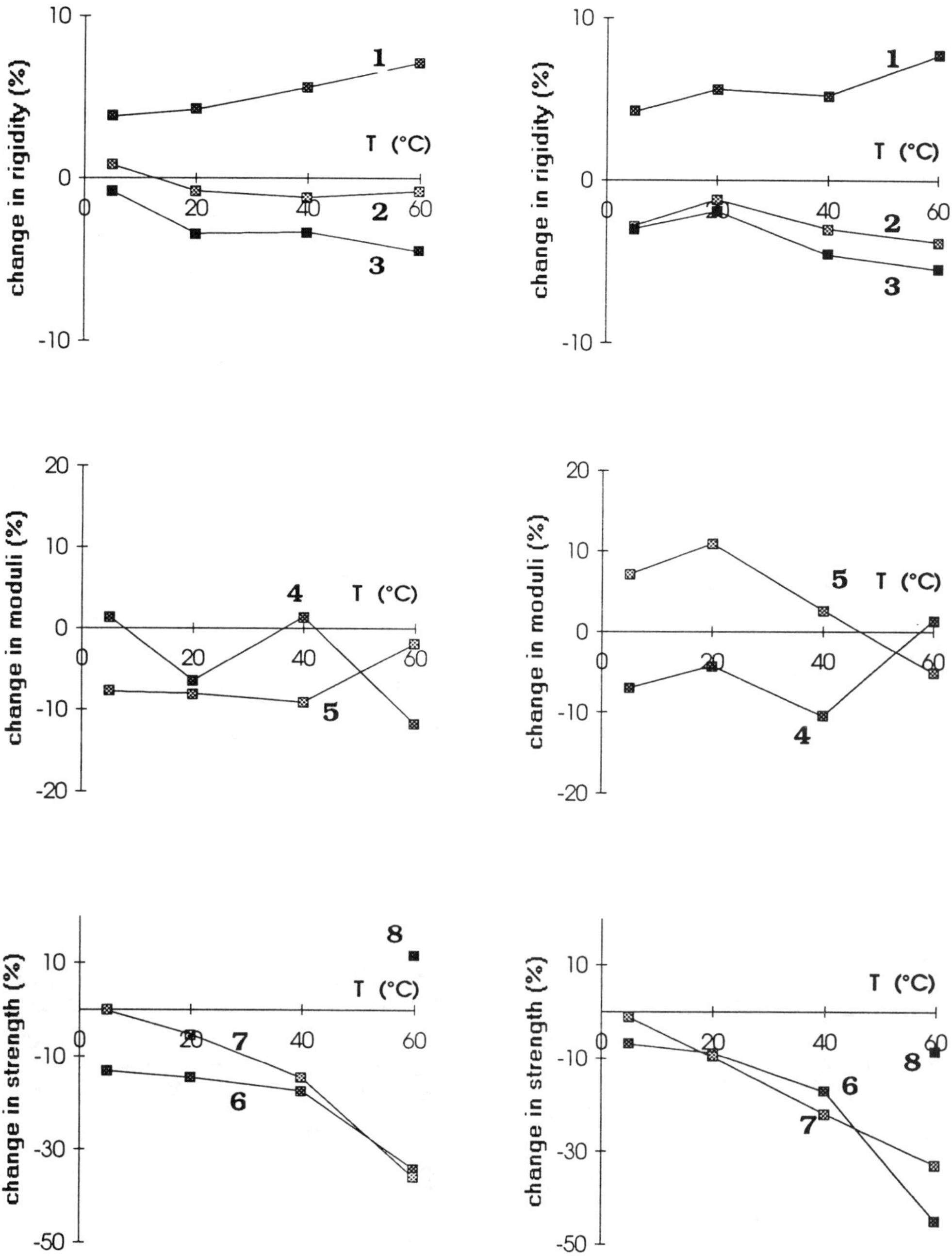

FIG. 2c—*Epoxy resin composite. Loss of mechanical properties after 2 years' immersion at 0.1 MPa* (left) *and 10 MPa* (right). *1*–D_{11}; *2*–D_{22}; *3*–D_{66}; *4*–E_{fl}; *5*–E_{ft}; *6*–σ_{fl}; *7*–σ_{ft}; *8*–τ_{13}.

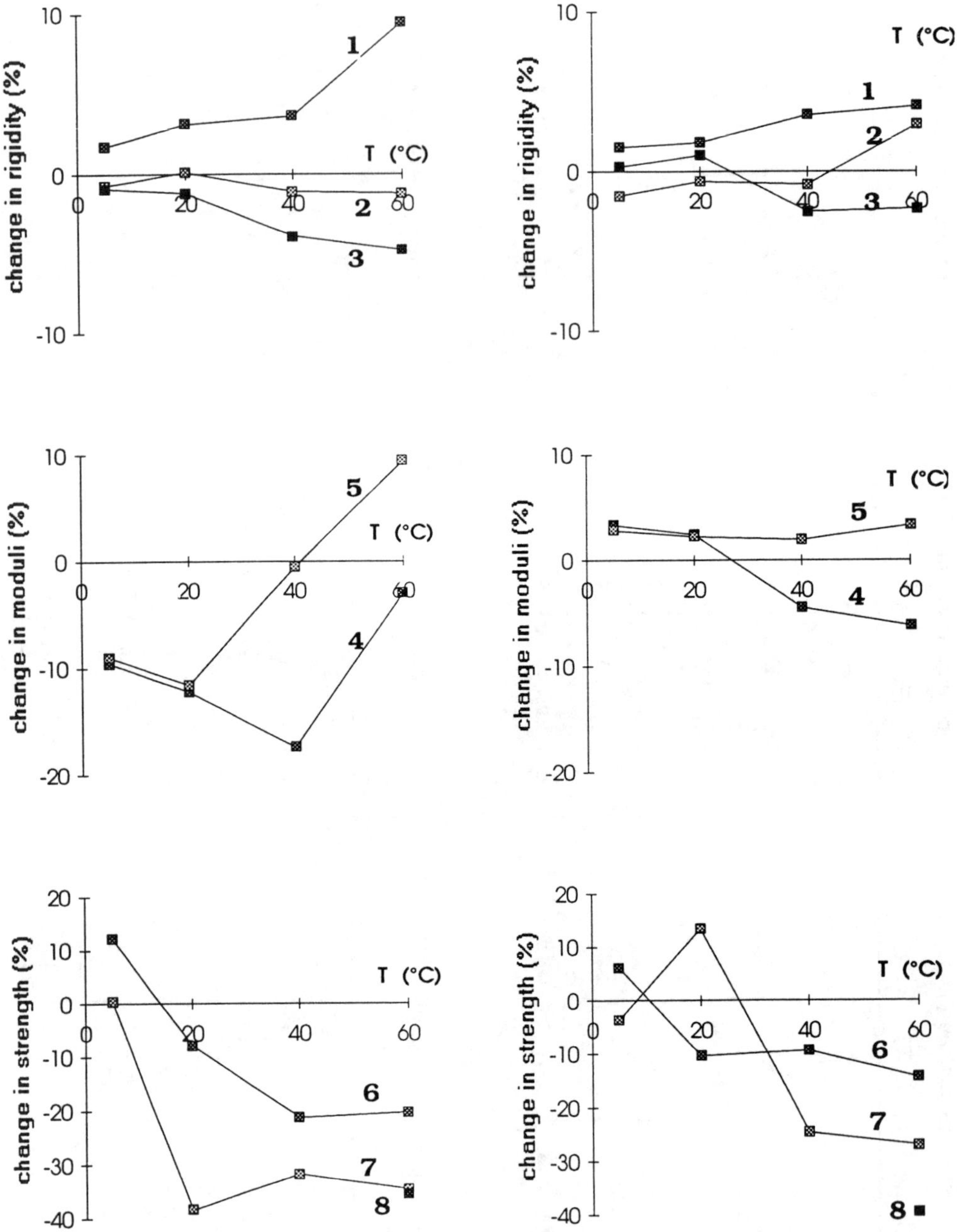

FIG. 2*d—Epoxy prepreg composite. Loss of mechanical properties after 2 years' immersion at 0.1 MPa* (left) *and 10 MPa* (right). *1*—D_{11}; *2*—D_{22}; *3*—D_{66}; *4*—E_{fl}; *5*—E_{ft}; *6*—σ_{fl}; *7*—σ_{ft}; *8*—τ_{13}.

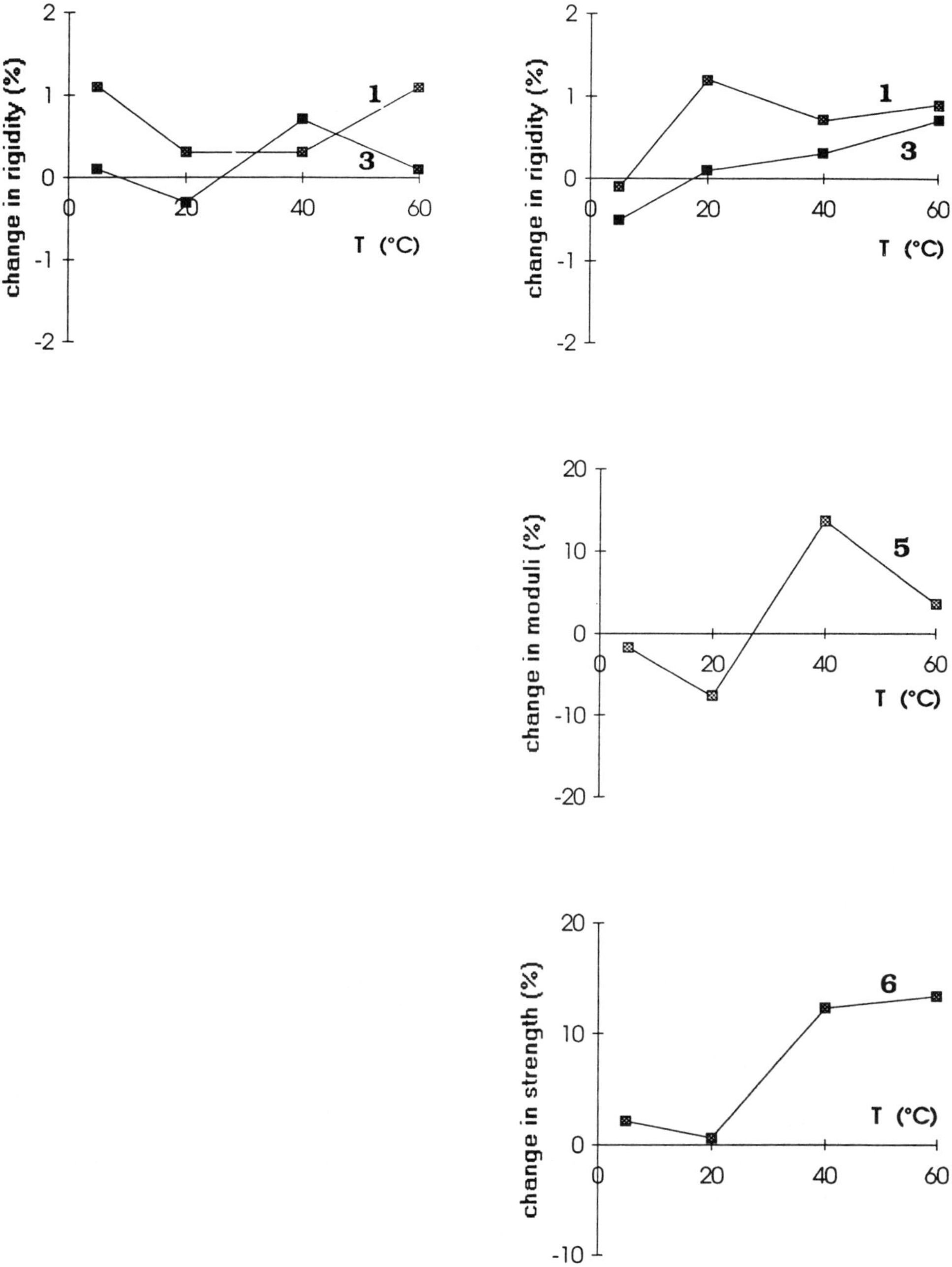

FIG. 2e—*PEEK composite. Loss of mechanical properties after 2 years' immersion at 0.1 MPa* (left) *and 10 MPa* (right). *1*—D_{11}; *2*—D_{22}; *3*—D_{66}; *4*—E_{fl}; *5*—E_{ft}; *6*—σ_{fl}; *7*—σ_{ft}; *8*—τ_{13}.

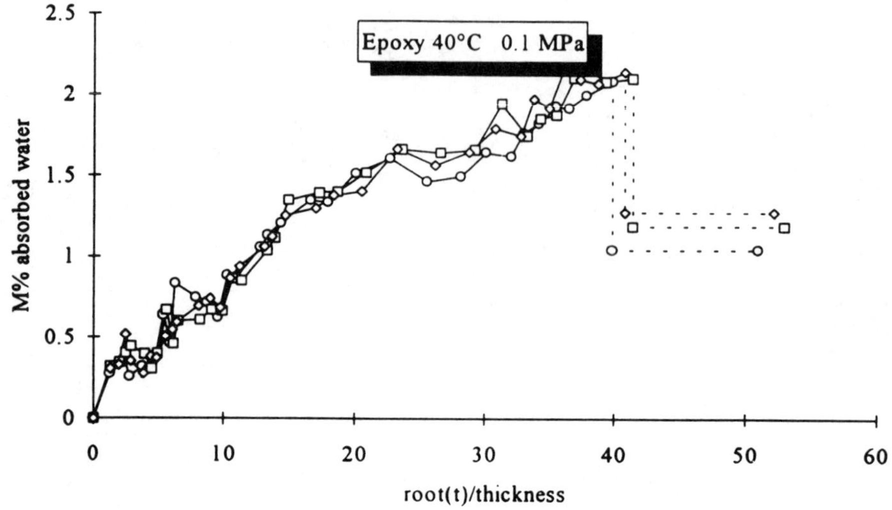

FIG. 3—*Epoxy resin composite. Absorption curve (% by $h^{1/2} \cdot mm^{-1}$) at 40°C and 0.1 MPa. Scatter between the three samples.*

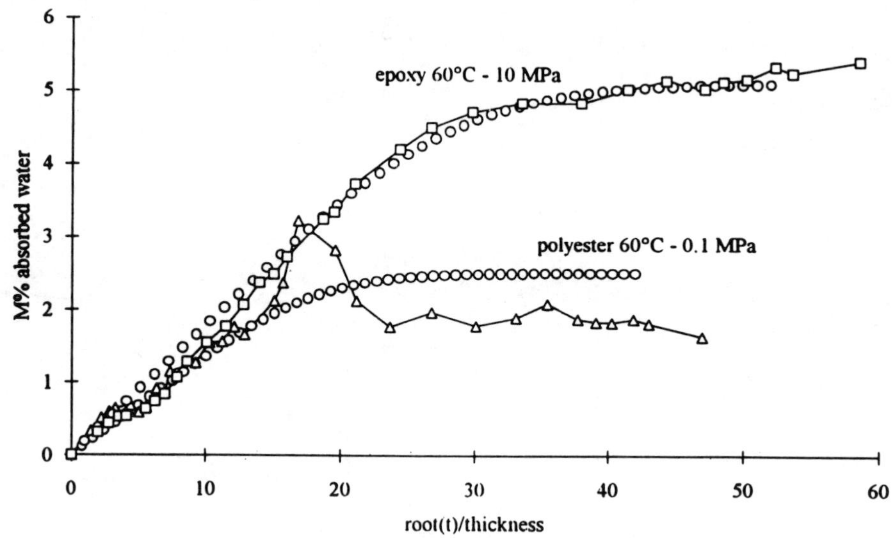

FIG. 4—*Absorption curve (% by $h^{1/2} \cdot mm^{-1}$). Ficks law fitting.*

TABLE 5—*Initial rate of diffusion* (10^{-2} *mm* · $h^{-1/2}$).

	Polyester		Vinyl Ester		Epoxy		Prepreg		PEEK	
	0.1 MPa	10 MPa	0.1 MPa	10 MPa	0.1 MPa	10 MPa	0.1 MPa	10 MPa	0.1 MPa	10 MPa
5°C	1.51	3.24	0.69	0.92	15.0	18.7	13.4	16.4	0.62	0.64
20°C	4.58	4.47	1.16	1.01	27.5	26.9	20.9	23.8	0.67	1.12
40°C	12.75	14.32	3.43	2.29	68.1	80.0	40.3	74.6	1.00	1.50
60°C	16.56	17.68	4.60	3.49	109.0	152.5	120.9	185.1	2.10	1.50

Following the properties by modal analysis appears to be very attractive. This technique allows the evolution of degradation to be determined rapidly in a nondestructive way (the same sample can be tested at different times during the period of test), giving an indication of the anisotropic global (large sample) degradation of the planar properties. The scatter in results is lower than that obtained by standard quasi-static tests.

Another method, based on ultrasonic measurement, has been used in a parallel study on these materials [*35*] and gives access to the through-thickness degradation of the material, in addition to planar properties. A preferential degradation of the properties through the thickness of the material has been observed. In addition, this study has shown that the evolution of the elastic properties of the composite material cannot be determined from the evolution of the elastic properties of the neat resin.

No simple correlation between the loss of mechanical properties and absorption parameters can be made. For example, a material that absorbs a very small amount of water (for example, vinylester) shows a loss of properties comparable to a material (epoxy composite) with a weight gain up to 5%. Scanning electron microscopy has been used to examine fracture surfaces before and after aging but clear indications of degradation mechanisms have not been determined. Some clean fibers are apparent in all thermoset materials suggesting that weakness of the fiber-matrix interface may be the preferential degradation mode as noted by other authors [*36*], but many different features are apparent on a single fracture surface in these materials.

Influence of Pressure

The influence of pressure on absorption is material-dependent and is clearly shown in Table 5. The weight gains of the two epoxy-based composites are strongly accelerated by a hydrostatic pressure of 10 MPa, while the three other materials appear far less sensitive to pressure. This is in contradiction with some published results for epoxy composites [*15,16*] but most previous work has been performed on higher temperature cure epoxies or carbon fiber-reinforced systems. In general, the hydrostatic pressure does not affect the loss in properties.

Accelerated Testing

The usefulness of results from tests over a long period of time is that they enable the validity of accelerated aging procedures to be evaluated. The initial slopes of the absorption curves and the loss of mechanical properties (after 2 years of exposure) are presented on log plots versus 1/T (°K) in Figs. 6 and 7, in order to validate an Arrhenius process. These figures indicate a very similar behavior of the two epoxy composites, for which the acceleration factor due to the temperature is identical. The thermoplastic material is more stable than epoxy composites that are in turn more stable than the vinyl ester and the polyester

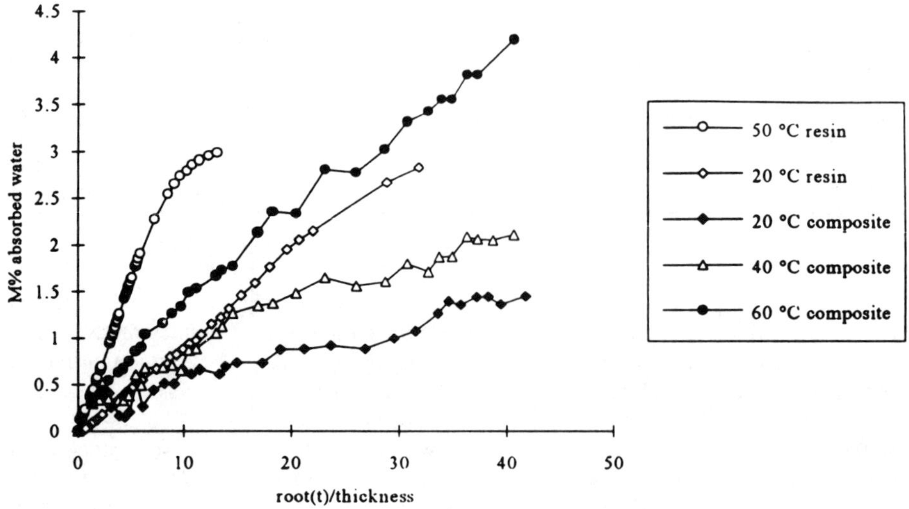

FIG. 5—*Epoxy resin. Absorption curve (% by $h^{1/2} \cdot mm^{-1}$) for composite and neat resin at 0.1 MPa.*

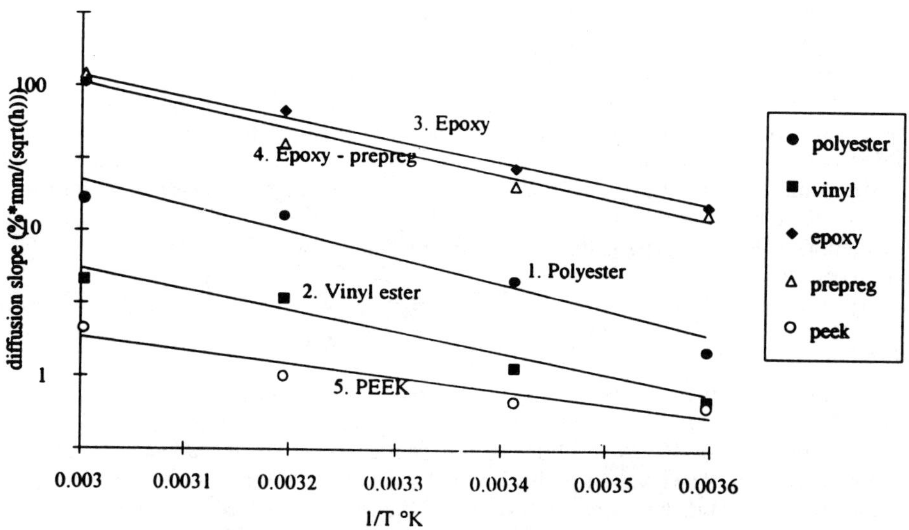

FIG. 6—*Initial diffusion slope ($10^{-2} \cdot mm \cdot h^{-1/2}$) versus $1/T$ (°K) for composite at 0.1 MPa.*

materials. The linearity of these curves is quite reasonable and some extrapolation can be made from results obtained at 60°C to predict long-term behavior at 5°C. This is one of the aims of the study and results are shown in Table 6 for some parameters. The accelerating factors differ significantly both between materials and for different parameters, underlining the importance of the validation process for each new material and environment considered.

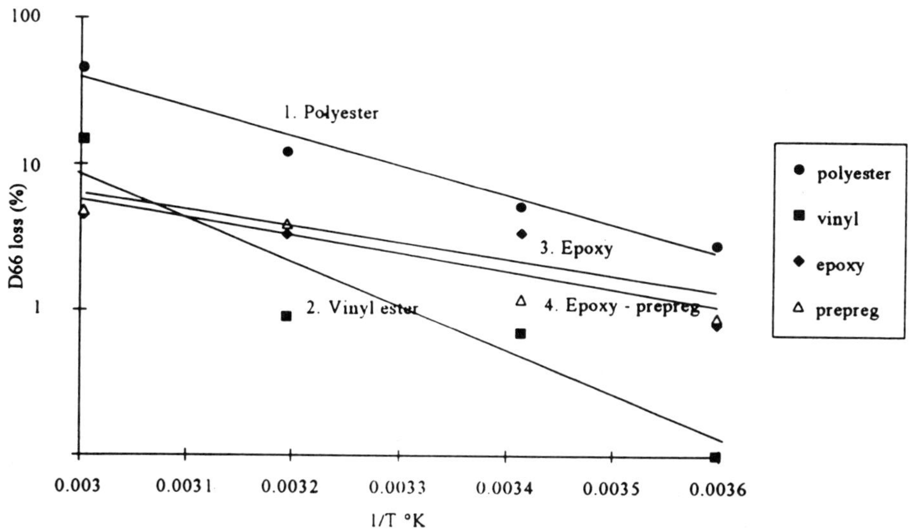

FIG. 7a—*Loss of properties* ΔD_{66} *(%) for composite at 0.1 MPa.*

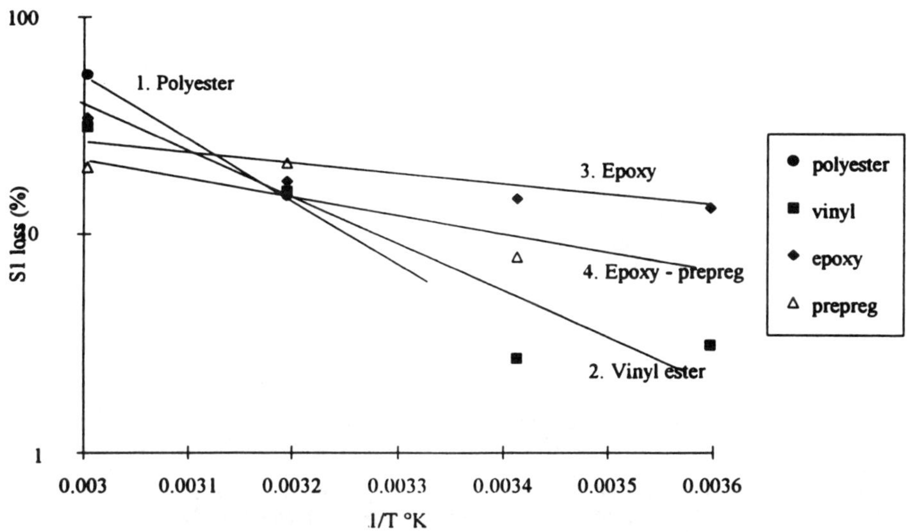

FIG. 7b—*Loss of properties* $\Delta\sigma_{fl}$ *(%) for composite at 0.1 MPa.*

Conclusions

This paper presents a large amount of data from a study of the wet aging of composites for marine applications over a 2-year period. The following conclusions may be drawn.

- Modeling of the absorption phenomena for marine composite materials by a single law is not possible.

TABLE 6—*Accelerating factors for rate of degradation and diffusion, resulting from raising temperature from 5 to 60°C.*

	Polyester	Vinyl Ester	Epoxy	Prepreg	PEEK
D_{66}	12	63	5	5	...
σ_{fl}	40	20	1.6	2.5	...
Diffusion	12	6	5	5	3

- Hydrostatic pressure (10 MPa) increases the rate of moisture absorption for some materials (for example, epoxies) but does not affect the evolution of mechanical properties.
- The mechanical behavior of these materials after immersion cannot be estimated from the absorption curve nor from the evolution of the properties of the neat resin.
- After two years' immersion at 60°C the loss of mechanical properties can be very significant (up to 50%) and the evaluation of this degradation by nondestructive tests (modal analysis and ultrasonic measurements) is very attractive. Standard test procedures could be developed for these techniques.
- The extrapolation of residual properties after immersion from accelerated tests, using time-temperature equivalence, can be estimated but the accelerating factors are quite small due to the low T_g's of these materials, and hence, the modest test temperatures permissible.

Acknowledgments

This work was performed in the "Materials for Deep Sea" project funded by the European Community Marine Science and Technology (MAST) program. The contributions of H. Loaec, D. Petton, and C. Simon of the IFREMER Brest Center are gratefully acknowledged.

References

[1] Smith, C. S., *Design of Marine Structures in Composite Materials*, Elsevier Applied Science, London & New York, 1990.

[2] Le Lan, J. Y., Parneix, P., and Gueguen, P. L., "Composite Material Superstructures," *Nautical Construction with Composite Materials*, P. Davies and L. Lemoine, Eds., IFREMER, France, 1992, pp. 339–411.

[3] Gibson, A. G., "Composite Structures in Offshore Applications," *Composite Materials in Marine Structures*, Chapter 3, R. A. Shenoi, Ed., Cambridge Ocean Technology Series, 1993.

[4] Lemiere, Y., "The Evolution of Composite Materials in Submarine Construction," *Nautical Construction with Composite Materials*, P. Davies and L. Lemoine, Eds., IFREMER, France, 1992, pp. 441–449.

[5] Chauchot, P. and Guillermin, O., "On the Use of Composite Materials for 6000 m Containers," *Proceedings*, 8th International Conference of the SAMPE European Chapter, De Bossu et al., Eds., 1987, pp. 312–328.

[6] Stachiw, J. D. and Frame, B., "Graphite-Fiber-Reinforced Plastic Pressure Hull Mod 2 for the Advanced Unmanned Search System Vehicle, Resistant Ceramic Housings for Deep Submergence Unmanned Vehicles," *Technical Report 1245*, Naval Ocean System Center, San Diego, CA, 1988.

[7] Collar, P., Rabb, R. J., Michel, J.-L., Brissett, L., and Kilpatrick, I. M., "Systems Research for Unmanned Autonomous Underwater Vehicles," *Proceedings, Oceans 94*, Brest Sept. 1994, Vol. 1, pp. 158–163.

[8] Graham, D., "Composite Pressure Hulls for Deep Ocean Submersibles," presented at the Eighth International Conference on Composite Structures in Paisley, Scotland, 1995.
[9] Springer, G. S., "Moisture Absorption in Fibre-Resin Composites," *Developments in Reinforced Plastics*, G. Pritchard, Ed., Vol. 2, 1982, pp. 43–66.
[10] Crank, J., *The Mathematical Theory of Diffusion*, Oxford University Press, 2nd edition, 1975.
[11] Loos, A. and Springer, G. S., "Moisture Absorption of Polyester – E glass Composite," *Environmental Effects on Composite Materials*, Vol. 1, G. S. Springer, Ed., 1981, Chapter 5, pp. 51–62.
[12] Weitsman, Y., "Effect of Fluids on Polymeric Composites — A Review," *Report MAES 95-1.0 CM*, July 1995, University of Tennessee.
[13] De Wilde, W. P. and Frolkovic, P., "The Modelling of Moisture Absorption in Epoxies, Effect at the Boundaries," *Composites*, Vol. 25, No. 2, 1994.
[14] Bonniau, P. and Bunsell, A. R., "A Comparative Study of Water Absorption Theories Applied to Glass Epoxy Composites," *Environmental Effects on Composite Materials*, Vol. 2, Chapter 18 G. S. Springer, Eds., 1981, pp. 209–229.
[15] Tucker, W. C., Lee, S-B., and Rockett, T. J., "The Effects of Pressure on Water Transport in Polymers," *Journal of Composite Materials,* Vol. 27, No. 8, 1993, pp. 756–763.
[16] Avena, A. and Bunsell, A. R., "Effect of Hydrostatic Pressure on the Water Absorption of Glass Fibre Reinforced Epoxy Resin," *Composites*, Vol. 19, No. 5, Sept. 1988.
[17] Pollard, A., Baggott, R., Wolstenholm, G. H., Yates, B., and George, A. P., *Journal of Materials Science,* Vol. 24, 1989, pp. 1665–1669.
[18] Grant, T. S., "Seawater Degradation of Composites," MSc Thesis, Texas A&M University, May 1991.
[19] Wolf, E. G., "Moisture Effects on Polymer Matrix Composites," *SAMPE Journal*, Vol. 29, No. 3, May/June 1993, p. 11.
[20] Weitsman, Y., "Moisture in Composites: Sorption and Damage," *Fatigue of Composite Materials*, K. L. Reifsnider, Ed., Elsevier, 1991, pp. 385–429.
[21] Pomies, F., Carlsson, L. A., Choqueuse, D., and Davies, P., "Degradation of Composite Materials in a Marine Environment," *Nautical Construction with Composite Materials*, P. Davies and L. Lemoine, Eds., IFREMER, France, 1992, pp. 375–385.
[22] Gutierrez, J., Le Lay, F., and Hoarau, P., "A Study of Aging of Glass Fibre-Resin Composite in a Marine Environment," *Nautical Construction with Composite Materials*, P. Davies and L. Lemoine, Eds., IFREMER, France, 1992, pp. 338–346.
[23] Grant, T. S. and Bradley, W. L., "In Situ Observation in SEM of Degradation of Graphite/Epoxy Composite Materials Due to Seawater Immersion," *Journal of Composite Materials*, Vol. 29, No. 7, 1995, pp. 852–867.
[24] Ellis, B. and Found, M. S., "The Effect of Water Absorption on a Polyester/Chopped Strand Mat Laminate," *Composites*, Vol. 14, No. 3, July 1983.
[25] Pritchard, G. and Speake, S. D., "The Use of Water Absorption Kinetic Data to Predict Laminate Property Changes," *Composites*, Vol. 18, No. 3, July 1987.
[26] Marom, G., "Environmental Effects on Fracture Mechanical Properties of Polymer Composites," Chapter 10, *Application of Fracture Mechanics to Composite Materials*, K. Friedrich, Ed., Elsevier, 1989.
[27] Dillard, D. "Viscoelastic Behavior of Laminated Composite Materials," Chapter 8, *Fatigue of Composite Materials*, K. L. Reifsnider, Ed., Elsevier, 1990.
[28] Kasamori, M., Funada, Y., Awazu, K., Watanabe, Y., Nakada, M., and Miyano, Y., "Accelerated Evaluation of Mechanical Degradation Behavior of GFRP in Hot Water," *Proceedings*, DURACOSYS 1995, Balkema publishers, 1996, pp. 225–231.
[29] Prevosto, M., Olagnon, M., Benveniste, A., Basseville, M., and Le Vey, G., "State Space Formulation, a Solution to Modal Parameter Estimation," *Journal of Sound and Vibration*, Vol. 148, No. 2, July 1991.
[30] Blevins, R. D., *Formulas for Natural Frequency and Modeshape*, Van Nostrand Reinhold Co., 1979.
[31] Sol, H., "Identification of Anisotropic Plate Rigidities Using Free Vibration Data," Ph.D. Thesis, Vrije Universiteit Brussel, Oct. 1986.
[32] Tsouvalis, N. G., "A Non-Destructive Method for the Identification of Anisotropic Plate Elastic Properties," IFREMER Internal Report D1T1 91.17 GO/MM, 1991.
[33] Demonet, C. M., "Interaction of Moisture with Resin Matrix Composites," *Proceedings*, 21st International SAMPE, 25–28 Sept. 1989, pp. 213–238.
[34] Jacquemet, F., "Ageing Effects of Sea Water on Composites Materials," *Proceedings*, European SAMPE, La Baule, 1987.

[*35*] Hosten, B., Baste, S., and Choqueuse, D., "Ultrasonic Evaluation of Hydrolytic Aging of Organic Matrix Composite Materials," *Annale des Composites*, AMAC, France, 1992 1/2, (in French), pp. 65–76.

[*36*] Pomies, F., Carlsson, L. A., and Gillespie, J. W., "Marine Environmental Effects on Polymer Matrix Composites," *Composite Materials: Fatigue and Fracture, ASTM STP 1230*, American Society for Testing and Materials, West Conshohocken, PA, 1995.

Giorgio Zaffaroni[1]

Two-Dimensional Moisture Diffusion in Hybrid Composite Components

REFERENCE: Zaffaroni, G., **"Two-Dimensional Moisture Diffusion in Hybrid Composite Components,"** *High Temperature and Environmental Effects on Polymeric Composites: 2nd Volume, ASTM STP 1302,* Thomas S. Gates and Abdul-Hamid Zureick, Eds., American Society for Testing and Materials, 1997, pp. 97–109.

ABSTRACT: A thermodynamics formulation for two-dimensional (2-D) diffusion in hybrid composite is discussed. The model had been transferred in a computer code and the validation is assessed here through the comparison with the analytical solution in the case where Fick's law is adequate (2-D, only 1 material) and with the experimental measurements in some simple cases (1-D, more than 1 material). An example of the use of the computer code is presented.

KEY WORDS: hybrid composites, moisture diffusion, environmental effects

Hybrid lay-up of fiber reinforced polymers plies are often used in aeronautical applications. In addition, it may be worth remembering that the hybrid situation must be often applied even to a single fiber/resin system as a consequence of different resin contents in different plies.

It is well known that absorption of a low molecular weight penetrant in a polymer reduces the mechanical (especially the "static") performances [*1*]. Then, when we deal with thick components where the moisture gain process continues during all the operative life and equilibrium cannot be achieved, the kinetic of diffusion into a hybrid is a problem of practical importance.

The special case in which we are interested is the moisture diffusion in fiber-reinforced thermosetting resins.

A popular approach to the problem is the Springer's model [2]. However, this model requires the knowledge of the dependence of the equilibrium moisture gain (M_∞%) from the relative humidity (RH%): M_∞%(RH%) in the range from zero to the actual external condition. This determination needs several experimental measurements.

In this paper, the computer code DIFF2D that is able to simulate the two-dimensional diffusion in a hybrid composite laminate will be presented. This computer program needs only the saturation levels of the materials interested in the external hygrothermal condition, the diffusion coefficients (in the two principal diffusion directions), and the geometrical description of the body through which the diffusion takes place.

[1] Material science researcher, AGUSTA S.p.A., via G. Agusta 520, 21017, Cascina Costa di Samarate (VA), Italy.

Chemical Equilibrium and Fick's Law in a Multi-phase System

Initially, it will be necessary to examine the (thermodynamics) equilibrium conditions in a multi-phase system. In the following context, such a system may be defined as the physical situation in which different materials lie in contact.

In this work, polymer relaxation will be neglected as second order phenomena and we shall restrict our interest to the isothermal case.

We start our analysis considering a one-dimensional system, such as is presented in Fig. 1. The chemical potential μ generally is a function of temperature, concentration, and so forth [3]. In the system of interest, μ can be considered to be dependent only on the concentration or partial pressure of H_2O.

If x is the spatial coordinate, the chemical equilibrium conditions for H_2O in the phases interested are

- (1) $\mu_{ext}(H_2O) = \mu_A(H_2O, x) \quad x = x_{EA}, x = x_{AE}$
- (2) $\mu_A(H_2O, x) = \mu_B(H_2O, x) \quad x = x_{AB}, x = x_{BA}$
- (3) $\mu_A(H_2O, x) = \mu_A(H_2O, x + dx) \quad x_{EA} \leq x \leq x_{AB}, x_{BA} \leq x \leq x_{AE}$
- (4) $\mu_B(H_2O, x) = \mu_B(H_2O, x + dx) \quad x_{AB} \leq x \leq x_{BA}$

that is, the chemical potential μ of the water is the same across the interfaces (Conditions 1 and 2) and μ is constant through the material (Conditions 3 and 4). These conditions may be summarized as follows: the equilibrium requires that μ is constant throughout the body. Then it is trivial to verify that the saturation concentration (from which μ depends) of the Material B is the same whether the Material A is present or not.

The application of Fick's first law is, on the other hand, subject to difficulties in the case of hybrid lay-up. In fact it states [4]

$$\text{flux} = -D \cdot dC/dx \tag{1}$$

where C is the concentration of the penetrant (H_2O).

Now if $C_{\infty,B} > C_{\infty,A}$, where C_∞ is the concentration at equilibrium, the process cannot be described by Law 1 (with D constant); in fact, the equilibrium gain requires that, after a certain time, diffusion occurs against the concentration gradient.

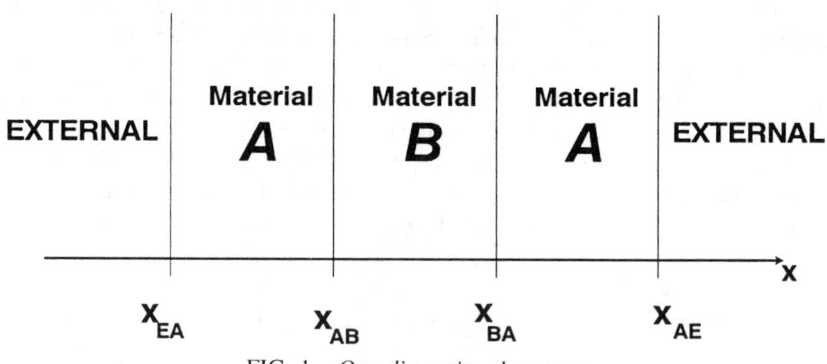

FIG. 1—*One-dimensional system.*

To override the problem Springer [2] in his computer code W8GAIN proposes that at the interfaces, equilibrium is immediately gained (that is, the moisture concentrations at the surfaces of two adjacent layers correspond to the same relative humidity).

It is trivial to extend these considerations to a system with continuous gradient of composition.

Linear Irreversible Thermodynamics Formulation of Diffusion in a System with a Continuous Composition Gradient and Correlation with Easily Measurable Parameters

Consider a media p in which, through the principal diffusion direction x ($0 \leq x \leq 1$) of a penetrant q, there exists a continuous composition gradient.

Using thermodynamics of irreversible processes in the linear region and neglecting cross terms we have for the flux J_q of q along with x

$$J_q = -L \cdot \frac{d}{dx}(\mu_p - \mu_q) \tag{2}$$

where L is the Onsager coefficient (see Refs *5* and *6*).

If $C = C(x) =$ local concentration of q:

$\mu_p = \mu_p(C,x)$ therefore: $d\mu_p/dx = \partial\mu_p/\partial x + (\partial\mu_p/\delta C)\cdot dC/dx)$

$\mu_q = \mu_q(c)$ therefore: $d\mu_q/dx = (\partial\mu_q/\partial C)\cdot(dC/dx)$

and

$$J_q = -L\cdot[(\partial\mu_q/\partial C + \partial\mu_p/\partial C)\cdot(dC/dx) + \partial\mu_p/\partial x] \tag{3}$$

Here $\partial\mu_p/\delta x$ takes into account the "permanent gradient" of composition. As expected, Eq 3 reduces itself to the usual Fick's law [*1*] if the composition is homogeneous, that is, $\partial\mu_p/\partial x = 0$. Now the problem is to find a function $f(C,x)$ such that

$$J_q = -D(C,x)\cdot(dC/dx) + f(C,x) \tag{4}$$

where $D = -L\cdot(\partial\mu_q/\partial C + \partial\mu_p/\partial C)$ and $f(C,x) = -L\cdot\partial\mu_p/\partial x$.

We shall suppose in the first approximation that D is position dependent: $D = D(x)$. The extension $D = D(C,x)$ is trivial. In this case $J_q = -D(x)\cdot(dC/dx) + f(C,x)$.

In addition, we shall suppose that f is linearly dependent on C and we shall use the following Conditions 1 and 2.

1. In an absorption experiment when the materials are in equilibrium we have $C(x) = C_\infty(x)$ and $J_q = 0$ then $J_q = -D(x)\cdot(dC_\infty/dx) + f(C_\infty,x) = 0$ and $f(C_\infty,x) = D(x)\cdot(dC_\infty/dx)$;
2. Before the diffusion process takes place (time ≤ 0, that is, the material is "dry") it is $J_q = 0$ and $(dC/dx) = 0$; then $f(C = 0,x) = 0$.

Therefore, it can be shown that $f(C,x) = C(x)\cdot[D(x)/C_\infty(x)]\cdot(dC_\infty/dx)$ and the flux expression becomes

$$J_q = -D(x)\cdot[dC/dx - (C(x)/C_\infty(x))\cdot dC_\infty//dx] \quad (5)$$

From the continuity equation $\partial C/\delta t = -\partial J/\partial x$ and if we define $g(x) = [dD/dx - D\cdot d(lnC_\infty)/dx)]$ and $h(x) = [dD/dx\cdot d(lnC_\infty)/dx - D\cdot d^2(lnC_\infty)/dx^2]$ then

$$\partial C/\partial dt = D(x)\cdot\partial^2 C/\partial x^2 + g(x)\cdot\partial C/\partial x + h(x)\cdot C \quad (6)$$

This expression is easily resolvable if $D = D(x)$ and $C_\infty = C_\infty(x)$ are known. If we define a "reduced concentration" as $C^*(x) = C(x)/C_\infty(x)$, Eq 5 becomes more simple in a form that resembles Eq 1

$$J_q = -D(x)\cdot[d(C^*\cdot C_\infty)/dx - C^*\cdot(dC_\infty/dx)] = -D(x)\cdot C_\infty(x)\cdot dC^*/dx = -D^*(x)\cdot dC^*/dx \quad (7)$$

where $D^*(x) = D(x)\cdot C_\infty(x)$.

The term dC^*/dx can be seen as the driving force of the process and the term $D^*(x)$ is a form of diffusion coefficient that depends on the position.

Equations 5 and 6 reduce themselves to the usual first and second Fick's law if $dC_\infty/dx = 0$ and $D(x) =$ constant.

Two-Dimensional Finite Difference Approximation of the Problem for a (Thermosetting) Polymer Reinforced Hybrid Laminate

In the following analysis we take into account a hybrid composite parallel piped for which a dimension (say z) is greater than the other two (x and y) where a penetrant diffuses along two axes (x and y). We divide the (rectangular) x-y section in rectangular cells with the constraint that the interfaces among different materials lie always in the wall(s) of a cell. With reference to Fig. 2, every generic elementary cell, which is characterized by two indices (i,j), is bounded by four walls: west, east, north, and south. The kinetic and the saturation constants of each cell are:

1. four Diffusion coefficients D_N, D_S, D_W, D_E and so forth, defined on the walls (two D_x for the W and E walls and two D_y for the others);
2. one $C_\infty(i,j)$ in the center of the cell;
3. one wall (two for the four angular cells) of the cells which lie on the borders of the body have a penetrant concentration which correspond to the equilibrium value (C_∞).

At every time t we have a value of the moisture concentration $C_{i,j}(t)$ in the center of the cell. If two different prepregs (say A and B) are brought in contact in the lay-up it is expected that during the cure there is a little interpenetration of A and B. The easiest way to model the interface is to consider a linear gradient of distribution of A into B and vice-versa.

If the material of the cell $(i-1,j)$:$m_{i-1,j}$, is the same of the cell (i,j), $m_{i,j}$ then we define a diffusion coefficient on a cell wall as D_x or D_y of the material; for example D_w on the west direction is equal to D_x of the material $m_{i,j}$. If $m_{i-1,j}$ is different from $m_{i,j}$, we calculate D_w from $D_x(m_{i-1,j})$ and $D_x(m_{i,j})$ using a series formula

$$(\delta x_{w,i}/D_W) = [(\delta x_{c,i-1}/2)/D_x(m_{i-1,j})] + [(\delta x_{c,i}/2)/D_x(m_{i,j})] \quad (8)$$

The finite difference form of the continuity equation in the cell (i,j) is (Fig. 2)

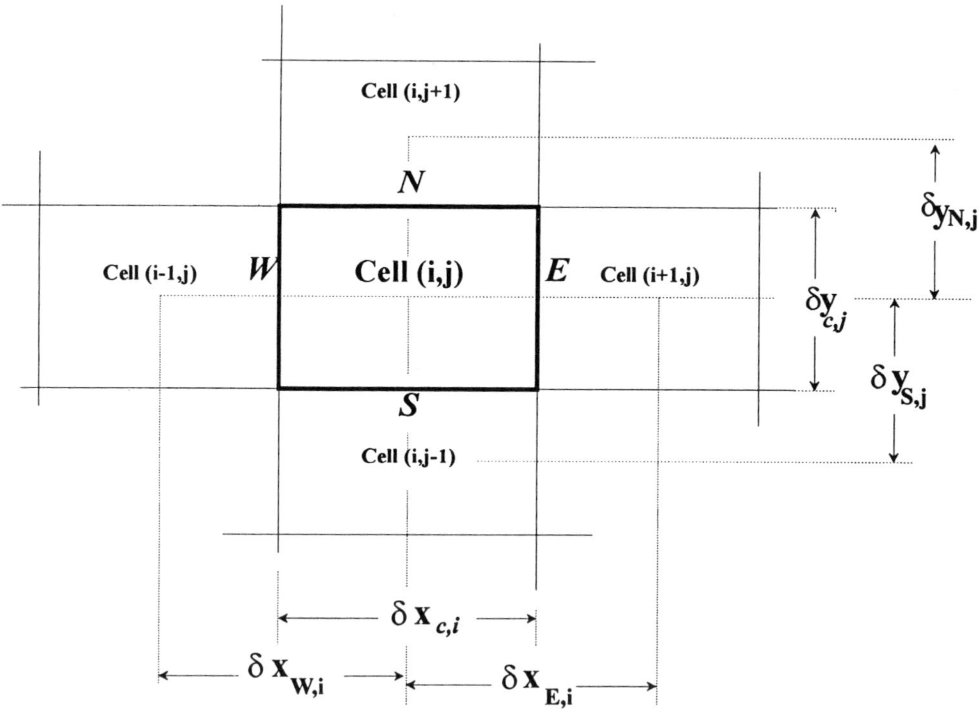

FIG. 2

$$\frac{C^{(t)}_{i,j} - C^{(t-\delta)}_{i,j}}{\delta t} = \frac{Jw(t) - J_E(t)}{\delta x_{c,i}} - \frac{Js(t) - J_N(t)}{\delta y_{c,j}} \tag{9}$$

where $J_{k(t)}$ = flux through the k wall at time t and $C^{(t)}_{i,j}$ = moisture concentration in the cell i,j at time t.

With the substitution of the finite difference expression for the fluxes (Eq 5), we shall obtain a system of linear equations which is resolvable by a computer. The finite difference expression is shown for J_W as example

$$J_W = -D_W \cdot \left\{ \frac{C^{(t)}_{i,j} - C^{(t)}_{i-1,j}}{\delta xw,i} - \frac{C^{(t)}_{i,j}}{C\infty_{i,j}} \cdot \frac{C\infty_{i,j} - C\infty_{i-1,j}}{\delta xw,i} \right\} \tag{10}$$

The stability restriction [*4*]

$$(D_x/\delta x^2)+(D_y/\delta y^2) \leq 1/(2 \cdot \delta t) \tag{11}$$

is easily satisfied. In fact D_x and D_y for composites are $10^{-6} \div 10^{-8}$ mm²/s, while a reasonable value for is 0.1 mm (the order of magnitude of a composite ply). Then the maximum δt value is $\delta t \leq \frac{1}{2}\,[D_x/\delta x^2 + D_y/\delta y^2]^{-1} \leq \frac{1}{2}\,[2 \cdot 10^{-7}/0.1^2]^{-1} = 2.5 \cdot 10^4$ s = 0.3 days which is a satisfactory value.

Validation of DIFF2D Code

The general validation of the code requires that the discretization of the transport Eq 5 and the diffusion equation be correct for the range of diffusion velocity in which we are interested. Then we must verify

1. the convergence and the stability of the 2-D discretization and
2. that the diffusion equation approximation in hybrid laminates is consistent with the experimental results.

The first condition will be checked comparing the DIFF2D results with the analytical solution for known boundary conditions. In fact for a homogeneous and isotropic parallelpiped with the z dimension much larger than the x and y dimensions, the equation of diffusion (Second Fick's law) is

$$\delta^2 C/\delta t^2 = D\cdot(\delta^2 C/\delta x^2 + \delta^2 C/\delta y^2) \tag{12}$$

with $-lx \leq x \leq lx$ and $-ly \leq y \leq ly$.

The boundary conditions in which we are interested are $C(x = \pm\, 1x,y,t) = C_\infty$ and $C(x,y = \pm\, 1y,t) = C_\infty$. In this case it is easy to verify that the solution of the differential equation (Eq 12) is a combination of the two one-dimensional solutions $G(x,t)$ and $H(y,t)$ of the diffusion equation for infinite plates of thickness, $21x$ and $21y$, respectively. These expressions are known from the literature [*4*]:

$$C(x,y,t)/C_\infty = 1 - (1-G(x,t)/C_\infty)\cdot(1-H(y,t)/C_\infty) \tag{13}$$

Integrating Eq 13 with respect to the space coordinates we have

$$Mt\% = m_x\%(t) + m_y\%(t) - [m_x\%(t)\cdot m_y\%(t)/M_\infty\%] \tag{14}$$

where $Mt\%$ is the moisture percent weight gain, $m_{x(y)}\%(t)$ is the (one-dimensional) moisture percent weight gain in an infinite plate of thickness, $21x$ ($21y$) and $M_\infty\%$ is the saturation percent weight gain. To demonstrate the equivalence of the results we have compared the $Mt\%$ computed from DIFF2D with the correspondent $Mt\%$ calculated using Eq 14 from the relative one-dimensional solutions. The comparison using Eq 14 validates the integration routine of DIFF2D too.

The test case is characterized by the following parameters, (Fig. 2): $lx = 2.5$mm, $\delta x = 0.0625$mm, $ly = 5.0$mm, $\delta y = 0.125$mm, $\delta t = 0.25$ days, $D_x = D_y = 1.0\cdot 10^{-7}$ mm^2/s, $C_\infty = 1.848\cdot 10-5$ (g/1) ($Mt\% = 1.2\%$).

For example, after 36 days (= 144 time steps) the relative error $r_{i,j}$ in each i,j position is calculated as follows

$$r_{i,j} = [C2(i,j) - c1(i,j)]/c1(i,j) \tag{15}$$

where $C2(i,j)$ is the DIFF2D results and $c1(i,j)$ is computed from Eq 13 using the one-dimensional solutions. The error is $10^{-11} \leq r_{i,j} \leq 10^{-4}$ while the mean relative error is less than 10^{-5}.

The comparison of the total moisture absorption using DIFF2D and using Eq 14 are shown in Table 1. The little percent absolute relative error are due to the integration algorithm of

TABLE 1—*Comparison of DIFF2D simulation with analytical results for 2-D diffusion.*

Time, days	$m_x\%(t)$	$m_y\%(t)$	$Mt\%$		% Absolute Relative Error
			From (14)	From DIFF2D	
4	0.1078	0.0573	0.1599	0.1607	0.5
9	0.1583	0.0825	0.2300	0.2305	0.2
16	0.2087	0.1079	0.2978	0.2982	0.1
25	0.2591	0.1331	0.3635	0.3637	0.05
36	0.3095	0.1583	0.4270	0.4272	0.04

NOTE—% absolute relative error = 100 · absolute value ($Mt\%$(DIFF2D)–$Mt\%$((13))/$Mt\%$(13)).

DIFF2D. In fact, difficulties arise when the humidity content is low and the moisture difference between contiguous points (center of the cells) is very high. If one is interested in this range of moisture content, it is necessary to increase the numbers of spatial divisions. In fact, we had verified that the increase of the number of the cells near the borders generates a reduction in the relative error.

The validation of the transport model in the case of hybrid lay up will be done following the absorption by weighing two kinds of specimens manufactured using carbon fiber-reinforced epoxy resin (CFRP) and glass fiber-reinforced epoxy resin (GFRP) (Fig. 3). The BIMAT specimens were built laying up 3 plies of GFRP prepreg + 3 plies of CFRP prepreg while the TRIMAT specimens lay-up were 4 piles of GFRP prepreg + 8 piles of CFRP prepreg + 4 plies of GFRP prepreg. The test condition was temperature = 45°C and relative humidity = 84%. In addition, the specimens BIMAT were tested also at 26°C/85%.

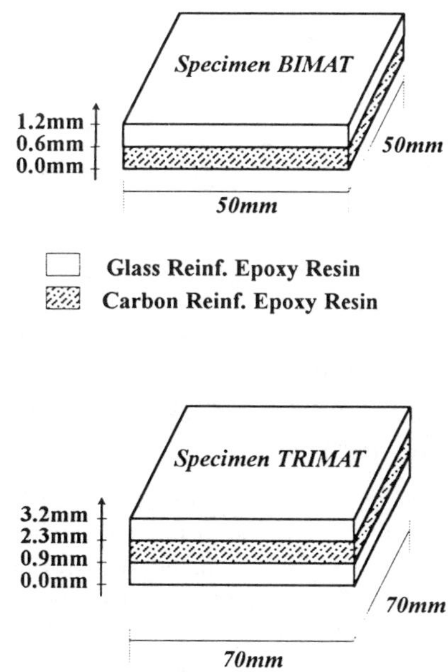

FIG. 3

The diffusion can be approximated as occurring through the thickness of an infinite (hybrid) plate because of the higher ratio (lateral dimension/thickness).

The diffusion coefficients and saturation weight gain for the moisture diffusion of the materials of interest are:

- Diffusion coefficients

$$\text{GFRP: } D_x = D_y = 0.592\,991 \cdot \exp(-5170/T(K))\ \text{mm}^2/\text{s}$$

$$\text{CFRP: } D_x = D_y = 5.077\,110 \cdot \exp(-5513/T(K))\ \text{mm}^2/\text{s}$$

- (saturation % weight gain/$_{\text{resin content\%}}$)

GFRP:$2.767 \cdot 10^{-2}$

CFRP:$2.830 \cdot 10^{-2}$

The moisture diffusion parameters (diffusion coefficients and saturation levels) were measured from appropriate test specimens as reported in Ref *7*.

The position of the interface(s) that divides different materials had been assessed observing the edge of the specimen(s) under a microscope. The resin content had been estimated taking into account

(GFRP ply thickness (mm)/fiber content (%W/W)) = $2.847 \cdot 10^{-3}$ and
(CFRP ply thickness (mm)/fiber content (%W/W)) = $2.401 \cdot 10^{-3}$.

The geometrical input data for the two kinds of specimens was

- **BIMAT**
  total thickness: 1.2 mm
  interfaces position: 0.61 mm
  number of spatial division: 30 (GFRP) + 30 (CFRP) time step: 0.25 day
- **TRIMAT**
  total thickness: 3.2 mm
  interfaces position: 0.9 mm and 2.3 mm
  number of spatial division: 40 (GFRP) + 40 (CFRP) + 40 (GFRP)
  time step: 0.25 day

The good agreement among the DIFF2D simulations and the experimental data can be seen in Figs. 4 and 5. The very little disagreement which can be found in Figs. 4 and 5 are probably due to the uncertainty of the interface positions (we have made only local measurements) and to the resin content estimation.

Example of Applications of the DIFF2D Code

DIFF2D code can be used to evaluate the equivalent time at 26°C/84% relative humidity needed to reach the same moisture distribution of the accelerated laboratory conditioning of the composite parts. In fact, to assess the (mechanical) performances of a thick composite element at the end of life it is usual to submit it to an accelerated laboratory hot/wet conditioning. Then it is necessary to know what period of "real life" does represent the accelerated aging. Temperature = 26°C and relative humidity = 84% is recognized to be the "natural" average worst world-wide hot/wet environmental condition [*8*].

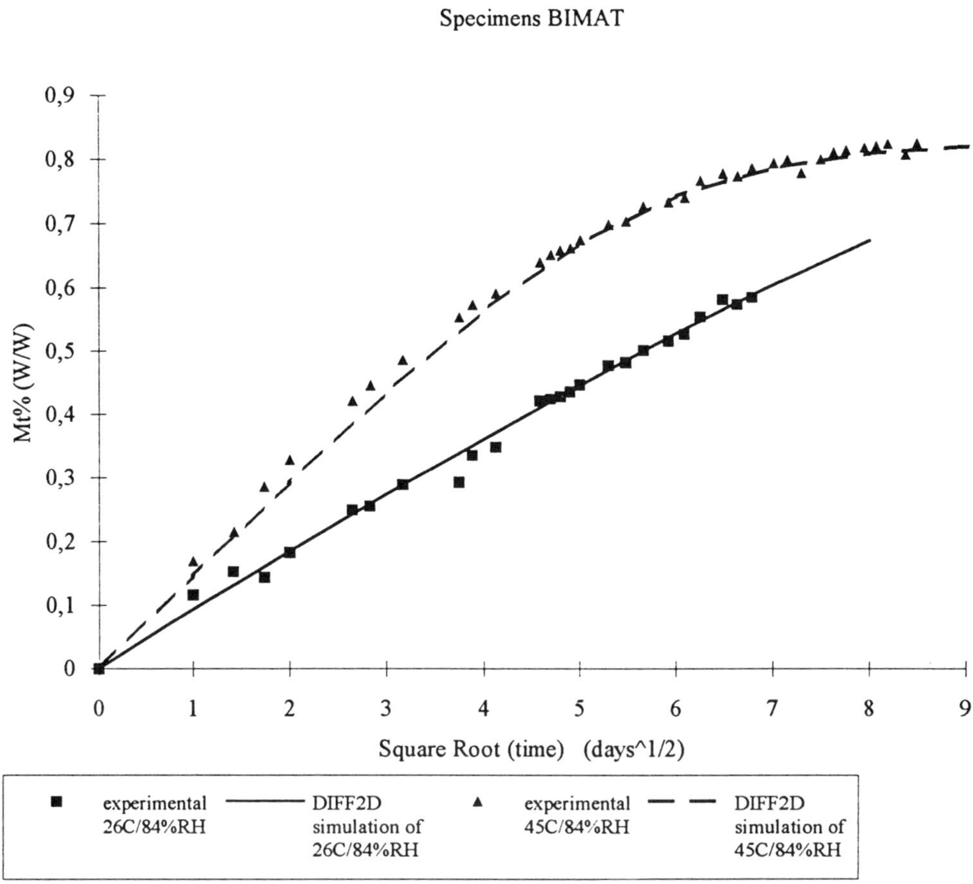

FIG. 4

We have aged, using the cycle shown in Table 2, the structural element sketched in Fig. 6. The choice of the three steps cycle had been in some aspect arbitrary, although some data [9] supports the suitability of this kind of accelerated aging to reproduce the much smoother natural low temperature moisture uptake. To reduce the aging time, we had increased the conditioning temperature and probably had overcome the conditions in which the Fick diffusion occurs. Therefore, to assess the moisture distribution it had been necessary to "slice" the composite. Every slice was dried (in vacuum at high temperature) and from the weight decrease we had been able to determine the through the thickness moisture content.

Before computing the equivalent time at 26°C/84% relative humidity we had tested the model and hypothesis made on the elements. In fact, the first step is the comparison of the moisture gain (measured by weighing some travellers) versus the simulated by DIFF2D in the first step of the aging cycle (at 50°C/84% relative humidity). The last curve agrees satisfactorily with the experimental point (Fig. 7). After doing this we had simulated the moisture absorption of the element using as input data the same geometry as above and the appropriate diffusion coefficients and moisture saturation values. Therefore, we had assessed

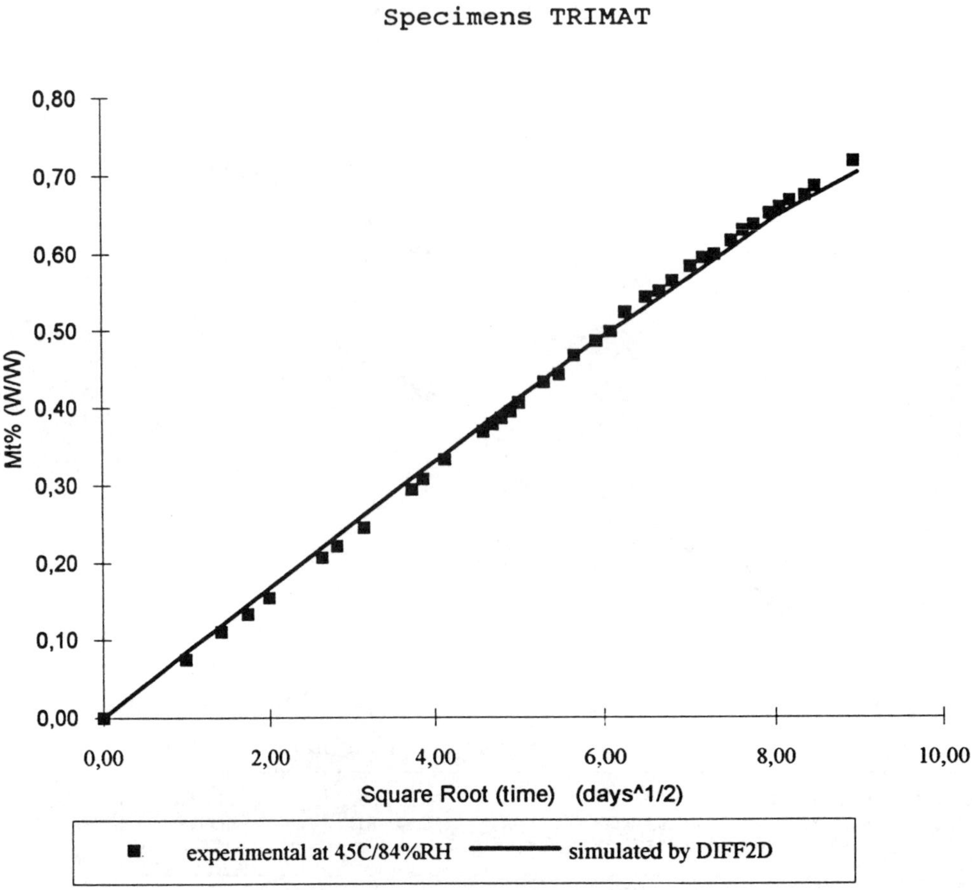

FIG. 5

the time needed at 26°C/84% to reach the same through the thickness situation as determined by slicing. In the particular case shown here, this time corresponds to 6 years. In fact, Fig. 8 shows that the two profiles are very similar [*7*].

Conclusion

By applying the principles of irreversible processes thermodynamics it might be concluded that the numerical simulation of moisture diffusion in a 2-D hybrid laminate is possible. A

TABLE 2—*Laboratory accelerated environmental aging of the test structural element.*

STEP	From, month	To, month	Temperature, °C	Relative Humidity, %
I	0	6	50	84
II	6	8	60	84
III	8	12	50	84
dwelling	12	test	26	84

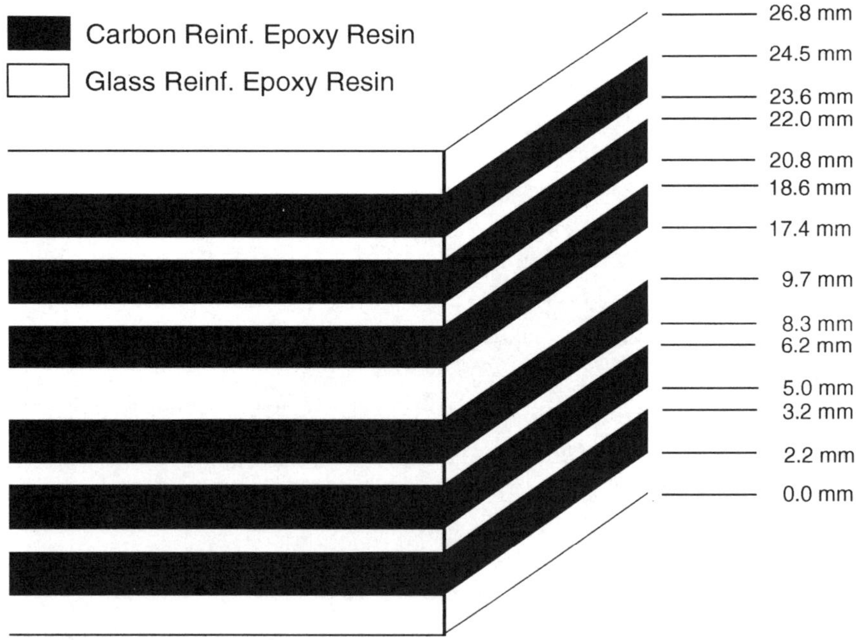
Carbon Reinf. Epoxy Resin
Glass Reinf. Epoxy Resin
26.8 mm
24.5 mm
23.6 mm
22.0 mm
20.8 mm
18.6 mm
17.4 mm
9.7 mm
8.3 mm
6.2 mm
5.0 mm
3.2 mm
2.2 mm
0.0 mm

FIG. 6

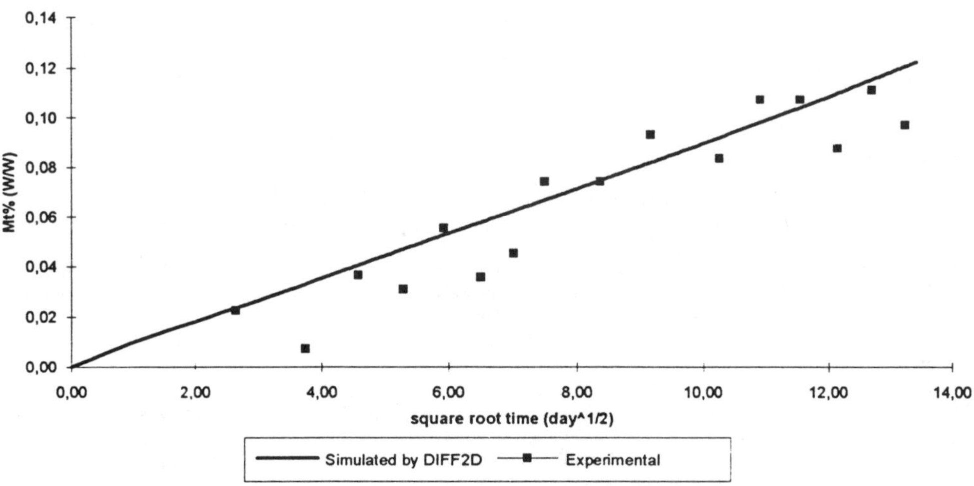
Mt% (W/W)
0,14
0,12
0,10
0,08
0,06
0,04
0,02
0,00
0,00
2,00
4,00
6,00
8,00
10,00
12,00
14,00
square root time (day^1/2)
Simulated by DIFF2D
Experimental

FIG. 7

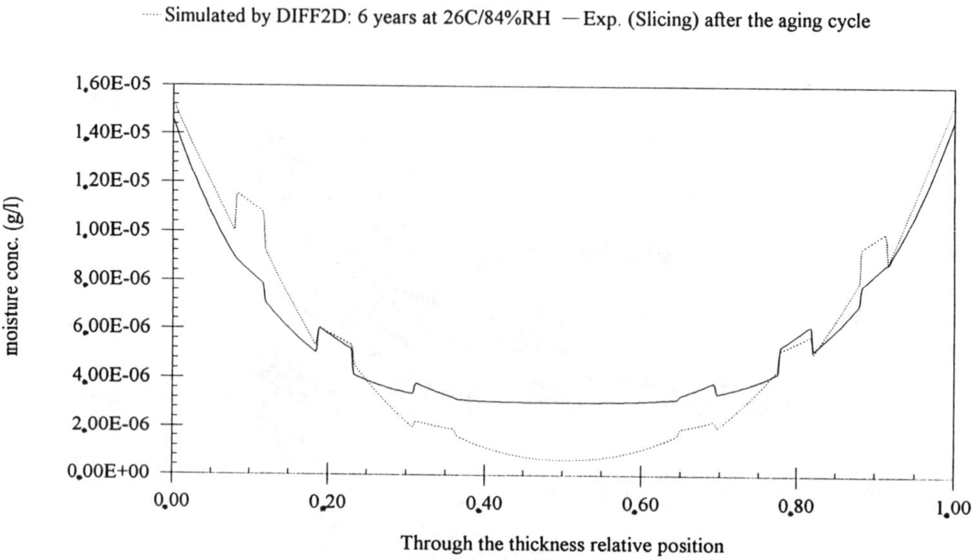

FIG. 8

limited number of experimental data regarding the materials in the test condition and the actual geometry (including the interfaces position) are necessary.

The DIFF2D code can simulate the moisture diffusion only in rectangular sections of hybrid. The application of DIFF2D to a too fast process (diffusion coefficients higher than 10^{-6} mm^2/s) can lead to unacceptable time step and cell size or both. In this last case, a complete implicit difference code is required.

Acknowledgments

The author wishes to thank Mr. T. A. Collings (formerly of the Defence Research Agency of Farnborough U.K.) for the slicing analysis, Mrs. S. M. Guerra for the experimental measurements, and Mr. C. A. Zanotti for the helpful comments and the support.

References

[1] Springer, G. S., "Moisture and Temperature Induced Degradation of Graphite Epoxy Composites," *Environmental Effects on Composite Materials*, Vol. 2, G. S. Springer, Ed., Technomic Publishing Co., Westport, CT, 1981, pp. 6–19.

[2] Springer, G. S., "Numerical Procedures for the Solution of One Dimensional Fickian Diffusion Problems," *Environmental Effects on Composite Materials*, Vol. 1, G. S. Springer, Ed., Technomic Publishing Co., Westport, CT, 1981, pp. 166–199.

[3] Denbigh, K., *The Principles of Chemical Equilibrium*, 1st Edition, Casa, Ed., Cambridge University Press, Italian Translation, Ambrosiana, Milan, 1977.

[4] Crank, J., *The Mathematics of Diffusion*, 2nd Edition, Oxford University Press, London, 1975.

[5] Sternberg, S. and Rogers, C. E., "Sorption and Diffusion in Asymmetric Membranes," *Journal of Applied Polymer Science*, Vol. 12, 1968, pp. 1017–1029.

[6] Frisch, H. L., "Multicomponent Non-Fickian Diffusion Through Inhomogeneous Polymer Membranes," *Journal of Polymer Science*, Part A, Vol. 2, 1964, pp. 1115–1119.

[7] Cappelletti, C., Rivolta, A., and Zaffaroni, G., "Environmental Effects on Mechanical Properties of Thick Composite Structural Elements," *Journal of Composite Technology & Research*, JCTRER, Vol. 17, No. 2, April 1995, pp. 107–114.

[8] Collings, T. A., "The Effect of Observed Climatic Conditions on the Moisture Equilibrium Level of Fibre Reinforced Plastic," *Composites*, Vol. 17, Jan. 1986, pp. 33–41.

[9] Marks, N. and Dalziel, A., "Artificial Ageing of Fiber Reinforced Composite Materials — Three Stage Method," Proceedings of the Third European Conference on Composite Materials, A. R. Bunsel, P. Lamicq, and A. Massiah, Eds., 20–23 March 1989, Bordeaux, pp. 587–594.

Y. Uematsu,[1] *T. Kitamura,*[2] *and R. Ohtani*[2]

Creep-Fatigue Interaction in Delamination Crack Propagation of Advanced CFRPs at High Temperatures

REFERENCE: Uematsu, Y., Kitamura, T., and Ohtani, R., **"Creep-Fatigue Interaction in Delamination Crack Propagation of Advanced CFRPs at High Temperatures,"** *High Temperature and Environmental Effects on Polymeric Composites: 2nd Volume, ASTM STP 1302,* Thomas S. Gates and Abdul-Hamid Zureick, Eds., American Society for Testing and Materials, 1997, pp. 110–132.

ABSTRACT: The objective of this study is to elucidate creep-fatigue interaction in Mode I delamination crack propagation of polymers reinforced by carbon fibers at high temperatures. The materials tested are two unidirectionally reinforced laminates, AS4/PEEK (carbon fiber: AS4, matrix: poly-ether-ether-ketone) and T800H/PMR-15 (carbon fiber: T800H, matrix: polyimide). Crack propagation tests are conducted in cyclic loading conditions with and without hold(s) at maximum tension and zero load at 473K for AS4/PEEK laminates and 573K for T800H/PMR-15 laminates, respectively. In fatigue with high frequency, the crack propagation rate per unit cycle da/dN is correlated well with the stress intensity factor range ΔK. However, the crack propagation in AS4/PEEK laminates depends strongly on the load waveform, while that in T800H/PMR-15 laminates is independent of it. The crack propagation in AS4/PEEK laminates is accelerated by the tensile load hold due to the creep deformation of matrix and it is purely time-dependent. In the time-dependent crack propagation, the rate per unit time da/dt is correlated well with the stress intensity factor K at the hold. On the other hand, the time-dependent crack propagation is decelerated by the zero-load-hold because of the creep recovery of matrix.

KEY WORDS: delamination, crack propagation, high temperature, CFRP, poly-ether-ether-ketone, polyimide, creep, fatigue, creep-fatigue interaction, creep recovery

Because of excellent performance such as high stiffness and superior relative strength, carbon fiber-reinforced polymers (CFRPs) are being widely used in many commercial fields. In terms of application to the primary aircraft structures, which is one of the weight-critical applications, the evaluation of their reliability has been of great concern in recent years. Some of the CFRPs are also expected to be used as the structural material of advanced aerospace planes of which components are exposed to the high-temperature environment. As the failure of such structures brings about serious accident, there is a strong demand for structural reliability.

[1] Assistant, Department of Mechanical Engineering and Systems, Graduate School of Engineering, Osaka University, Yamadaoka, Suitashi, Osaka 565, Japan.

[2] Associate professor and professor, respectively, Department of Engineering Physics and Mechanics, Graduate School of Engineering, Kyoto University, Yoshida-Honmachi, Sakyo-ku, Kyoto 606-01, Japan.

Delamination is a common failure mode in CFRPs and great attention is paid to the mechanics in the propagation of delamination cracks. ASTM recently published a test method, ASTM D 5528, Mode I Interlaminar Fracture Toughness of Continuous Fiber-Reinforced Composite Materials. Although the fracture toughness and the fatigue crack propagation at room temperature have been intensively investigated on the basis of the fracture mechanics concept [*1-7*], few research works have been conducted on the behavior at elevated temperatures [*8-14*]. Since the failure of polymer at high temperatures is strongly affected by the time-dependent deformation, namely creep [*15*], it is important to evaluate the effect of creep-fatigue interaction on the durability of CFRPs used for high-temperature components. Moreover, it should be noted that the creep takes place even at room temperature for some CFRPs. Considering long-term use as a component, this also signifies the importance of understanding the coupling effect between time-dependent (creep) and cycle-dependent (fatigue) delamination mode (fatigue).

In this study, the authors investigate the delamination crack propagation under Mode I cyclic loading of CFRPs in creep-fatigue interaction at high temperatures. The effects of load hold at tension and zero load on the propagation are examined in order to elucidate the influence of creep and creep recovery of matrix.

Experimental Procedure

The materials tested are two kinds of unidirectionally reinforced CFRPs. One of them consists of a carbon fiber, AS4, and a thermoplastic polymer, poly-ether-ether-ketone (PEEK) and is denoted by AS4/PEEK. The other comprises a carbon fiber, T800H, and a thermoset polymer, polyimide (PMR-15), and is labeled as T800H/PMR-15. Both laminates have excellent heat-resisting durability and are used for high-temperature applications. Table 1 summarizes the constitutions of both laminates.

The crack propagation tests are conducted using double cantilever beam (DCB) specimens where the fiber direction is aligned along the direction of crack propagation, as shown in Fig. 1. Upilex film or aluminum foil, which is treated with release agent on both sides, is inserted at the edge of the specimen during fabrication as the starter of a delamination crack. The detail of DCB specimens is summarized in Table 2. The creep-fatigue tests are conducted by an electro-mechanical or an electro-servo hydraulic fatigue testing machine with an infrared radiation furnace. The load is applied to the specimen through hinges at the specimen edges (Fig. 1). The temperature is monitored by thermo-couples that are attached to the DCB specimen and is controlled within ±3K during the tests.

For the AS4/PEEK laminate, tests are conducted at 473 K (200°C) under the load ratio R ($=P_{min}/P_{max}$) of 0.1 or 0, and the following three kinds of load waveforms: (1) sinusoidal or rectangular load waveform with high frequency (R = 0.1), (2) trapezoidal with tensile

TABLE 1—*Constitution of test materials, AS4/PEEK and T800H/PMR-15.*

CFRP	AS4/PEEK	T800H/PMR-15
Volume fraction of fiber, V_f	62%	60%
Constitution of laminate	$(0)_{48}$	$(0)_{40}$
Crystallinity	34%	...
Glass transition point, T_g	413K	613K
Melting point, T_m	603K	...

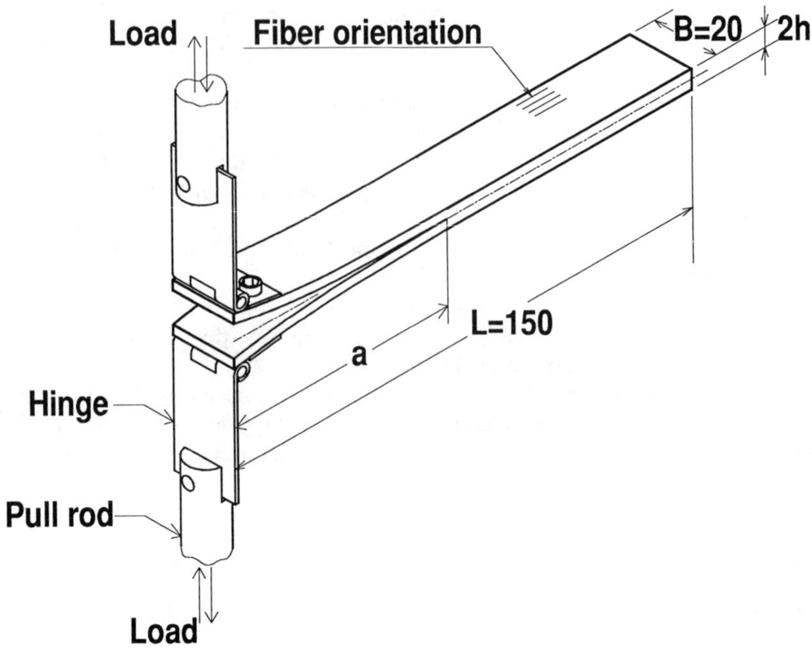

FIG. 1—*Double cantilever beam (DCB) specimen.*

load hold (R = 0.1), and (3) trapezoidal one with zero load hold as well as tensile load hold (R = 0). The detailed loading conditions are listed in Table 3.

For the T800H/PMR-15 laminate, the tests are conducted at 573 K (300°C) under Conditions 1 and 2 with R = 0.1. In the test with the load frequency of ν = 0.01 Hz, the crack is extended to about 50 mm from the initial slit by the fatigue with 1 Hz before the test in order to shorten the experimental time. The test conditions for the T800H/PMR-15 laminate are summarized in Table 4.

Fracture surface is examined by means of a scanning electron microscope (SEM) after the test.

Fracture Mechanics Parameter

It has been reported that the relationship between the elastic compliance λ and the crack length a in the DCB specimen can be formulated as follows [*16*].

TABLE 2—*DCB specimen.*

CFRP	AS4/PEEK	T800H/PMR-15
Thickness of specimen	6 mm	5.6 mm
Inserted film	poly-imide (or aluminum foil)	poly-imide
Thickness of film	13 μm (or 15 by 2 μm)	7.5 μm
Initial crack length	45 ~ 50 mm	45 mm

TABLE 3—*Test conditions of AS4/PEEK.*

Load Waveform	Test No.	R= P_{min}/P_{max}	P_{max}, N	ν, Hz	t_{HH}, s	t_{HL}, s	Symbol in figures
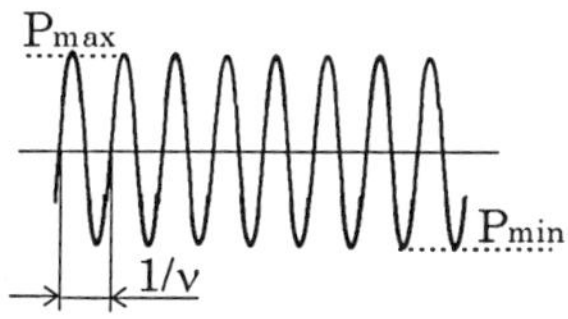	1	0.1	228	2	...	...	○
	2		190	2	...	...	△
	3		150	2	...	...	□
	4		190	0.5	...	...	▽
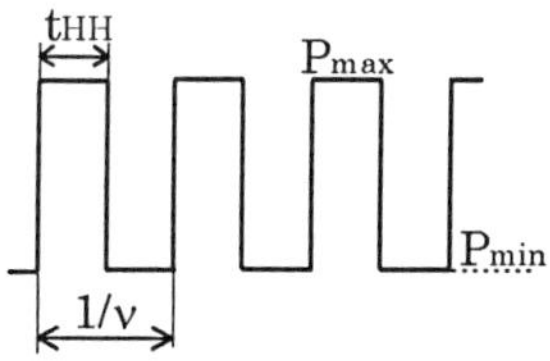	5	0.1	228	0.5	1	1	◇
	6		228	0.05	10	10	◩
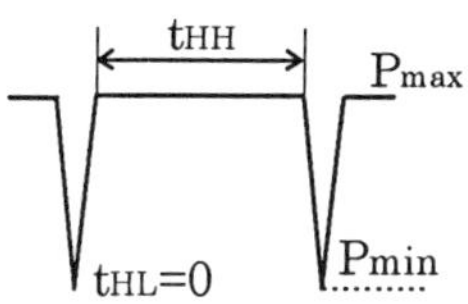	7	0.1	228	...	360	0	●
	8		190	...	3600		▲
	9		228	...	3600		■
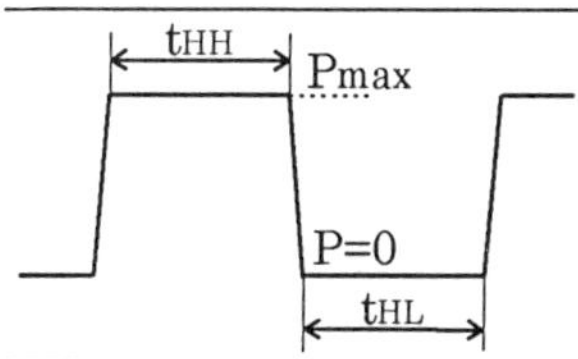	10	0	228	...	3600	360	⊕
	11		228	...	3600	3600	⊞
	12		249	...	3600	10800	◈
Static creep	13	...	249	...	...	...	+
	14	...	228	...	...	...	×

$$\lambda^{1/3} = C_0 + C_1 \cdot \frac{a}{h'} \qquad (1)$$

where

C_0, C_1 = material constants and
h = half thickness of specimen.

The elastic compliance of the DCB specimen used in this study was experimentally examined at the test temperature by measuring the crack opening displacement, and the values of C_0 and C_1 were determined as listed in Table 5 [*10,12*]. On the other hand, the energy release rate G is evaluated by the following equation on the basis of the concept of linear elastic fracture mechanics

TABLE 4—*Test conditions of T800H/PMR-15.*

Load Waveform	Test No.	R= P_{min}/P_{max}	P_{max}, N	ν, Hz	t_H, s	Symbol in figures
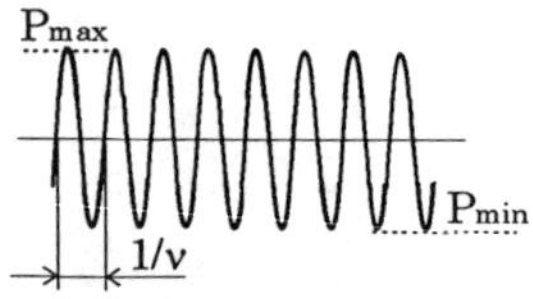	15		80		...	⊙
	16	0.1	75	1	...	△
	17		65		...	⊡
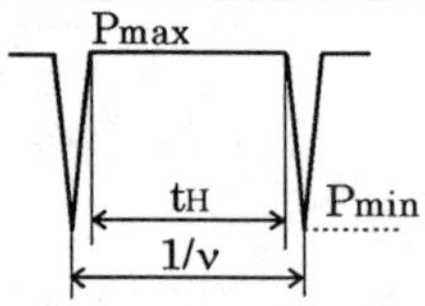	18	0.1	80	1 ↓ 0.01	99	◇ ↓ ◆

$$G = \frac{P^2}{2B} \cdot \frac{d\lambda}{da} \tag{2}$$

where

P = load and
B = width of specimen.

Then, the elastic stress intensity factor K is evaluated as follows [*17*].

$$K = \sqrt{\frac{G}{H}} \tag{3}$$

where

H = a constant calculated from the elastic constants.

The detailed form of this constant H is described in Ref *10*.

Fatigue Crack Propagation

Figure 2 shows the relationship between the crack propagation rate per unit cycle da/dN and the elastic stress intensity factor range ΔK in AS4/PEEK laminates under fatigue conditions with the high frequency (sinusoidal or rectangular load waveform). The small symbols

TABLE 5—*Constants in Eq 1.*

CFRP	Temperature, K	Constants	
		C_0, mm/N$^{1/3}$	C_1, mm/N$^{1/3}$
AS4/PEEK	473	0.600	0.139
T800H/PMR-15	573	0.394	0.133

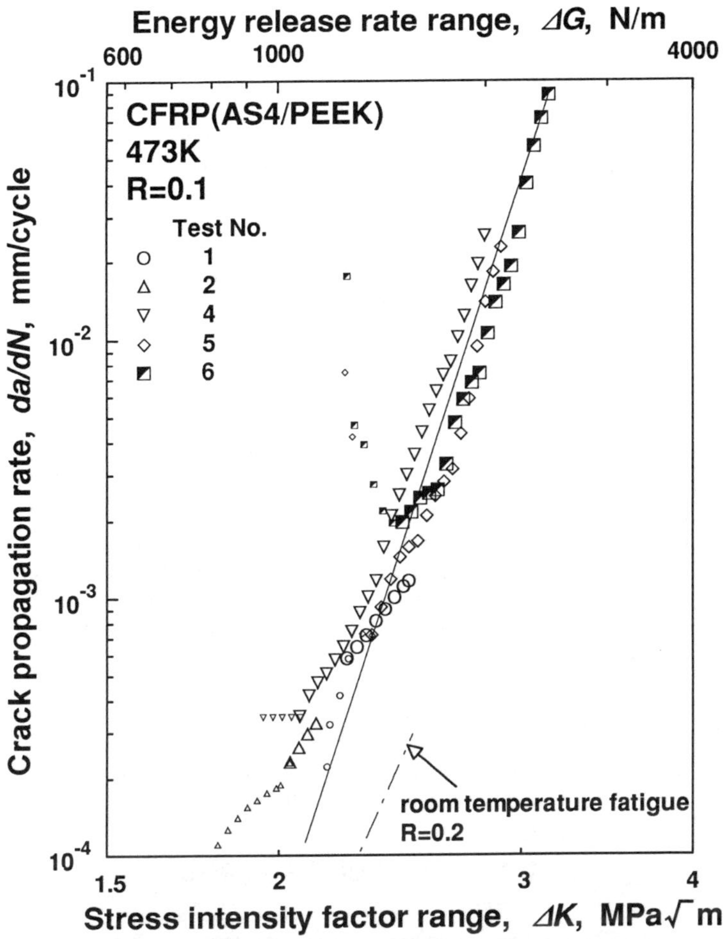

FIG. 2—*Relationship of AS4/PEEK between crack propagation rate per unit cycle* da/dN *and stress intensity factor range* ΔK.

in Fig. 2 point out the results obtained at the very early stage of propagation, where the propagation rate decreases as the crack propagates due to the formation of fiber bridging [*18,19*]. After the crack propagates to the length of about 5 mm from the initial slit, the effect of fiber bridging tends to saturate and the propagation is gradually accelerated due to the increase of the stress intensity. In this paper, the focus is put on the crack propagation at the latter stage, as the effect of fiber bridging was discussed in other papers in detail [*11,13*].

Figure 2 reveals that the $da/dN-\Delta K$ relation is scarcely dependent on the period of cycle, indicating that the crack propagation is purely cycle-dependent. Thus, da/dN is correlated well with ΔK and the relation can be indicated by the straight line. The K-governed correlation implies that, even for such high propagation rate and a ductile matrix, the crack propagates under the small scale yielding (SSY) condition. It may be attributed to the constraint of the plastic deformation of the matrix by fibers. The dotted line in the figure represents the $da/dN-\Delta K$ relation in fatigue at room temperature [*7*]. The difference points out that the resistance against the delamination becomes weaker as the temperature goes up. On the other hand, since the fracture toughness is improved at high temperatures [*9,10*], the fatigue crack propagates steadily until $\Delta K = 3.2$ MPa·m$^{1/2}$ at 473 K while it becomes unstable at ΔK of about 2.5 MPa·m$^{1/2}$ at room temperature [*7*].

The SEM fractograph is shown in Fig. 3. Most of the fracture surface is covered by the thin polymer and characterized by the matrix fracture near fibers. It has been reported that matrix cracking is dominant in the fatigue of AS4/PEEK laminate at room temperature [*7*]. This signifies the similarity in the failure mechanism at the crack tip between high-temperature fatigue and room-temperature fatigue.

The relationship between *da*/*dN* and ΔK of T800H/PMR-15 laminates is shown in Fig. 4. The fluctuation of the *da*/*dN*-*K* relation in the lower ΔK region is larger than that of AS4/PEEK laminates. This is because the fiber bridging in T800H/PMR-15 laminates is more pronounced than that in AS4/PEEK laminates so that the effect does not saturate at the lower ΔK region [*12*]. Comparing AS4/PEEK with T800H/PMR-15 laminates, the former has higher resistance for the crack propagation than the latter.

Figure 5*a* shows the SEM fractograph of T800H/PMR-15 laminate in high-temperature fatigue. It clearly points out that the fracture occurs in the brittle manner along the interface between fibers and matrices. The fracture toughness tests at high temperatures [*9*] bring the fracture of similar morphology as indicated in Fig. 5*b*. The resemblance of cracking morphology between fracture toughness and fatigue is often reported for CFRPs with brittle matrices, for example, CF/Epoxy laminates [*7*]. This suggests that the crack propagation under cyclic loading is governed by the static failure near the crack tip while it is subjected to the tensile stress in each cycle.

Effect of Tensile Load Hold on Crack Propagation

Figure 6 shows the effect of tensile load hold on the crack propagation rate *da*/*dN* of AS4/PEEK laminates for equal values of stress intensity factor at the hold *K*. The crack

FIG. 3—*SEM fractograph of AS4/PEEK in the cycle-dependent fatigue. Test 2,* P_{max} = *190 N,* ν = *2 Hz.*

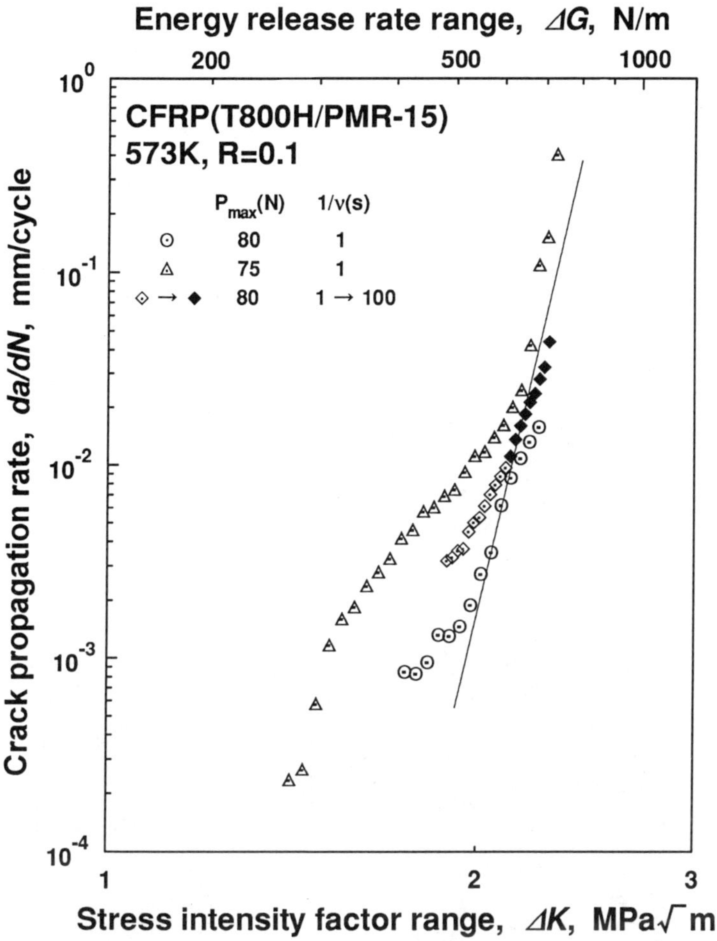

FIG. 4—*Relationship of T800H/PMR-15 between crack propagation rate per unit cycle* da/dN *and stress intensity factor range,* ΔK.

propagation is accelerated by the hold and da/dN is approximately in proportion to the hold time t_{HH}. Figures 7*a* and 7*b* show the relationships between the crack propagation rate per unit cycle da/dN (mm/cycle), and the frequency of load cycle, ν (Hz), and the crack propagation rate per unit time, da/dt (mm/h) and ν, respectively, including the results in fatigue without the load hold. The results of static creep tests are also shown at the left end in Fig. 7*b*. Figure 7*a* shows that da/dN is independent of the period in the high frequency area ($\nu \geq 0.05$ Hz) so that the crack propagation is purely cycle-dependent. On the contrary, from Fig. 7b, da/dt is independent of ν in the longer load cycle. This means that the crack propagation in the fatigue with low frequency is purely time-dependent and the rate da/dt is nearly constant during the load hold.

Figure 8 shows the relationship between the rate per unit time da/dt and the elastic stress intensity factor at the load hold K in the time-dependent fatigue with $t_{HH} > 10$ s. The crack propagation rate is correlated well with K and the relationship coincides with that in static creep [*8*]. It implies that the time-dependence of crack propagation originates from the matrix

FIG. 5—*SEM fractographs of T800H/PMR-15. (a) Test 16,* P_{max} *= 75 N, ν = 1 Hz. (b) Fracture toughness test at 573 K.*

creep. Moreover, it is well-known [*20*] that the stress singular field in large-scale creep (LSC) conditions, where the creep zone largely extends around the crack tip, is represented by creep *J*-integral J^* instead of *K*. Thus, the *K*-governed correlation in Fig. 8 points out that the crack propagation is characterized by small-scale creep (SSC) conditions, where the matrix creep is confined in the zone near the crack tip as schematically illustrated in Fig. 9 because the carbon fibers strongly constrain the deformation in the matrix. The detail of crack propagation in the SSC of the CFRP was discussed elsewhere [*8*]. The relation between da/dt [mm/h] and K (MPa·m$^{1/2}$) is formulated as

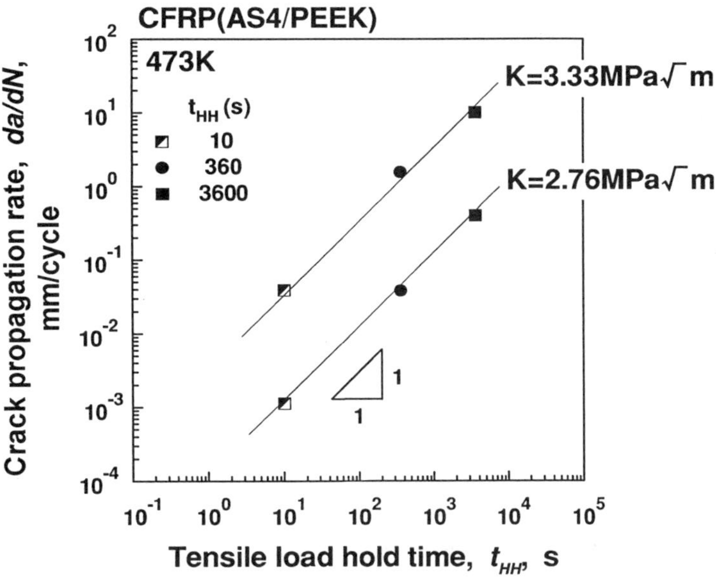

FIG. 6—*Relationship of AS4/PEEK between crack propagation rate per unit cycle* da/dN *and tensile load hold time* t_{HH} *under constant stress intensity factor range.*

$$\frac{da}{dt} = 1.60 \times 10^{-11} K^{22.9} \tag{4}$$

Since the exponent of K in Eq 4 is large (22.9), the crack propagation during the stress change in the trapezoidal waveform is negligible.

The SEM fractograph in the time-dependent fatigue (t_{HH} = 3600 s), which is shown in Fig. 10, is characterized by interfacial failure between fibers and matrices with the elongated polymer torn in pieces. This is similar to the fractograph obtained in static creep [*8*].

Figure 11 shows the relationship between da/dN and ν in T800H/PMR-15 laminates at the equal value of ΔK. It is clear that da/dN is independent of the frequency and the tensile load hold. This implies that the creep scarcely occurs at this temperature (573 K) and the crack propagation is purely cycle-dependent, which is attributed to the characteristics of the thermoset matrix, PMR-15. The SEM observation reveals that the fracture surface of Test 18 (ν = 0.01 Hz, Table 4) is similar to that of Test 16 (ν = 1 Hz, Fig. 5*a*), which also proves that the crack propagation is little affected by the load hold. However, the time-dependence might appear in long-term use at high temperatures.

Effect of Zero-Load-Hold on Time-Dependent Crack Propagation

Deceleration of Crack Propagation

Figure 12 shows the relationship of AS4/PEEK laminates between time-dependent da/dt and K under cyclic loading conditions with the zero-load-hold as well as the tensile load hold. The solid line in Fig. 12 represents Eq 4 in the condition without the zero-load-hold discussed in the previous section (Fig. 8), and the comparison with the dotted line for the

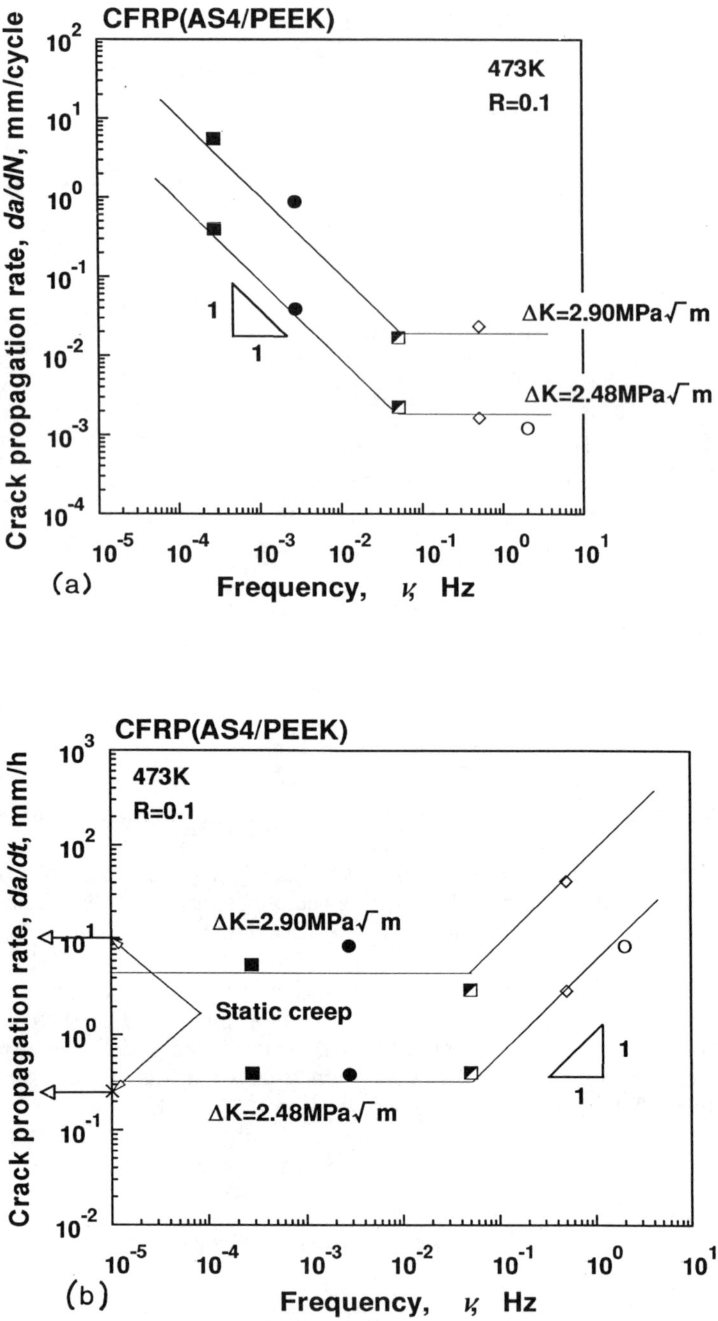

FIG. 7—*Relationship of AS4/PEEK between crack propagation rate and frequency. (a) Relationship between crack propagation rate per unit cycle* da/dN *and frequency* ν. *(b) Relationship between crack propagation rate per unit time* da/dt *and frequency* ν.

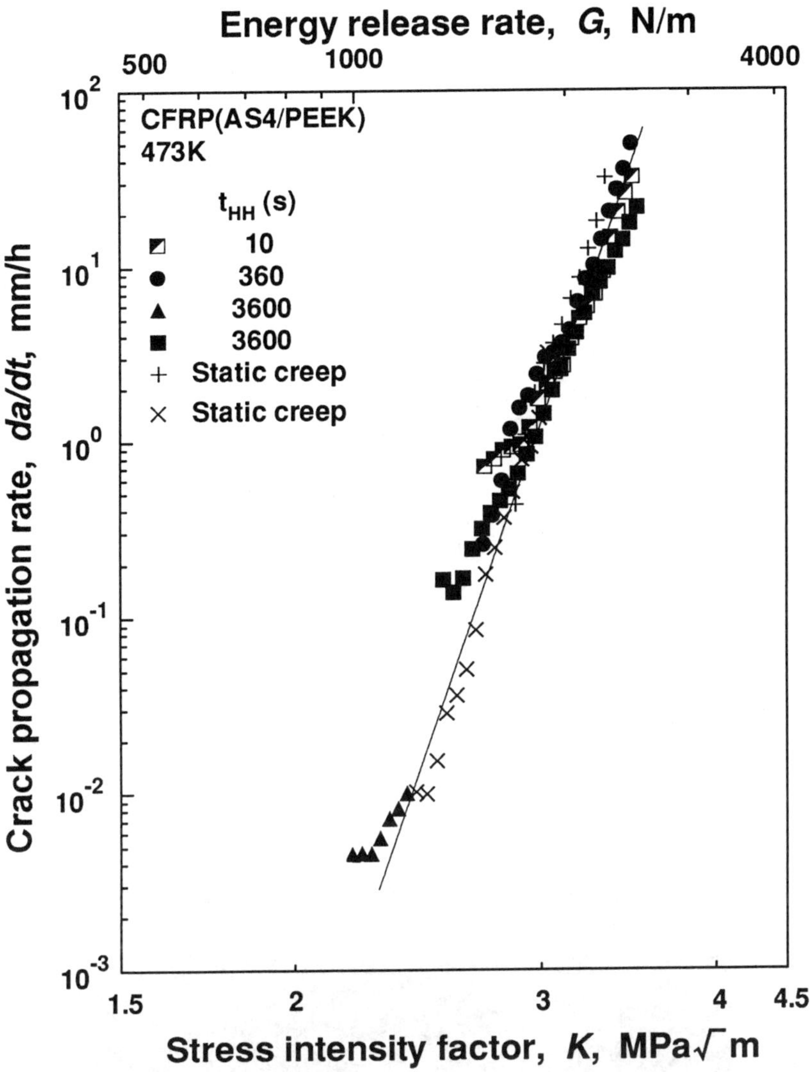

FIG. 8—*Relationship of AS4/PEEK between crack propagation rate per unit time* da/dt *and stress intensity factor* K.

zero-load-hold condition. The zero-load-hold eminently decreases *da/dt* for an equal value of *K* at the tensile load hold. This implies that the crack propagation in the time-dependent fatigue is strongly dependent on the load waveform. Figure 13 shows the deceleration of crack propagation against the zero-load-hold period. As the hold time is longer, the crack propagation becomes slower and reaches a saturated value. Figure 13 also reveals that *da/dt* saturates faster at a higher stress intensity.

The fractograph in Test 10 with the zero-load-hold is shown in Fig. 14. This is similar to that of Test 9 without the zero-load-hold (Fig. 10), suggesting that the crack propagates due to the matrix creep as well. The morphology near the crack tip is examined by means of an

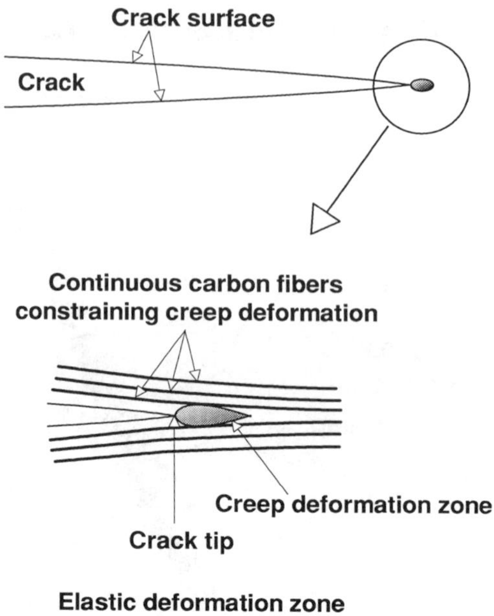

FIG. 9—*Schematic illustration of small-scale creep (SSC) condition.*

FIG. 10—*SEM fractograph of AS4/PEEK in the time-dependent fatigue. Test 9,* P_{max} *= 228 N,* t_{HH} *= 3600 s,* t_{HL} *= 0 s.*

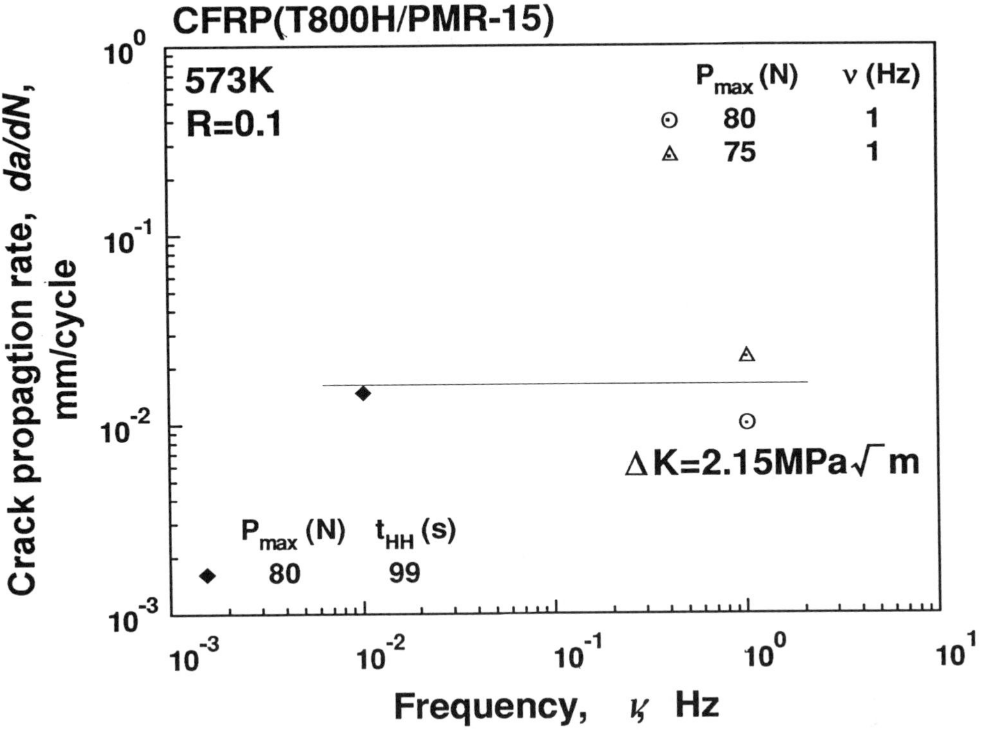

FIG. 11—*Relationship of T800H/PMR-15 between crack propagation rate per unit cycle* da/dN *and frequency ν.*

optical microscope [*12*] and it is found that the bridging state, for example, density of bridged fibers along the crack front, is scarcely dependent on the zero-load-hold. Moreover, the crack propagation rate at the initial slit, which can not be affected by the fiber bridging, is compared in both cases in order to eliminate the effect of bridging. As listed in Table 6, the crack propagation rate under the zero-load-hold is much slower than that without the hold. Therefore, it can be estimated that the deceleration does not originate from the fiber bridging.

Figure 15 shows the relation between da/dN and ΔK in the tests with the zero-load-hold (larger symbols) in comparison with the crack propagation in the cycle-dependent fatigue (Fig. 2). Since da/dN under the zero-load-hold is much higher than da/dN in the cycle-dependent fatigue, the effect of zero-load-hold does not completely cancel out the creep damage introduced during the tensile load hold.

Creep Recovery of Matrix during Zero Load

In order to identify the cause of the above-mentioned deceleration, creep deformation tests of PEEK are conducted in cyclic load conditions such as (1) the loading with only tensile stress hold and (2) the loading with tensile stress hold and zero-stress-hold, as shown in Table 7. Plate specimens of PEEK without carbon fibers, of which width, gage length, and thickness are 8, 20, and 4.8 mm, respectively, are subjected to the cyclic stress. The tests are carried out by means of an electro-hydraulic fatigue testing machine with a radiation furnace at 473 K in an air after 5 h exposure in a high-temperature environment. Because

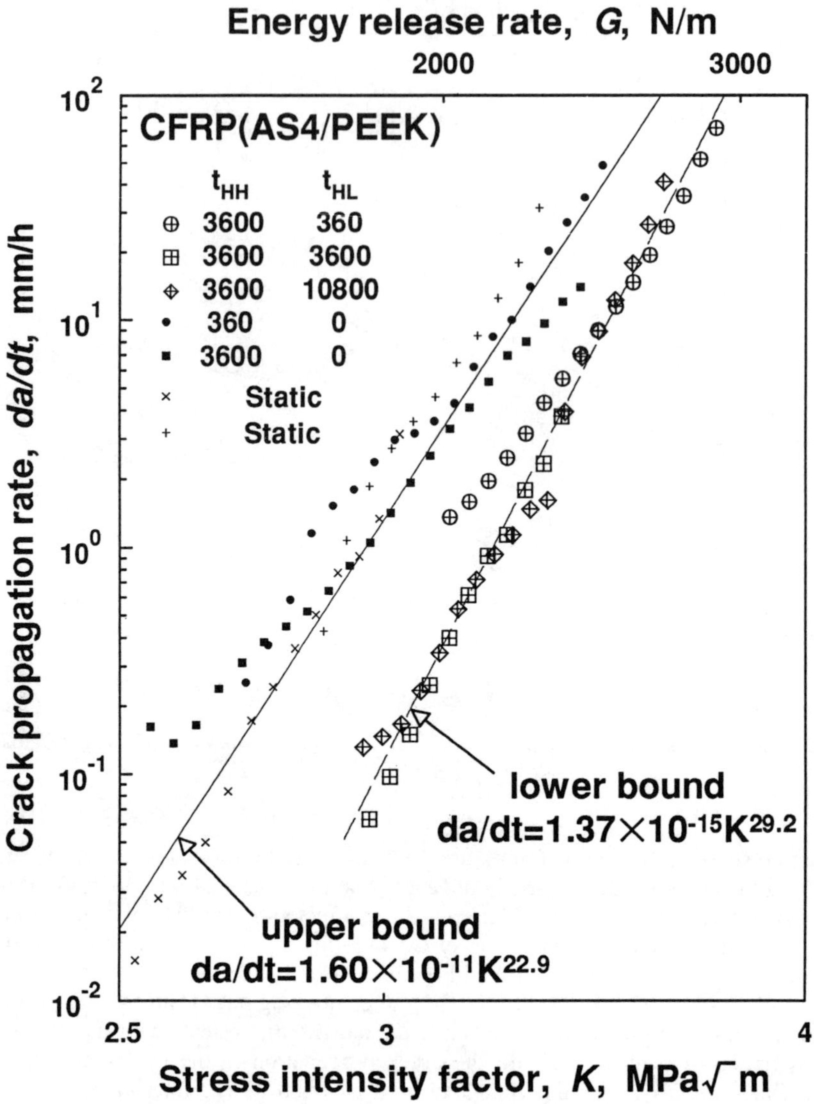

FIG. 12—*Relationship between crack propagation rate per unit time* da/dt *and stress intensity factor* K *in the time-dependent fatigue with and without the zero-load-hold.*

of differences in the fabrication process, the material cannot be identical to the matrix of the AS4/PEEK used in the delamination crack propagation tests. The purpose of these experiments, however, is put on the qualitative inquiry of the deformation behavior of matrix under cyclic creep loadings.

Figures 16*a* and 16*b* show the change in the strain during a cycle in the tests with and without the zero-stress-hold, respectively. In both cases, creep takes place during the tensile stress hold after the instantaneous strain and the transient creep are observed at the early stage of hold period. Figure 16*a* shows that a greater part of the creep strain remains in the

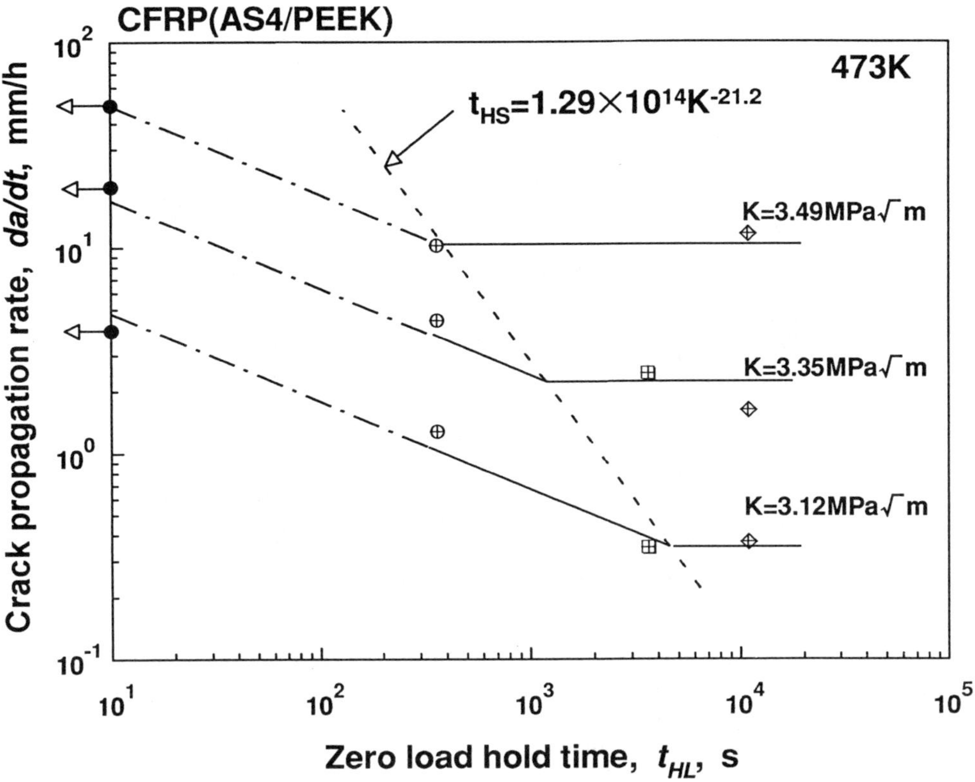

FIG. 13—*Relationship between crack propagation rate* da/dt *and zero-load-hold time* t_{HL} *under constant* K.

succeeding cycle. Therefore, the strain ratchetting takes place due to the accumulation of the creep strain. On the other hand, in the test with the zero stress, the strain decreases during the zero-stress-hold after the instantaneous strain at the unloading as shown in Fig. 16*b*. In other words, the creep strain introduced in the tensile hold is recovered during the zero hold. Although the reduction of strain during the zero-stress-hold, which is termed "recovery of creep strain," is also observed in metallic materials, the amount is very small as compared with that in polymers. Due to the extensive recovery, the accumulation of creep strain or the creep ratchetting is held back. The curves in Fig. 17 show the change in the strain measured at the end of tensile hold in each cycle. The zero-stress-hold represses the creep ratchetting due to the recovery. The deceleration of crack propagation in AS4/PEEK laminates, therefore, might be attributed to the repression of creep ratchetting of matrix in the vicinity of the crack tip due to the recovery during the zero-load-hold.

Crack Propagation in Time-Dependent Fatigue

Since the creep recovery influences the crack propagation of CFRP, the *da/dt-K* relation depends on the load waveform in the time-dependent fatigue. The dependence can be summarized as a schematic diagram shown in Fig. 18. The creep cracking without the zero-load-hold give the upper bound of the *da/dt-K* relation in the time-dependent fatigue, which is

FIG. 14—*SEM fractograph in the time-dependent fatigue of AS4/PEEK with zero-load-hold. Test 10,* P_{max} *= 228 N,* t_{HH} *= 3600 s,* t_{HL} *= 360 s.*

formulated by Eq 4. On the other hand, the lower bound in the time-dependent fatigue is given by the conditions with long zero-load-hold, where the effect of creep recovery is saturated. The lower bound is formulated as

$$\frac{da}{dt} = 1.37 \times 10^{-15} K^{29.2} \tag{5}$$

which is obtained from the relation represented by the dotted line in Fig. 12. The zero-load-hold period to reach the lower bound, t_{HS}, is given by the dotted line in Fig. 13 and it is correlated with K as

$$t_{HS} = 1.29 \times 10^{14} K^{-21.2} \tag{6}$$

In the case of $t_{HS} > 1.29 \times 10^{14} K^{-21.2}$, the crack propagation is governed by the Eq 5. In

TABLE 6—*Crack propagation rate per unit time* da/dt *at the initial slit of AS4/PEEK in the time-dependent fatigue with and without the zero load hold.*

	Test	t_{HH}, s	t_{HL}, s	Crack propagation rate da/dt at initial slit (K = 2.50 MPa·m$^{1/2}$), mm/h
Without zero load hold	8	360	0	2.50
	9	3600	0	0.161
With zero load hold	17	3600	3 600	0.0348
	18	3600	10 800	0.0915

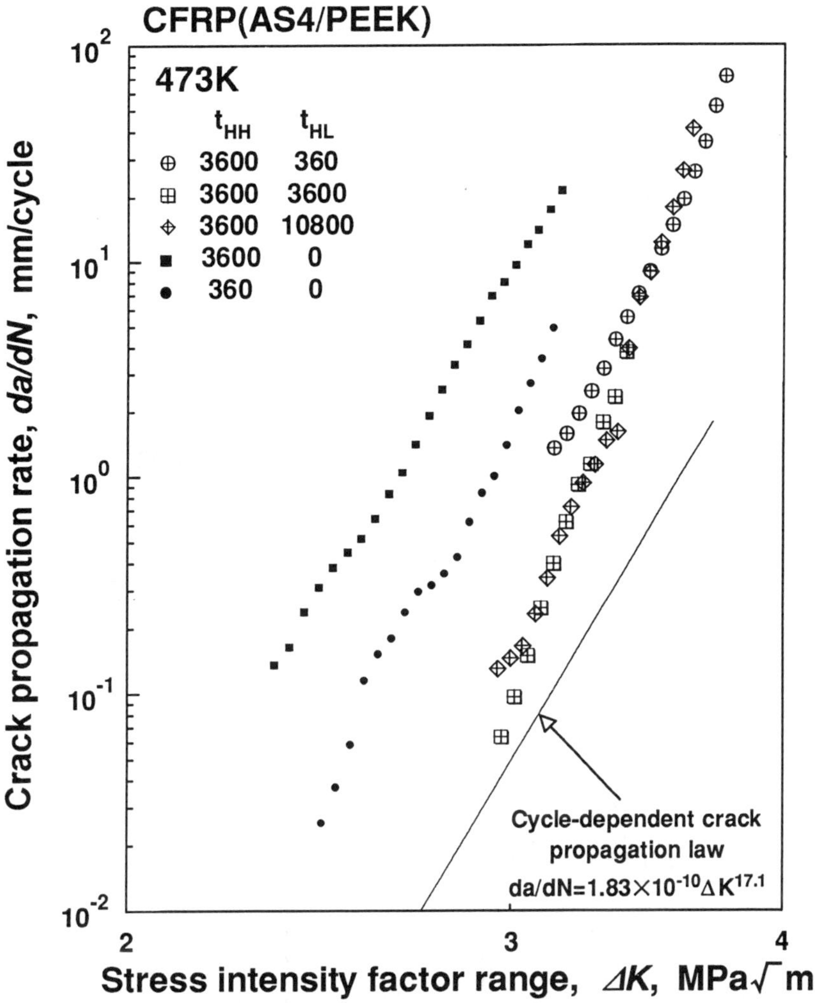

FIG. 15—*Relationship between crack propagation rate per unit cycle* da/dN *and stress intensity factor range* ΔK, *in the time-dependent fatigue without zero-load-hold.*

the case of $t_{HS} < 1.29 \times 10^{14} K^{-21.2}$, the crack propagation behavior is characterized by the following equation interpolating the upper and lower bounds

$$\frac{da}{dt} = F\left(\frac{da}{dt}\right)_{upper\ bound} + (1 - F)\left(\frac{da}{dt}\right)_{lower\ bound} = F \times 1.60 \times 10^{-11} K^{22.9} + (1 - F) \times 1.37 \times 10^{-15} K^{29.2} \quad (7)$$

where $(da/dt)_{\text{upper bound}}$ and $(da/dt)_{\text{lower bound}}$ are the crack propagation rate given by Eqs 4 and 5, respectively. Here, F is a function of zero-load-hold period t_{HL} and monotonically decreases from unity to zero as t_{HL} changes from zero to t_{HS}. For example

TABLE 7—*Test conditions of PEEK.*

<table>
<tr><th>Load Waveform</th><th>Test No.</th><th>t_{HH}, s</th><th>t_{HL}, s</th><th>σ_{max}, MPa</th></tr>
<tr><td rowspan="2">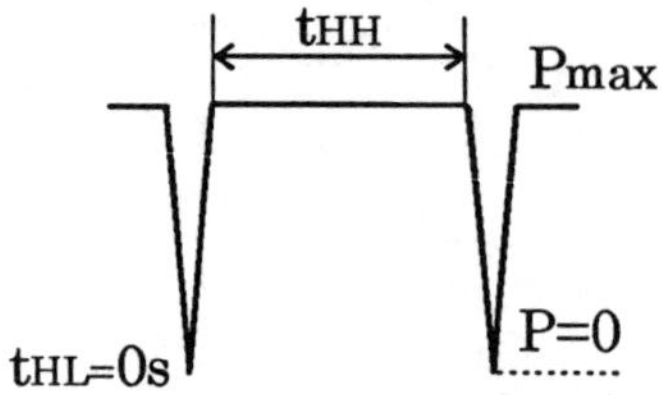</td><td>19</td><td rowspan="2">3600</td><td rowspan="2">0</td><td>9</td></tr>
<tr><td>20</td><td>21</td></tr>
<tr><td rowspan="2">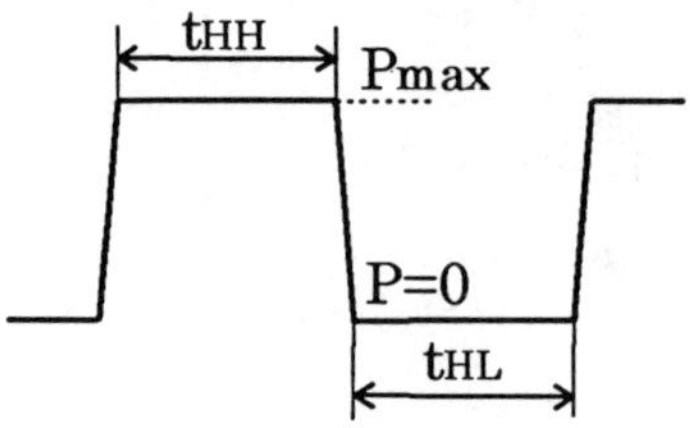</td><td>21</td><td rowspan="2">3600</td><td rowspan="2">3600</td><td>9</td></tr>
<tr><td>22</td><td>21</td></tr>
</table>

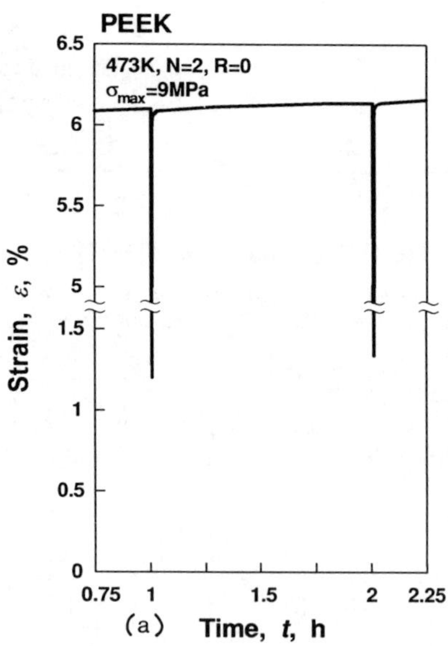

FIG. 16—*Creep curves of PEEK in a cycle. (a) Without zero-load-hold. (b) With zero-load-hold.*

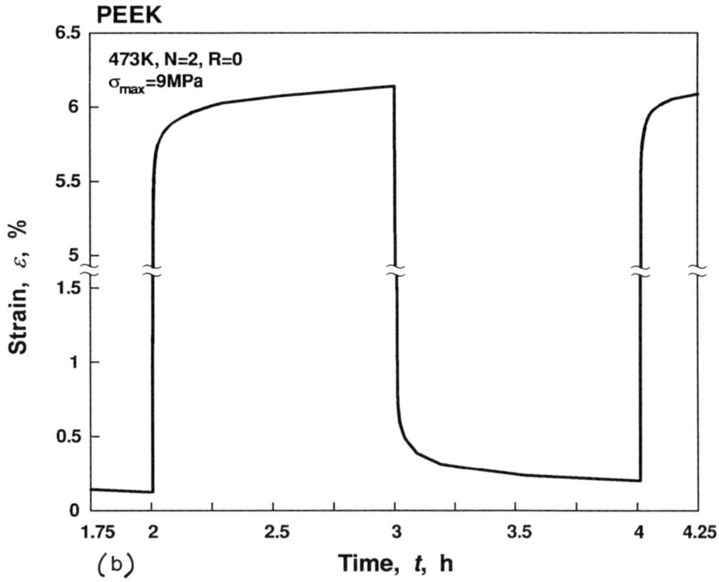

FIG. 16—*Continued*

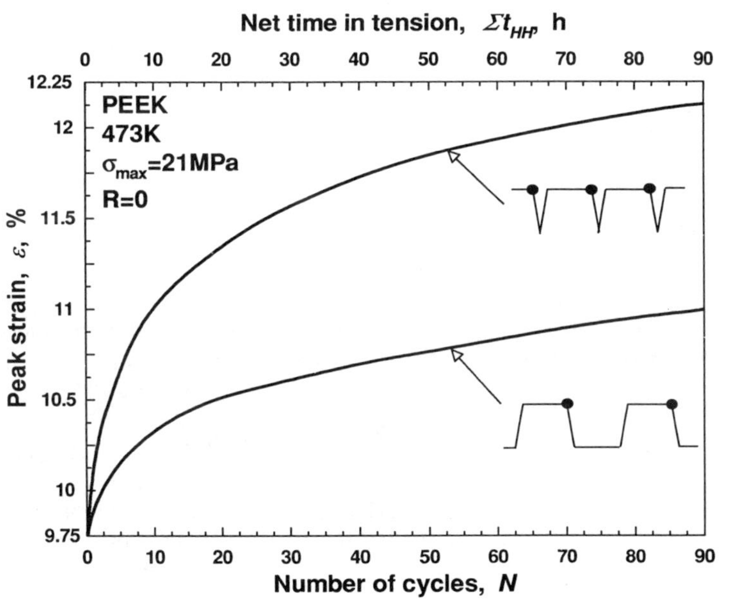

FIG. 17—*Creep ratchetting curves showing the peak strain against number of cycles in cyclic loading tests with and without zero-stress-hold.*

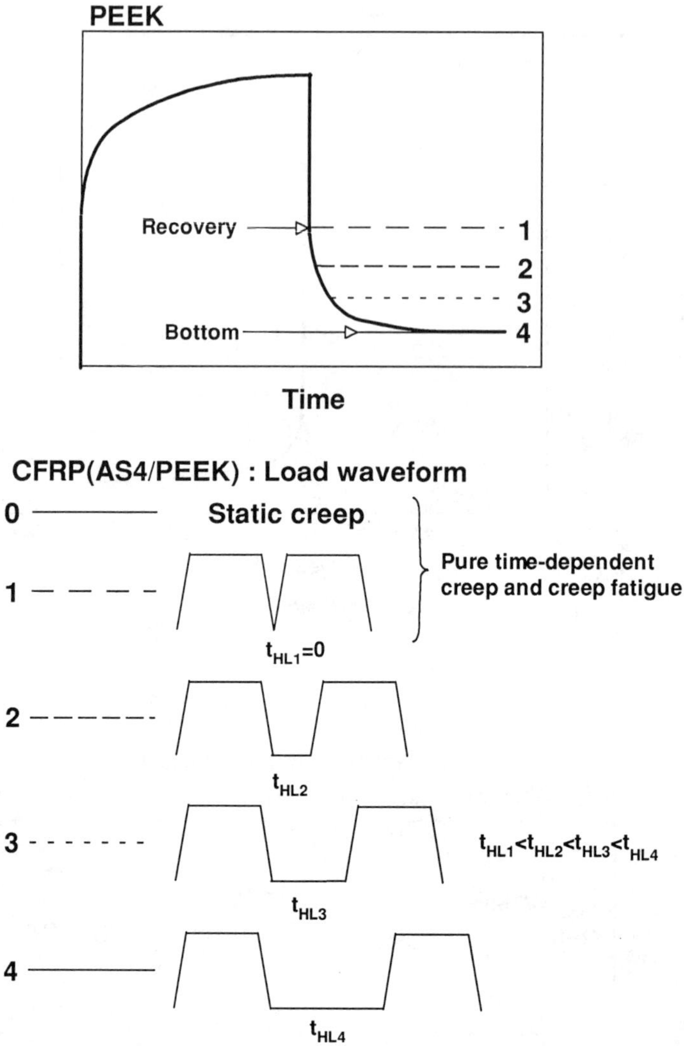

FIG. 18—*Schematic diagrams explaining the crack propagation law with the zero-load-hold as well as the tensile load hold in the time-dependent fatigue.*

$$F = 1 - \left(\frac{t_{HL}}{t_{HS}}\right)^{m}, \; m\text{: constant} \tag{8}$$

is a candidate for the function of F. However, because of lacking experimental data in this study, the determination of exact form of the function, F remains for future work.

Conclusions

The delamination crack propagation behavior in creep-fatigue interaction at high temperature is investigated using DCB specimens of AS4/PEEK and T800H/PMR-15 laminates. The results are summarized as follows.

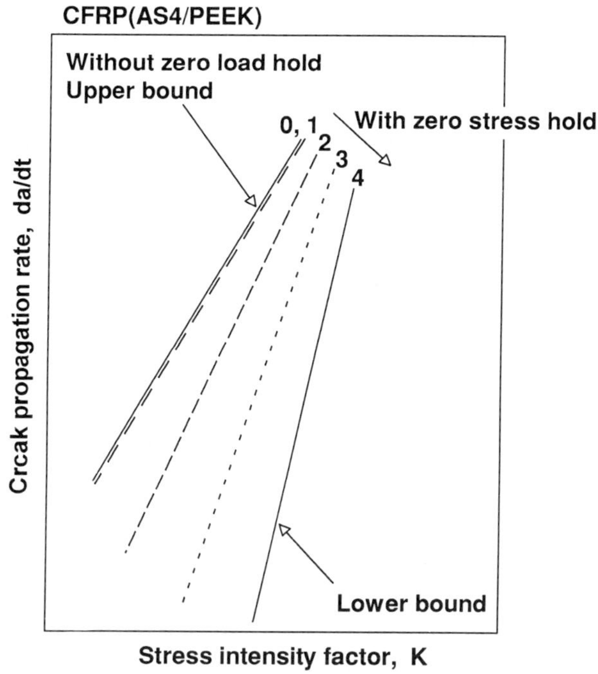

FIG. 18—*Continued*

1. Crack propagation of both laminates under fatigue loading with high frequency is purely cycle-dependent and its rate per unit cycle da/dN is correlated well with stress intensity factor range ΔK. This implies that the crack propagates under small-scale yielding (SSY) conditions where the plastic deformation of matrix is confined in the vicinity of the crack tip.
2. Tensile load hold accelerates the crack propagation of AS4/PEEK laminates. It is caused by the creep deformation of matrix and the crack propagation shows pure time-dependence. The crack propagation rate per unit time da/dt is correlated well with stress intensity factor K and the relation coincides the results in static creep. This signifies that the crack propagates under small scale creep (SSC) conditions where the creep of matrix is confined near the crack tip due to the constraint of fibers.
3. Time-dependent crack propagation is not observed in T800H/PMR-15 laminates even under the cyclic loading with the tensile load hold. This is because the thermoset polyimide, PMR-15, scarcely creeps whereas the thermoplastic matrix, PEEK, shows eminent time-dependent deformation at high temperatures.
4. The time-dependent crack propagation of AS4/PEEK laminate is decelerated by the zero-load-hold. It is revealed from the deformation tests of PEEK without fibers in creep-fatigue conditions that the deceleration is attributed to the recovery of creep strain in the matrix at the crack tip during the zero-load-hold.

References

[1] Keary, P. E. and Ilcewicz, L. B., "Mode I Interlaminar Fracture Toughness of Composites Using Slender Double Cantilevered Beam Specimens," *Journal of Composite Materials*, Vol. 9, 1985, pp. 154–177.

[2] Jar, P. Y. and Kausch, H. H., "The Influence of Moulding Conditions on Delamination Behaviour of Carbon-Fiber/PEEK Composites," *Composites Science and Technology*, Vol. 52, 1994, pp. 349–359.

[3] Zheng, S. and Sun, C. T., "A Double-Plate Finite-Element Model for the Impact-Induced Delamination Problem," *Composites Science and Technology*, Vol. 53, 1995, pp. 111–118.

[4] Wilkins, D. J., Eisenmann, J. R., Camin, R. A., and Benson, R. A., "Characterizing Delamination Growth in Graphite-Epoxy," *Damage in Composite Materials, ASTM STP 775*, American Society for Testing and Materials, West Conshohocken, PA, 1982, pp. 168–183.

[5] Hojo, M., Tanaka, K., Gustafson, C. G., and Hayashi, R., "Effect of Stress Ratio on Near-Threshold Propagation of Delamination Fatigue Cracks in Unidirectional CFRP," *Composites Science and Technology*, Vol. 29, 1987, pp. 273–292.

[6] Perl, Y. J., Davies, P., Benaeggagh, M. L., and de Charentenay, F. X., "Mode I and Mode II Delamination of Thermosetting and Thermoplastic Composites," *Composite Materials: Fatigue and Fracture, Vol. 2, ASTM STP 1012*, American Society for Testing and Materials, West Conshohocken, PA, 1989, pp. 251–269.

[7] Hojo, M. and Tanaka, K., "Fractography of the Delamination Fatigue Crack Growth in CFRP Laminates," *Advances in Fiber Composite Materials*, T. Fukuda, Z., Maekawa, and T. Fujii, Eds., Elsevier, 1994, pp. 123–155.

[8] Ohtani, R., Kitamura, T., Uematsu, Y., and Hojo, M., "Mode I Delamination of Carbon Fiber Reinforced Thermoplastic Polymer Under Static and Cyclic Creep at Elevated Temperatures," *High Temperature and Environmental Effects on Polymeric Composites, ASTM STP 1174*, American Society for Testing and Materials, West Conshohocken, PA, 1993, pp. 23–38.

[9] Uematsu, Y., Ohtani, R., Kitamura, T., and Tatsumi, T., "Temperature Dependence of Delamination Fracture Toughness of CF/PEEK and CF/PMR-15 Laminates," *Transactions of the Japan Society of Mechanical Engineers*, Ser. A, Vol. 56, No. 566, 1993, pp. 2286–2291 (in Japanese).

[10] Uematsu, Y., Kitamura, T., and Ohtani, R., "Delamination Behavior of a Carbon-Fiber-Reinforced Thermoplastic Polymer at High Temperatures," *Composites Science and Technology*, Vol. 53, 1995, pp. 333–341.

[11] Uematsu, Y., Ohtani, R., Kitamaru, T., and Oka, K., "Effect of Crack Shielding due to Fiber Bridging on Delamination Propagation of AS4/PEEK Laminates in High Temperature Fatigue," *Journal of the Society of Materials Science*, Japan, Vol. 44, No. 501, 1995, pp. 762–768 (in Japanese).

[12] Uematsu, Y., "Mode I Delamination Crack Propagation of Unidirectionally Reinforced CFRPs at Elevated Temperatures," Doctoral dissertation, 1995 (in Japanese).

[13] Kitamura, T., Ohtani, R., and Uematsu, Y., "Time Dependent Behavior of Mode I Delamination in Unidirectionally Reinforced CF/PEEK Under Cyclic Loading at Elevated Temperature," 8th International Congress on Experimental Mechanics, 10-13 June 1996, Nashville, Society for Experimental Mechanics, pp. 435–436.

[14] Newaz, G. M., Lustiger, A., and Yung, J. Y., "Delamination Growth Under Cyclic Loading at Elevated Temperature in Thermoplastic Composites," *Advances in Thermoplastic Matrix Composite Materials, ASTM STP 1040*, American Society for Testing and Materials, West Conshohocken, PA, 1989, pp. 264–278.

[15] Ohtani, R., Ohnami, M., and Inove, T., Eds., *High Temperature Creep-Fatigue, Current Japanese Materials Research*, JSME, Vol. 3, Elsevier, 1988.

[16] Kageyama, K. and Hojo, M., "Proposed Methods for Interlaminar Fracture Toughness Tests of Composite Laminates," Proceedings of the Fifth U.S./Japan Conference on Composite Materials (CCM-5), 1990, pp. 227–234.

[17] Shi, G. C., Paris, P. C., and Irwin, G. R., "On Cracks in Rectilinearly Anisotropic Bodies," *International Journal of Fracture Mechanics*, Vol. 1, 1965, pp. 189–203.

[18] Johnson, W. S. and Mangalgari, P. D., "Investigation of Fiber Bridging in Double Cantilever Beam Specimens," *Journal of Composite Technology and Research*, Vol. 9, Spring, 1987, pp. 10–13.

[19] O'Brien, T. K. and Martin, R. H., "Results of ASTM Round Robin Testing for Mode I Interlaminar Fracture Toughness of Composite Materials," *ASTM Journal of Composites Technology and Research*, Vol. 15, No. 4, Winter, 1993.

[20] Taira, S., Ohtani, R., and Kitamura, T., "Application of J-Integral to High-Temperature Crack Propagation," Part I, Creep Crack Propagation, *Transactions of ASME*, Ser. H, Vol. 101, 1979, pp. 154–161.

Mahmood M. Shokrieh,[1] *Olivia P. Eilers,*[1] *and Larry B. Lessard*[1]

Characterization of a Graphite/Epoxy Composite Under In-Plane Shear Fatigue Loading

REFERENCE: Shokrieh, M. M., Eilers, O. P., and Lessard, L. B., **"Characterization of a Graphite/Epoxy Composite Under In-Plane Shear Fatigue Loading,"** *High Temperature and Environmental Effects on Polymeric Composites: 2nd Volume, ASTM STP 1302,* Thomas S. Gates and Abdul-Hamid Zureick, Eds., American Society for Testing and Materials, 1997, pp. 133–148.

ABSTRACT: To examine the in-plane shear behavior of graphite/epoxy material in fatigue, the three-rail shear test specimen, normally used for static testing, is modified by inserting notches on the edge of the specimen. The performance of the notched specimen shows great improvement in both static and fatigue loading conditions. Therefore, by using the notched specimen, a more reliable experimental data base can be obtained.

The main objective of this research is to fully characterize a graphite/epoxy composite under in-plane shear fatigue loading by using the modified notched specimen. To fully characterize the material under in-plane shear, both static and fatigue tests are performed. Fatigue life, residual strength, and residual stiffness of AS4/3501-6 are measured. Experimental results from both static and fatigue loading show a very clear mode of failure under in-plane shear in the gage area of test specimens. To measure the fatigue life, a cyclic load is applied to the specimen and tests are continued until catastrophic failure is achieved. To measure the residual strength and residual stiffness, the fatigue loading is stopped after a fixed number of cycles (for example, 100, 1000, and so forth) and the residual properties are measured by applying a static load and destroying the specimen. Nonlinear stress-strain behavior of the material under static loading conditions is observed and the effect of fatigue loading on the nonlinearity is studied and discussed. To study the effects of mean stress on the fatigue behavior of the material, fatigue tests are performed under different stress ratios ($\kappa = \sigma_{min}/\sigma_{max}$). The fatigue life, residual strength, and residual stiffness of the material are presented in normalized forms that can be used to predict the fatigue behavior of the composite material, under arbitrary state of stress and stress ratio, based on the presented experimental data base. Thus, the behavior of AS4/3501-6 graphite/epoxy composite under in-plane shear fatigue loading is fully characterized.

KEY WORDS: shear testing, three-rail shear, test methods, fatigue, strength degradation, stiffness degradation, life, notched specimen

Although much research exists for static behavior of composite materials under shear loading conditions [*1*], only limited research exists for shear fatigue behavior of these materials [*2-10*]. In the design of composite materials under fatigue loading conditions, existence of a reliable experimental data base is a key point. The material properties of a com-

[1] Research assistant, graduate student, and assistant professor, respectively, Mechanical Engineering Department of McGill University, 817 Sherbrooke St., West, Montreal, Canada, H3A 2K6.

posite structure subjected to fatigue loading degrade with repeated cycles. The degradation of the structure is due to material degradation in the fiber, matrix, and shear directions. Thus, shear behavior is an important component of any complete material model. Moreover, creation of a reliable experimental data base is influenced by the testing technique used by the investigator. The main objective of this research is to fully characterize AS4/3501-6 graphite/epoxy unidirectional composite under in-plane shear fatigue loading. Therefore, the testing technique should have satisfactory performance in both static and fatigue shear loading conditions.

There are about 30 different test methods [*1*] for characterizing the in-plane shear properties of composite material under static and fatigue loading conditions. Therefore, selecting a simple, suitable, and reliable testing technique is a tedious task. Nine of the in-plane shear testing techniques for advanced composite materials were examined by Lee and Munro [*10*]. The top four testing techniques evaluated by Lee and Munro [*10*] are Iosipescu, 45° off-axis tensile, 10° off-axis, and three-rail shear test methods. It should be mentioned that Lee and Munro [*10*] did not consider the fatigue performance of these testing techniques but merely studied the performance of the methods under static loading conditions. Although the Iosipescu method is a standard technique, to the best knowledge of the authors this method has not been used in fatigue loading conditions. Therefore, the performance of the Iosipescu technique is not known in fatigue loading conditions. Furthermore the Iosipescu fixture, which is suitable for static testing, would have to be modified for fatigue experiments. Even though the off-axis specimen testing techniques, such as 45° or 10°, are very simple methods for measuring the static shear properties of composite materials, nevertheless, due to existence of biaxial state of stress, the application of these methods under pure in-plane shear fatigue loading conditions is arguable. The three-rail shear test method (Fig. 1) introduced by ASTM (Guide for Testing In-Plane Shear Properties of Composite Laminates, D 4255) as a standard guide for testing in-plane shear properties of composite laminates can be adapted for fatigue testing of composite materials. However, the three-rail shear testing method suffers from the existence of stress concentrations that can create cracks on the free edges of the test sample. Therefore, using the traditional specimen results in a large scatter in the experimental data base.

Recently the authors modified the three-rail shear test sample by inserting notches on the free edges of the specimen [*11*], as shown in Fig. 2, and the fixture was modified to be suitable for fatigue loading conditions. The philosophy behind the insertion of notches at the locations of stress concentrations is to replace a very sharp crack with a blunt crack with much lower stress concentration factor. It should be mentioned here that such improvement was devised by Butler et al. [*8*] for the two-rail shear specimen. The performance of the three-rail notched specimen showed great improvement in both static and fatigue loading conditions. Furthermore, it is noticed that by using the modified specimen, scatter in experimental results is reduced drastically and a reliable experimental data base for a unidirectional ply under in-plane shear loading condition can be obtained.

It should be noted that this research is a part of a larger program for full fatigue characterization of a unidirectional graphite/epoxy composite material under tension, compression, in-plane and out-of-plane shear loading conditions. In this paper, the in-plane shear behavior (fatigue life, residual strength, and residual stiffness) of AS4/3501-6 graphite/epoxy composite material under fatigue loading conditions is studied. The results of experiments are presented as normalized master curves that are useful for complete characterization of shear fatigue. Note that the master curves obtained are for AS4/3501-6 material and experiments must be repeated for any other material.

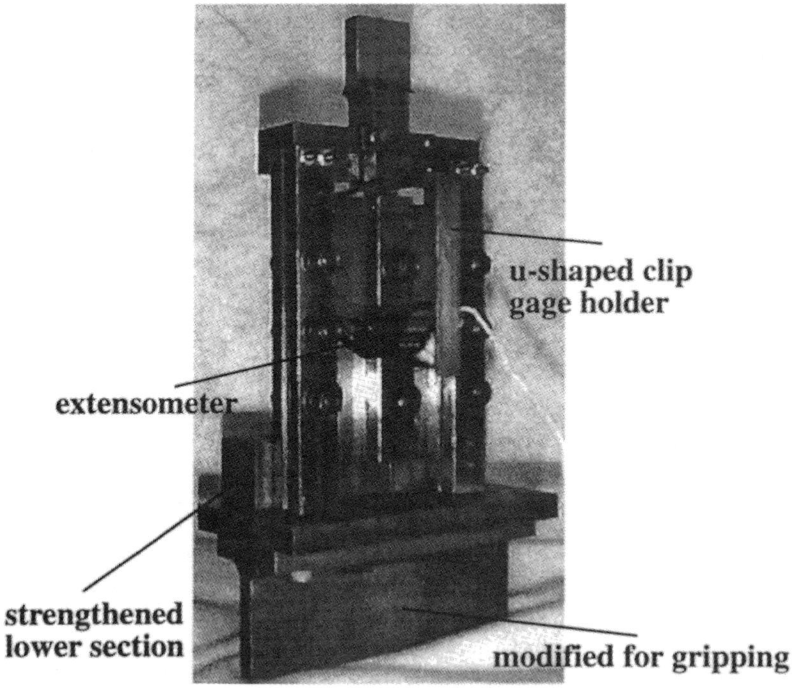

FIG. 1—*Three-rail shear fixture modified for gripping and equipped with extensometer.*

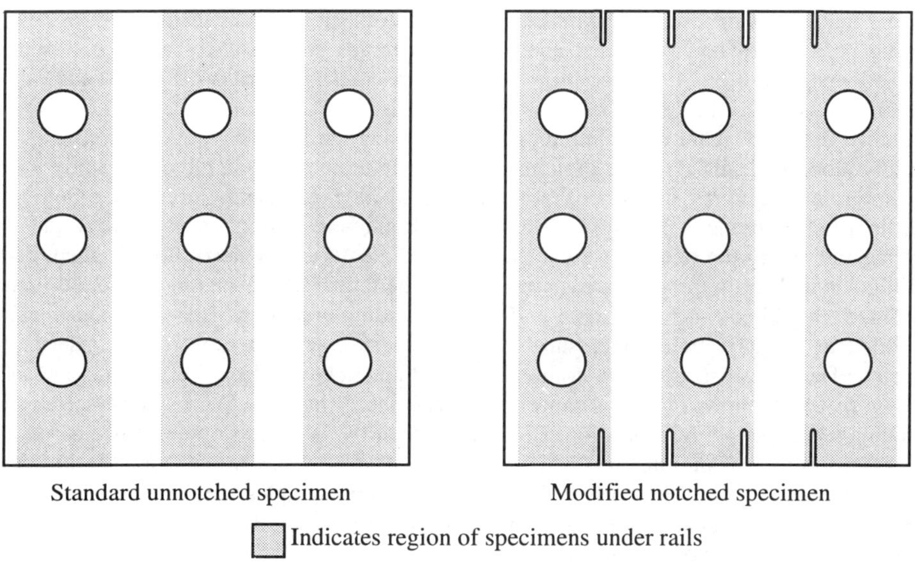

FIG. 2—*Geometry of unnotched and notched specimen.*

Modified Three-Rail Shear Fixture and Specimen

In this section, in order to show the improvement in performance of the modified specimen, a comparison between the results of experiments using the traditional and the modified three-rail shear specimen is presented. For more details concerning the modified specimen and the finite element analysis, the interested reader can refer to a paper recently published [*11*].

As shown in Fig. 1, the lower part of the standard fixture is redesigned for fatigue experiments in order to grip the fixture from the top as well as the bottom. The samples were fabricated from AS4/3501-6 graphite/epoxy based on the standard specimen geometry. Theoretically, the behavior of the 0° ply and the 90° ply is the same under pure shear conditions. In the case of a rail shear test, load is not really applied in pure shear (four sides of the specimen), but rather on only two sides of the specimen. Therefore, instead of using a unidirectional ply (0° or 90°) lay-up for the samples and to increase the stability of the sample during the shear test, the cross-ply lay-up $(0/90]_{4s}$ is selected for testing specimens. Note that by classical laminated plate theory, an applied shear load on a cross-ply lay-up corresponds to pure shear stresses on each of the 0° and 90° layers. For the notched sample, the locations of notches were found by finite element techniques [*11*]. The notches are manufactured by using a milling machine and the edges of the samples are polished for both the notched and unnotched specimens. For static and fatigue tests, stroke and load control conditions are used, respectively. To measure the strain for the static case, strain gages can be used; however, within 100 cycles for the fatigue cases, the strain gage separates from the specimen and the strain can not be monitored. To overcome this problem, instead of using strain gages an extensometer is used and a simple holder is designed to hold the extensometer, as shown in Fig. 1. The holder is a u-shaped device that clamps the extensometer to both sides of the specimen without touching the specimen anywhere else. The holder is spring-tensioned to keep the extensometer in place. The fatigue load is applied in a sinusoidal form. Testing frequency was changed by trial and error until a frequency was found that caused minimal temperature increase to the specimen, between 5 and 10°C, as observed during the fatigue experiments. The final frequency selected was 2 Hz for the fatigue tests.

The behavior of the unnotched and notched samples under static loading conditions is studied first. By increasing load from zero, sharp cracks appear on the free edge of the unnotched sample. The failure initiation mode of the unnotched sample under static loading is shown in Fig. 3. After crack initiation, by increasing the load level, cracks propagate very rapidly along the rails. The final failure mode of the unnotched sample under static loading is shown in Fig. 4. As shown in Fig. 4, instead of having shear failure in the gage areas, the final failure mode is a combination of a small amount of shear in gage areas and crack propagation along the rails. It is clear that the specimen does not absorb the applied load by failing under shear in the gage areas and it seems the failure along the rails is more dominant. In Fig. 5, the failed notched sample under static loading conditions is shown. The specimen is failed in the gage area under shear, and the sharp cracks are arrested by inserting the notches. The typical shear stress-strain behavior of unnotched and notched samples are shown in Fig. 6. The stiffness of both samples in the nonlinear elastic region is the same; however, the unnotched sample fails at a lower load level than the notched sample. It is clear that the premature failure of the unnotched sample is due to edge crack propagation. In Fig. 7, the results of static strength tests of unnotched and notched samples are shown and compared. The results show higher scatter and lower average strength for the unnotched samples in comparison with the notched specimens.

After studying the behavior of both unnotched and notched specimens under static loading, the two types of samples are studied under fatigue loading conditions. The failed unnotched

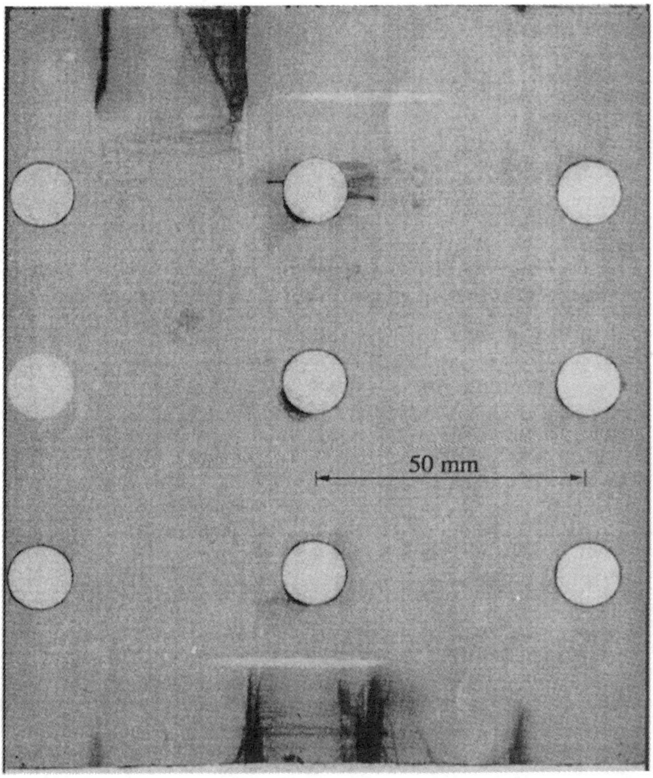

FIG. 3—*Failure initiation mode of unnotched sample under static loading.*

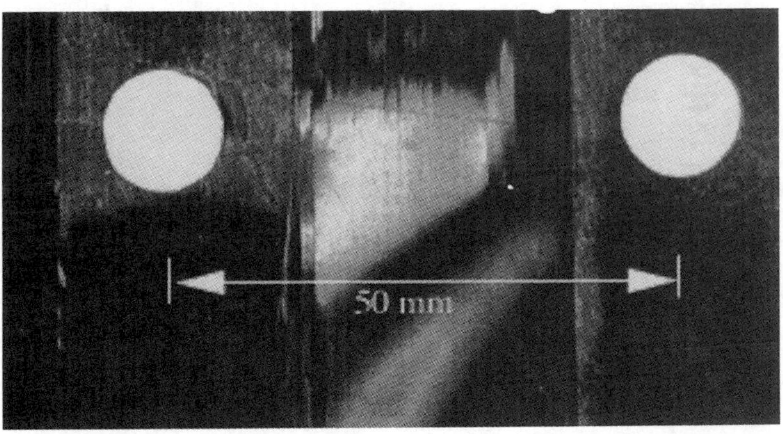

FIG. 4—*Final failure mode of unnotched sample under static loading (close-up view).*

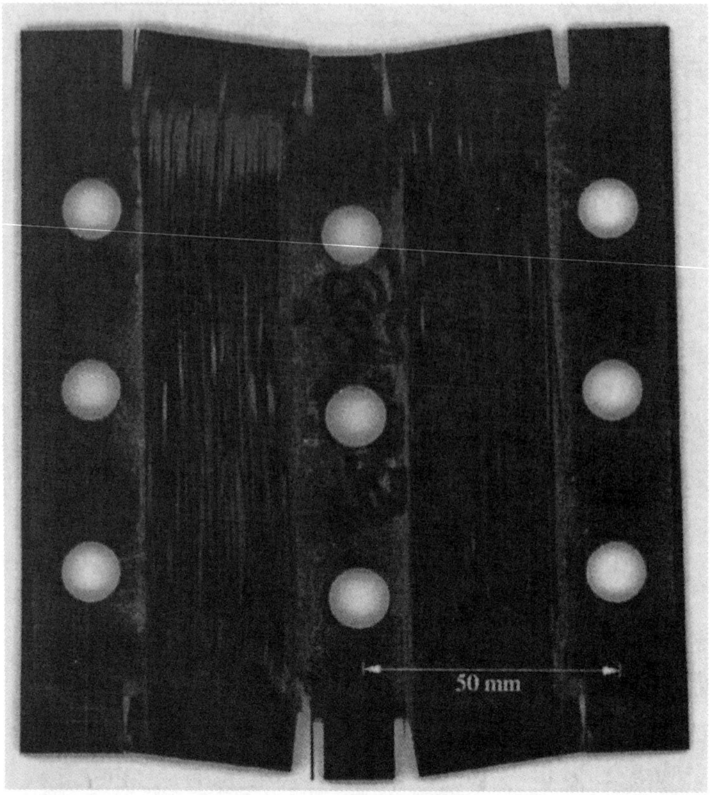

FIG. 5—*Final failure mode of notched sample under fatigue loading.*

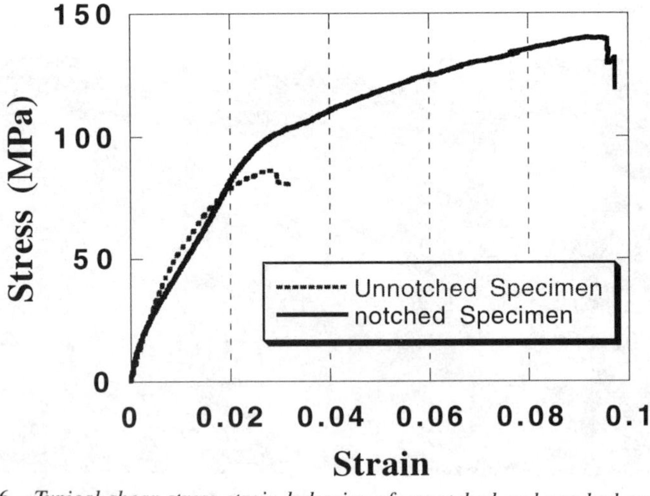

FIG. 6—*Typical shear stress-strain behavior of unnotched and notched samples.*

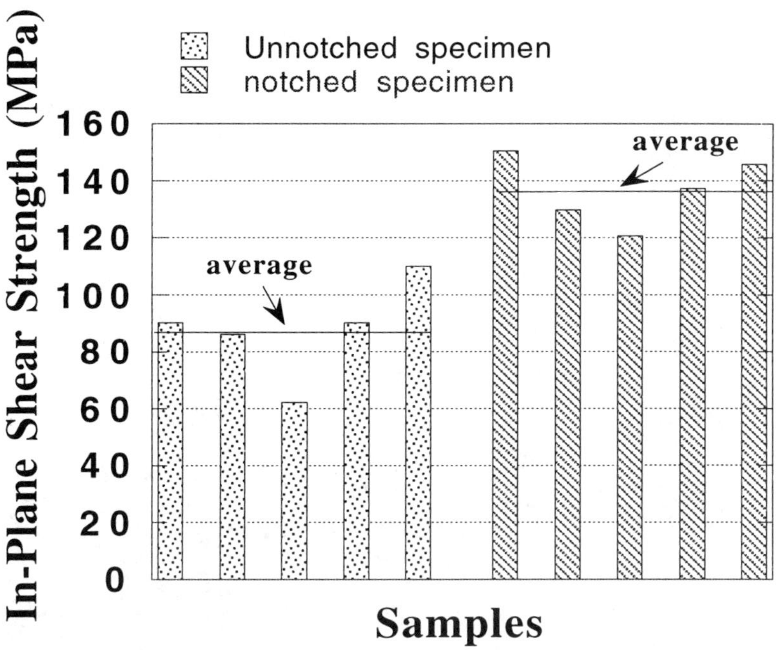

FIG. 7—*Comparison of static strength tests of unnotched and notched samples.*

sample under fatigue loading is shown in Fig. 8. By comparing Figs. 4 and 8, it is seen that the final mode of failure under fatigue and static loading conditions are the same. From Fig. 8, it is clear that the unnotched sample under fatigue loading could not absorb the external load by inducing shear in gage areas and the sample failed mostly due to crack propagation along the edge of the rails. The experimental results show a large amount of scatter in fatigue life of the unnotched sample, which is mostly due to existence of sharp cracks along the rails. In Fig. 9, the failed notched sample under fatigue loading is shown; it can be seen that the specimen failed under shear in the gage areas. Similar to the unnotched sample, the final mode of failure for the notched sample under fatigue and static loading conditions is similar. The fatigue life of both unnotched and notched samples under the same fatigue loading conditions (max load = 47.6 KN, and load ratio = 0.1), are shown in Fig. 10. The results show higher scatter and lower average life for the unnotched samples in comparison with the notched specimens. Note that all tests were performed at stress levels that were high enough to show materials degradation within 106 cycles. No attempt was made to test samples at stress levels below 40% of static failure stress in shear.

The improved performance of the notched samples, both in static and fatigue loading conditions, shows the reliability of the test method. Therefore, the test method can be used to characterize the material properties (such as, fatigue life, residual strength, and residual stiffness) under static and shear fatigue loading conditions.

Material Characterization Using Modified Specimens

In this section the result of experiments for full characterization of AS4/3501-6 graphite/epoxy composite material under in-plane shear fatigue loading conditions are summarized.

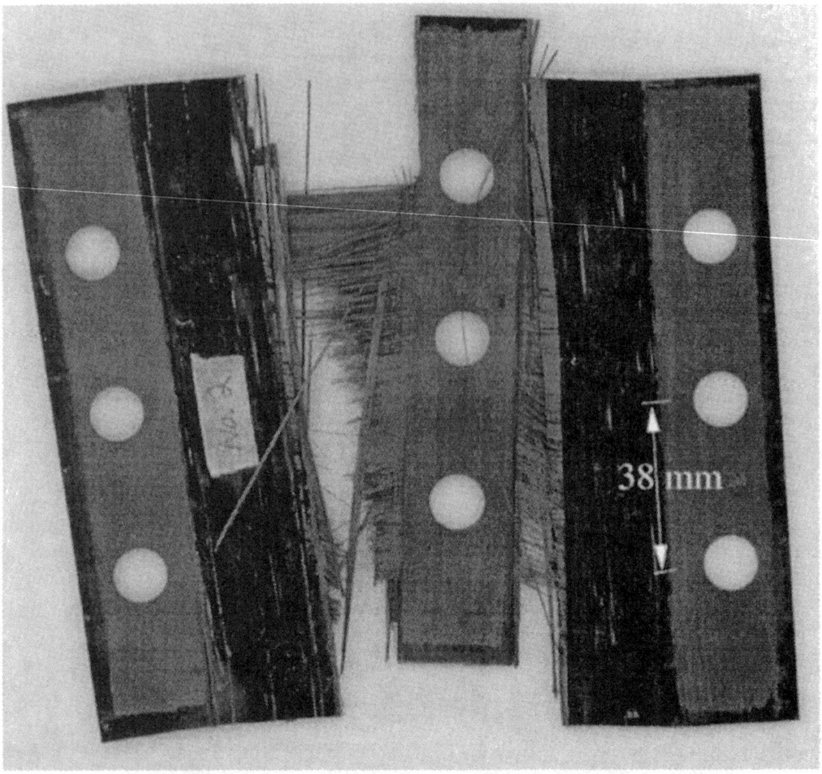

FIG. 8—*Failed unnotched sample under fatigue loading.*

The residual strength, residual stiffness, and fatigue life of the material under different stress levels and stress ratios are measured. To measure the residual properties of the material, the fatigue cycling is stopped after some limited number of cycles (for example, 100, 1000, and so forth) and the specimen is destroyed by applying static shear load. While for measuring the fatigue life of the material, the fatigue loading is continued until catastrophic failure is achieved. To present the experimental results in form of master curves, normalization techniques are used that are explained in the following. Note that these normalization techniques involve the establishment of certain curve-fitting parameters. These parameters are material dependent; consequently, the results obtained here will apply only to AS4/3501-6 material. However, the technique can be applied to any material.

Fatigue Life

In this section, the fatigue life of AS4/3501-6 graphite/epoxy composite material under in-plane shear fatigue loading is presented in a normalized form. The effect of mean stress $((\sigma_{max} + \sigma_{min})/2)$ on the fatigue life can be presented efficiently by using constant life, Goodman-type diagrams. Establishing the constant life diagrams for a unidirectional ply is time-consuming and expensive. However, by using the normalized fatigue life technique developed by Gathercole et al. [*12*], all different fatigue life curves for different mean stress

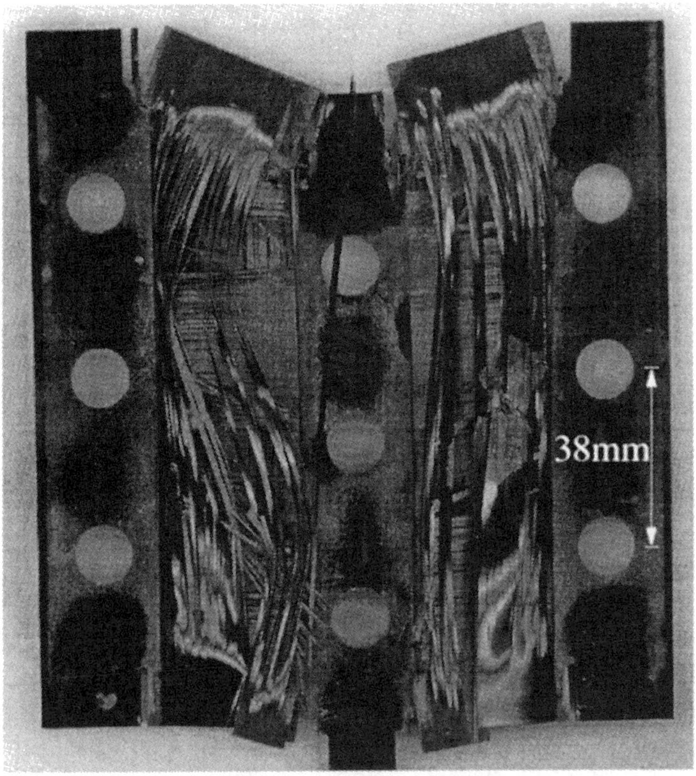

FIG. 9—*Failed notched sample under fatigue loading.*

conditions collapse to a single two-parameter master curve, which decreases the number of required experiments drastically. In the following, a brief summary of the method, developed by Gathercole et al. [*12*], for simulating the fatigue life of composite materials is presented. They applied their model [*12*] for simulating the behavior of composite laminates under tension/tension or compression/compression fatigue loading conditions successfully. In this study, their model [*12*] is modified and applied for simulating the fatigue life of a unidirectional ply under shear fatigue loading conditions.

Introducing non-dimensional stresses by division of the mean stress σ_m, the alternating stress σ_a, and the compressive strength σ_c by the static shear strength ($R(0)$), where $q = \sigma_m/R(0)$, and $a = \sigma_a/R(0)$, an empirical interaction curve may be derived.

$$u = \log\left(\frac{\ln(a/f)}{\ln[(1-q)(1+q)]}\right) = A + B\log N_f \tag{1}$$

where

u = master curve life parameter,
f, A, and B = curve fitting constants,
$\sigma_a = (\sigma_{max} - \sigma_{min})/2$ = alternating stress, and
$\sigma_m = (\sigma_{max} + \sigma_{min})/2$ = mean stress.

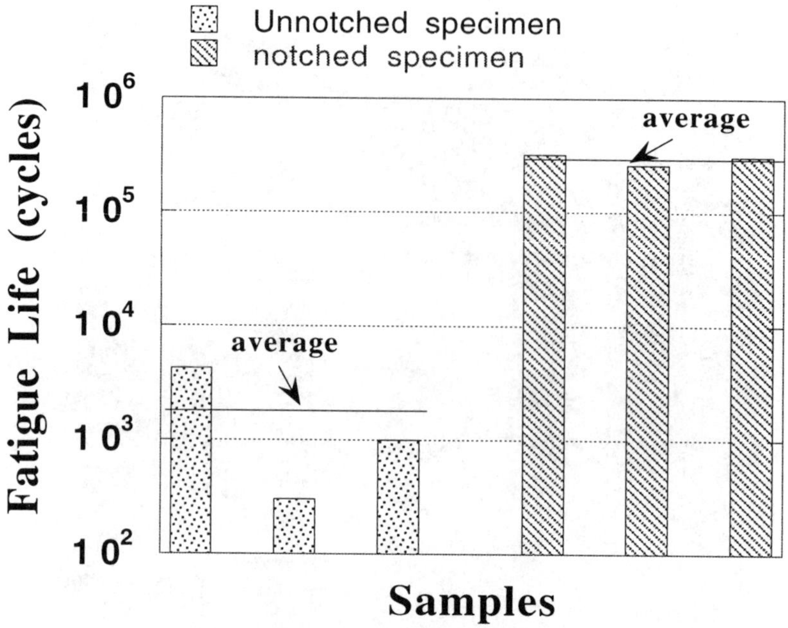

FIG. 10—*Fatigue life of unnotched and notched samples under the same fatigue loading conditions.*

Two different stress ratios ($\kappa = \sigma_{min}/\sigma_{max} = 0.1$ and 0.0) and different percentages of the static strength (selected as the maximum stress) are applied to find the fatigue life of the material under in-plane shear. In Fig. 11, the master curve for the fatigue life of the material under in-plane shear loading conditions is presented in a normalized form. It should be

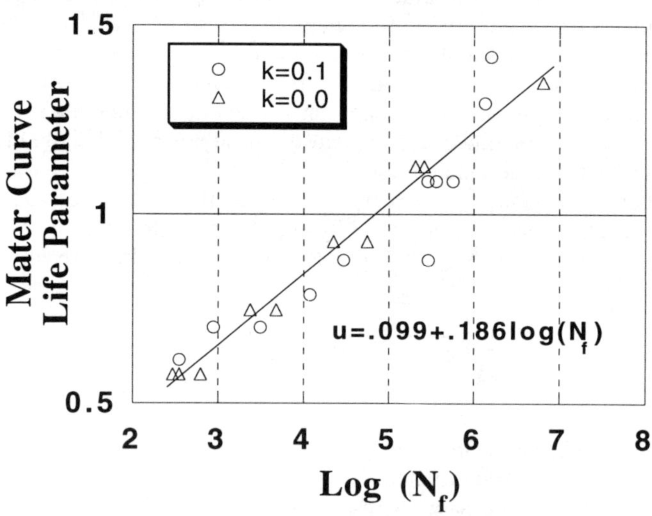

FIG. 11—*Master curve for the fatigue life of the material under in-plane shear loading conditions.*

mentioned that the left-hand side of Eq 1 is denoted by u. By using Eq 1, the curve fitting parameters are found and mentioned in Fig. 11. Although the parameter f seems to be a function of the strength of laminated composite materials under tension or compression fatigue loading,[2] in this study this parameter appears to be a constant equal to 1.06.

Residual Strength

It is observed that composite materials have different behavior under high and low states of stress in fatigue loading conditions [*13*]. Therefore, to measure the residual strength and residual stiffness of the material under fatigue loading, two different states of stress (59 and 40% of the maximum static shear strength of the material) are selected. By selecting these two different states of stress, the high and low load level is applied. Ideally, more tests should be performed at lower frequencies such as 20% of maximum static shear strength, but fatigue tests at such low load levels are too time consuming. A stress ratio equal to 0.1 and a frequency of 2 Hz is applied.

Different models have been presented in the literature to simulate the residual strength of composite materials under fatigue loading. The sudden death model [*14*] is suitable to show the behavior of the material under high state of fatigue stress, while the wear out model [*15*] simulates the behavior under a low-level state of stress. This implies that by plotting the residual strength versus the number of cycles, different curves are achieved, with each curve corresponding to one state of stress. Using the normalization technique proposed by Adam et al. [*13*], all different curves for different states of stress collapse to a single curve. This single curve can be used as the master curve, which is very useful for design purposes. In Fig. 12, the results of residual strength experiments are shown in normalized form. As shown in that fatigue, the material degradation, in a normalized form, in the very first cycles, is very fast. Then the material degrades gradually, and finally there is a sharp degradation at the end of life. To show the material degradation mathematically, the following equation is used and the experimental curve fitting parameters (α and β) are found [*13*].

$$\left(\frac{R(n) - \sigma}{R(0) - \sigma}\right)^{\alpha} = 1 - \left(\frac{\log(n) - \log(0.25)}{\log(N_f) - \log(0.25)}\right)^{\beta} \quad (2)$$

where

$R(n)$ = residual shear strength,
$R(0)$ = static shear strength,
σ = magnitude of applied maximum stress,
n = number of applied cycles,
N_f = fatigue life at σ, and
α and β = experimental curve fitting parameters.

By using the normalization equation (Eq 2), the experimental residual strength curve-fitting parameters α and β for the unidirectional AS4/3501-6 graphite/epoxy ply under in-plane shear loading conditions are found to be equal to 0.156 and 9.112, respectively. The static shear strength $R(0)$ for the unidirectional ply is equal to 136 MPa. The number of cycles to failure for a selected state of stress N_f can be found from the normalized fatigue life equation (Eq 1).

[2] Harris, B., private communication, 1995.

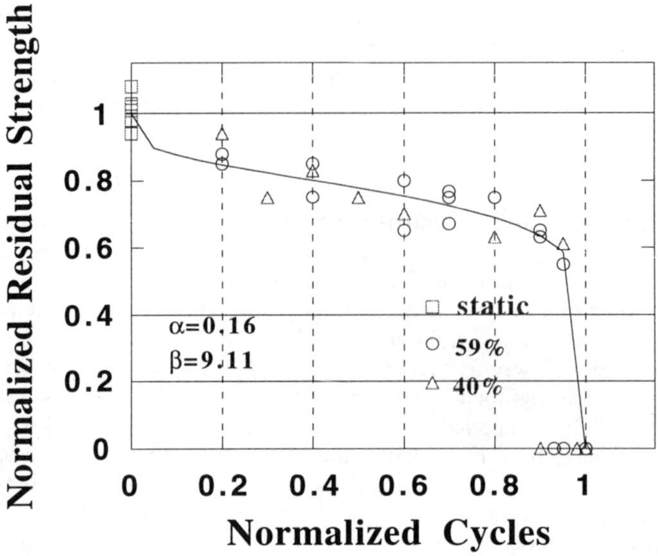

FIG. 12—*Master curve for the residual shear strength of the material under in-plane shear loading conditions.*

Residual Stiffness

The stress-strain behavior of AS4/3501-6 graphite/epoxy unidirectional ply under shear loading conditions is nonlinear (Fig. 6). The initial slope of the curve is defined as the stiffness of the material. The residual stiffness of the material is a function of the state of stress and number of cycles. Using a suitable normalization technique, all the different curves for different states of stress can be shown by a single master curve. To present the residual stiffness as a function of number of cycles in a normalized form, the following equation (Eq 3) is developed in this study. The same strategy used for developing Eq 2 is used to develop Eq 3.

$$\left(\frac{E(n) - \dfrac{\sigma}{\epsilon_f}}{E(0) - \dfrac{\sigma}{\epsilon_f}}\right)^{\eta} = 1 - \left(\frac{\log(n) - \log(0.25)}{\log(N_f) - \log(0.25)}\right)^{\zeta} \tag{3}$$

where

$E(n)$ = residual shear stiffness,
$E(0)$ = static shear stiffness,
σ = magnitude of applied maximum shear stress,
ϵ_f = average engineering shear strain to failure,
n = number of applied cycles,
N_f = fatigue life at σ, and
η and ζ = experimental curve fitting parameters.

It should be mentioned that the average engineering shear strain to failure ϵ_f is assumed to be a constant and independent on the state of stress and number of cycles. This assumption is used by many authors [*16–19*] and verified experimentally in this study. The magnitude of average strain to failure ϵ_f for the unidirectional AS4/3501-6 graphite/epoxy ply under in-plane shear loading is found equal to 0.101, experimentally. The static shear stiffness $E(0)$ for the unidirectional ply is equal to 5.03 GPa. The results of residual shear stiffness of the material is shown in a normalized form in Fig. 13. By using the experimental results and the curve-fitting equation (Eq 3), the experimental residual stiffness curve-fitting parameters η and ζ for the unidirectional AS4/3501-6 graphite/epoxy ply under in-plane shear loading conditions are found to be equal to 0.70 and 11, respectively.

The nonlinear elastic shear stress-strain behavior of a composite ply is a well-known phenomenon [*20*]. The nonlinear elastic part of the stress-strain curve can be expressed mathematically by the Hahn-Tsai equation [*20*], and the curve fitting parameters E_{xy}, and δ can be found experimentally.

$$\epsilon_{xy} = \frac{\sigma_{xy}}{E_{xy}} + \delta\sigma_{xy}^3 \tag{4}$$

where

ϵ_{xy} = engineering shear strain,
σ_{xy} = shear stress,
E_{xy} = in-plane shear stiffness (initial slope of $E(n)$), and
δ = parameter of nonlinearity.

Typical shear stress-strain behavior of the unidirectional AS4/3501-6 graphite/epoxy ply after 5000, 100 000, and 200 000 cycles under 59% of its maximum static shear strength is shown in Fig. 14. It should be noted that only the elastic part of the curves are shown. Using

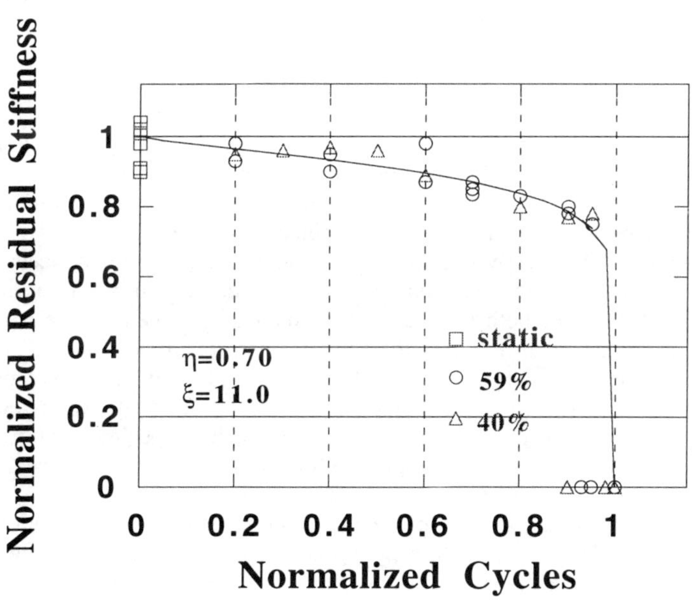

FIG. 13—*Master curve for the residual shear stiffness of the material under in-plane shear loading conditions.*

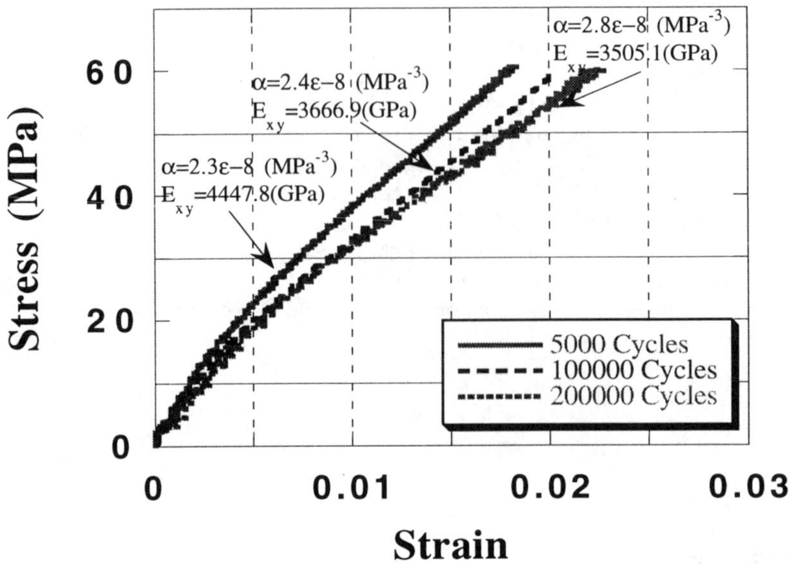

FIG. 14—*Typical shear stress-strain behavior of the material after shear fatigue loading.*

Eq 4, the experimental curve-fitting parameters are calculated for each case. As shown the in-plane shear stiffness of the material E_{xy} is decreasing by increasing the number of cycles, while the parameter of nonlinearity δ is increasing. This indicates that during fatigue cycling, the material is becoming softer and more nonlinear. Although it is observed that the parameter of the material nonlinearity is an increasing function of number of cycles, further research is needed to express this relationship in a normalized form.

Conclusions

A theoretical and experimental study is performed to characterize the behavior of unidirectional AS4/3501-6 graphite/epoxy composite ply under static and fatigue shear loading conditions. The three-rail shear test method is selected as the testing technique. The three-rail shear specimen is modified by inserting notches on the edges. It is shown that the unnotched specimen fails mostly due to propagation of sharp cracks induced near the edge of the rails, while the notched specimen absorbs the external load by failing in shear in the gage areas. The stiffness of both unnotched and notched specimens in the nonlinear elastic region are the same. However the strength of the notched specimen both in static and fatigue loading conditions and the fatigue life are higher than those of the unnotched specimen. Also it is noticed that by using the notched specimen, scatter in experimental results is reduced drastically. The better performance of the notched specimen, both in static and fatigue loading conditions, shows the capability of the newly developed specimen for more realistic measurement of shear properties of the material.

Using the notched specimen, the fatigue life, residual strength, and residual stiffness of the unidirectional AS4/3501-6 graphite/epoxy composite ply under different shear fatigue loading conditions are measured. The results of experiments are presented in normalized form curves that are useful for complete characterization of shear fatigue in terms of fatigue life, residual strength, and residual stiffness. The effect of fatigue loading on nonlinear shear stress-strain behavior of the material in elastic region is studied and the Hahn-Tsai equation is used to express this behavior. However, to express the relationship between the parameter of material nonlinearity and the number of cycles, further research is needed. The ability to accurately measure shear properties will play an important role in the three-dimensional modeling of composite materials, where all material properties must be precise for proper modeling.

References

[1] Purslow, D., "The Shear Properties of Unidirectional Carbon Fibre Reinforced Plastics and Their Experimental Determination," *Aeronautical Research Council*, Current Papers 1381, 1977.

[2] Sims, D. F. and Brogdon, V. H., "Fatigue Behavior of Composites Under Different Loading Modes," *Fatigue of Filamentary Material, ASTM STP 636*, K. L. Reifsnider and K. N. Lauraitis, Eds., American Society for Testing and Materials, West Conshohocken, PA, 1977, pp. 185–205.

[3] Smith, E. M. and Pascoe, K. J., "The Role of Shear Deformation in the Fatigue Failure of a Glass Fiber-Reinforced Composite," *Composites*, 1977, pp. 237–243.

[4] Phillips, D. C. and Scott, J. M., "The Shear Fatigue of Unidirectional Fibre Composite," *Composites*, Oct. 1977, pp. 233–236.

[5] Philips, D. C. and Scott, J. M., "Shear Fatigue of CFRP Under Low Frequency Torsion," *Fiber Science and Technology*, No. 11, 1978, pp. 23–54.

[6] Purslow, D., "Some Fundamental Aspects of Composite Fractography," *Composites*, Oct. 1981, pp. 241–247.

[7] Farve, J. P. and Vidal, G., "In-Plane and Interlaminar Shear Fatigue Characterization of Unidirectional GFRP and CFRP, Including Moisture Effects," *TEQC83, Proceedings of the International Conference on Testing, Evaluation and Quality Control of Composites*, Surrey University, Guildford, England, Sept. 1983, pp. 57–65.

[8] Butler, R. J., Barnard, P. M., and Curtis, P. T., "The Development of a Satisfactory, Simple, Shear Fatigue Test for Unidirectional E-Glass Epoxy," *Composite Materials: Testing and Design (Eighth Conference), ASTM STP 972*, J. D. Whitcomb, Ed., American Society for Testing and Materials, West Conshohocken, PA, 1988, pp. 227–240.

[9] Yaniv, G., Daniel, I. M., and Lee, J. W., "Method for Monitoring In-Plane Shear Modulus in Fatigue Testing of Composites," *Test Methods for Design Allowables for Fibrous Composites: 2nd Volume, ASTM STP 1003*, C. C. Chamis, Ed., American Society for Testing and Materials, West Conshohocken, PA, 1989, pp. 276–284.

[10] Lee, S. and Munro, M., "Evaluation of In-Place Shear Test Methods for Advanced Composite Materials by the Decision Analysis Technique," *Composites*, Vol. 17, No. 1, 1986, pp. 13–22.

[11] Lessard, L. B., Eilers, O. P., and Shokrieh, M. M., "Testing of In-Plane Shear Properties Under Fatigue Loading," *Journal of Reinforced Plastics and Composites*, Vol. 14, No. 9, 1995, pp. 965–987.

[12] Gathercole, N., Reiter, H., Adam, T., and Harris, B., "Life Prediction for Fatigue of T800/5245 Carbon-Fibre Composites: I. Constant Amplitude Loading," *Fatigue*, Vol. 16, Nov. 1994, pp. 523–532.

[13] Adam, T., Dickson, R. F., Jones, C. J., Reiter, H., and Harris, B., "A Power Law Fatigue Damage Model for Fiber-Reinforced Plastic Laminates," *Proceedings of the Institution of Mechanical Engineers, Part C: Mechanical Engineering Science*, Vol. 200, No. C3, 1986, pp. 155–166.

[14] Chou, P. C. and Croman, R., "Degradation and Sudden-Death Models of Fatigue of Graphite/Epoxy Composites," *Composite Materials: Testing and Design (Fifth Conference), ASTM STP 674*, S. W. Tsai, Ed., American Society for Testing and Materials, West Conshohocken, PA, 1979, pp. 431–454.

[15] Haplin, J. C., Jerina, K. L., and Johnson, T. A., "Characterization of Composites for the Purpose of Reliability Evaluation," *Analysis of the Test Methods for High Modulus Fibers and Composites, ASTM STP 521*, American Society for Testing and Materials, West Conshohocken, PA, 1973, pp. 5–64.

[16] Poursartip, A., Ashby, M. F., and Beaumont, P. W. R., "The Fatigue Damage Mechanics of a Carbon Fiber Composite Laminate: I — Development of the Model," *Composites Science and Technology*, Vol. 25, 1986, pp. 193–218.

[17] Hwang, W. and Han, K. S., "Fatigue of Composites — Fatigue Modulus Concept and Life Prediction," *Journal of Composite Materials*, Vol. 20, March 1986, pp. 154–165.
[18] Whitworth, H. W., "Modeling Stiffness Reduction of Graphite/Epoxy Composite Laminates," *Journal of Composite Materials*, Vol. 21, April 1987, pp. 362–372.
[19] Hwang, W. and Han, K. S., "Fatigue of Composite Materials — Damage Model and Life Prediction," *Composite Materials: Fatigue and Fracture, Second Volume, ASTM STP 1012*, P. A. Lagace, Ed., American Society for Testing and Materials, West Conshohocken, PA, 1989, pp. 87–102.
[20] Hahn, H. T. and Tsai, S. W., "Nonlinear Elastic Behavior of Unidirectional Composites Laminate," *Journal of Composite Materials*, Vol. 7, Jan. 1973, pp. 102–118.

H. Parvatareddy,[1] *A. Pasricha,*[2] *D. A. Dillard,*[3] *B. Holmes,*[4] *and J. G. Dillard*[5]

High Temperature and Environmental Effects on the Durability of Ti-6Al-4V/FM5 Adhesive Bonded System

REFERENCE: Parvatareddy, H., Pasricha, A., Dillard, D. A., Holmes, B., and Dillard, J. G., **"High Temperature and Environmental Effects on the Durability of Ti-6Al-4V/FM5 Adhesive Bonded System,"** *High Temperature and Environmental Effects on Polymeric Composites: 2nd Volume, ASTM STP 1302,* Thomas S. Gates and Abdul-Hamid Zureick, Eds., American Society for Testing and Materials, 1997, pp. 149–174.

ABSTRACT: A fracture mechanics based approach using wedge and double cantilever beam (DCB) specimens is being used to evaluate the durability of a titanium/adhesive system. The adhesive used was a polyimide, developed by NASA Langley Research Center, modified and supplied by Cytec Engineered Materials, Inc., and designated as FM5. Prior to bonding, the adherend surfaces were pretreated by one of two surface pretreatments: phenylethynyl trimethyl silane or chromic acid anodization.

Samples were aged at one of three different temperatures, 150, 177, and 204°C, all of which are well below the (250°C) glass transition temperature (T_g) of the adhesive. Aging was also carried out in one of three different environments, ambient atmospheric air and reduced air pressures of 2 psi air (13.8 KPa) and 0.2 psi air (1.38 KPa), for several months. Samples aged for different times were then tested to evaluate both static and fatigue properties. Results obtained thus far indicate that the greatest loss in strength occurs after aging in air at the highest aging temperature of 204°C. Strain energy release rate G values obtained from the static DCB tests dropped by 10 to 20% after aging for periods of up to 6 months.

Solvent uptake tests were conducted on neat FM5 resin in several common organic solvents followed by tensile tests to evaluate both saturated and residual properties. Strain energy release rate values were also computed from wedge tests of samples immersed in the different solvents for several days. Results showed the FM5 resin to be quite solvent resistant.

KEY WORDS: wedge test, double cantilever beam, fracture mechanics, environmental aging, adhesive bond, FM5 polyimide, strain energy release rate, solvent effects

Structural adhesive joints are being widely used and studied as alternatives to conventional fasteners in the automotive, aerospace, and other industries. Adhesive bonding offers advan-

[1] Graduate research assistant, Engineering Science and Mechanics Department, Virginia Polytechnic Institute and State University, Blacksburg, VA 24061-0219.

[2] Research associate, Engineering Science and Mechanics Department, Virginia Polytechnic Institute and State University, Blacksburg, VA 24061-0219.

[3] Professor, Engineering Science and Mechanics Department, Virginia Polytechnic Institute and State University, Blacksburg, VA 24061-0219.

[4] Graduate research assistant, Department of Chemistry, Virginia Polytechnic Institute and State University, Blacksburg, VA 24061-0219.

[5] Professor, Department of Chemistry, Virginia Polytechnic Institute and State University, Blacksburg, VA 24061-0219.

tages such as lower weight and lower manufacturing costs [*1*]. Furthermore, high-performance adhesives that are currently being synthesized (for example, epoxies, phenolics, acrylics, thermoplastic polyimides) offer other useful properties such as higher modulus, higher toughness, and stability at high temperatures. Also, recent advances made with surface pretreatments for the adherend surfaces have spurred an increased interest in using adhesive joints for aircraft and automotive structures [*2*].

In general, adhesive joints fail by progressive crack growth and the initiation and propagation of cracks in a bonded joint are directly related to the service conditions to which the joint is exposed [*3*]. Static loads, fatigue loads, elevated temperature, and humidity are some of the principal conditions that cause the formation and propagation of cracks in an adhesive bond [*4*]. Therefore, any study involving adhesive joints should include a detailed assessment of the durability of the bond system in the service conditions of that particular application. Durability study of bonded joints must involve, among other things, an approach based on fracture mechanics.

In the present study, the durability of the Ti-6Al-4V/FM5 adhesive bonded system is being evaluated using wedge and double cantilever beam (DCB) fracture specimens. The current study is part of a comprehensive study to develop a durable material system for application in the proposed mach 2.4 high speed civil transport (HSCT) aircraft. According to the design criteria, the material system to be used on the aircraft should be durable for over 60 000 h of flight encountering temperatures during flight in the range of 177°C. Physical aging and chemical aging of the adhesive material are some of the important issues that have to be evaluated and taken into consideration for predicting the bond durability. In order to simulate the service environment conditions of the HSCT, the Ti-6A1-4V/FM5 bonds were aged in one of three temperatures: 150, 177, and 204°C, at one of three different environments: atmospheric air, and reduced air pressures of 2 psi air (13.8 KPa) and 0.2 psi air (1.38 KPa). The 2 psi air environment corresponds to the service altitude air pressure of the HSCT aircraft.

Wedge tests are very popular for evaluating adhesive bond durability. These tests are very simple to conduct and very cost effective. References *5* through *8* are examples of literature describing the wedge test, the analysis of the wedge test, and some data collected on other adhesive systems. In the current study, the wedge test has been used extensively as a screening tool to evaluate several surface pretreatments for the titanium adherends. Based on these screening tests, two surface treatments, the phenylethynyl trimethyl silane (PS) treatment and the chromic acid anodization (CAA), were chosen for this study.

DCB tests have also been used very extensively to study the effects of static, fatigue and environmental loads on adhesive bonds and thereby the durability of the bonded joints [*9–24*]. Rakestraw et al. [*9,10*] used the DCB specimen to study steel/epoxy bonds, Kinloch and Osiyemi [*11*], Kodokian and Kinloch [*12*], and Blackman et al. [*13*] used this specimen to study graphite epoxy composite/adhesive bonded joints. Lefebvre et al. [*14*] used a modified DCB specimen to study rubber/metal bond durability.

It has been observed that several thermoplastic adhesives and polymers are susceptible to solvent-induced environmental stress cracking (ESC) phenomena [*25–30*]. Parvatareddy et al. [*31*] showed that ESC required a synergistic effect between solvent and stress, and further showed that high-performance composites showed ESC when exposed to common organic solvents used in the aircraft industry. In order to assess the solvent resistance of the FM5 resin system, solvent uptake tests were performed on neat FM5 resin samples by exposing the samples to several common organic solvents. Following the solvent uptake tests, tensile tests were conducted on both saturated and dry adhesive samples to study stress-strain characteristics. Wedge tests were also conducted on the Ti-6Al-4V/FM5 bonds by immersing the specimens in the various solvents for several hours and measuring the crack growth in the joints.

Experimental

Materials

The adhesive used in the study was designated as FM5, a polyimide originally developed at NASA Langley Research Center and modified and supplied by Cytec Engineering Materials, Inc., Havre de Grace, Maryland [*32*]. The adhesive was supplied in two forms; as polymer film supported on glass (scrim) cloth and as unsupported film. The supported film contained 85% polymer by weight. This supported film also contained approximately 4% by weight of the solvent NMP. The supported film was used to bond titanium adherends to conduct wedge and DCB tests. The unsupported film had less than 0.5% by weight of the solvent and was used principally to conduct the solvent resistance studies and to collect weight loss data. The glass transition temperature T_g of thc adhesive was 250°C using both dynamic mcchanical analyzer (DMA) and differential scanning calorimetry (DSC) measurements. The adhesive was cured at a temperature of 350°C for 60 min while maintaining a curing pressure of 75 psi (0.518 MPa).

The adherends used in the study were cut from Ti-6Al-4V plates, having dimensions of 200 by 25 by 3.175 mm, and supplied by President Titanium Company, Hanson, Massachusetts. The adherends were surface pretreated before bonding with the adhesive.

Surface Pretreatments

The steps involved in the phenylethynyl trimethyl silane (PS) treatment are as follows. The titanium adherends were cleaned by soaking in a 3% (w/w) solution of sodium hydroxide (NaOH), sulfuric acid (H_2SO_4), and water. The cleaned titanium was then soaked in a 5% (w/w) solution of phenylethynyl trimethyl silane in 95% ethanol at 50°C, followed by a 95% ethanol (C_2H_5OH) rinse.

The chromic acid anodization (CAA) treatment involves the following steps. First the titanium adherends were wiped with acetone and cleaned with a 3M Scotch-Brite pad and deionized water. After ensuring that most of the surface contaminants were removed, the titanium was then soaked in Isoprep 177 (Allied Kelite) for 10 to 15 min, followed by a 5-min warm water rinse in deionized water at a temperature greater than 43°C. The sample pieces were then soaked in a nitric acid (HNO_3) bath (114 mL H_2O, 151 mL HNO_3, 10.9 mL of 50% HF) for 1.5 min followed by a cold water rinse in deionized water. The anodization was accomplished at 5 V and a current of 0.11 to 0.14 A/m^2 for 20 min, followed again by a cold water rinse in deionized water. The composition of the bath was 45 g of CrO_3/L. Ti-6Al-4V plates were used as the cathodes. The samples were then dried in an oven for 1 to 1.5 h at 66°C.

Wedge and DCB Specimen Fabrication

Wedge and double cantilever beam specimens were produced by individually bonding 200- by 25- by 3.175-mm plates of pretreated Ti-6Al-4V. The supported film was cut to dimensions of 175 by 25 mm and placed in between the titanium adherends and bonded in a hot press using the recommended cure cycle (see below). The initial 25 mm of the specimen was left unbonded and this region was used to attach end blocks in the case of the DCB

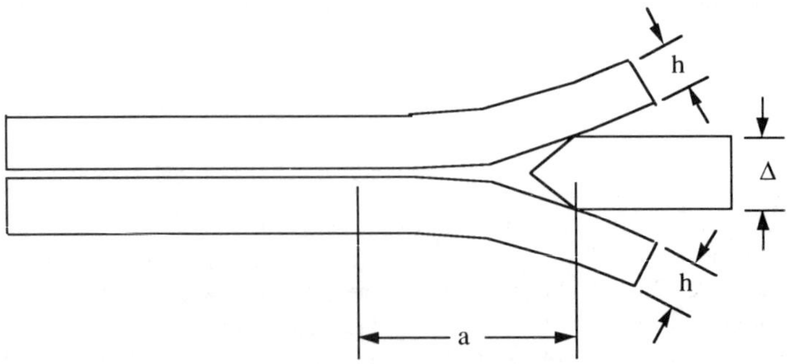

FIG. 1*a*—*Schematic of a wedge specimen with the related parameters for calculation of strain energy release rate* G.

specimens. The thickness of the adhesive film was measured before and after bonding and was found to be around 0.5 mm before bonding an around 0.125 mm after bonding. Typical schematics of the wedge and DCB specimen geometrics are shown in Figs. 1*a* and 1*b*.

The PS-treated titanium pieces were bonded together by using a cure cycle as follows. The adhesive system was heated up to a temperature of 350°C using a heating rate of around 5°C/min and maintaining a pressure of 75 psi (0.518 MPa) at all times. The desired temperature of 350°C was reached and maintained for 60 min. Following the 60-min curing at the high temperature and pressure, the specimens were cooled down to room temperature at a cooling rate of around 15°C/min. The above curing procedure was later modified for bonding the CAA titanium samples. In the modified procedure the samples were first heated up to 250°C under pressure of 75 psi (0.518 MPa) at the same heating rate of 5°C/min, held there for 30 min and then heated up to 350°C and maintained there for an additional 60 min. Then the samples were brought down to room temperature using the same cooling rate as used previously. This modified cure was used to eliminate air bubbles and voiding which was seen in the bondline with the PS treated samples. The solvent present in the adhesive, which in this case is NMP has a boiling point of 202°C. In the original curing process, the NMP did not have enough time to escape from the bond. By holding the temperature at 250°C for an additional 30 min, most of the solvent apparently evaporated from the system. This modified curing procedure seemed to work well for the CAA samples and it is being currently used for making the bonds for on-going work. Figures 2 and 3 are scanning electron

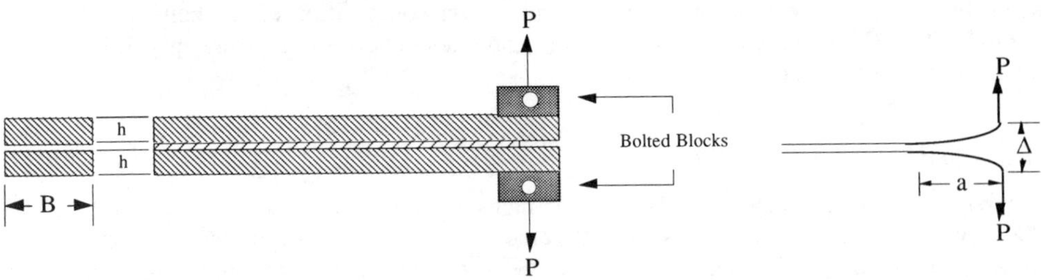

FIG. 1*b*—*Schematic of a DCB specimen with the related parameters for calculation of strain energy release rate* G.

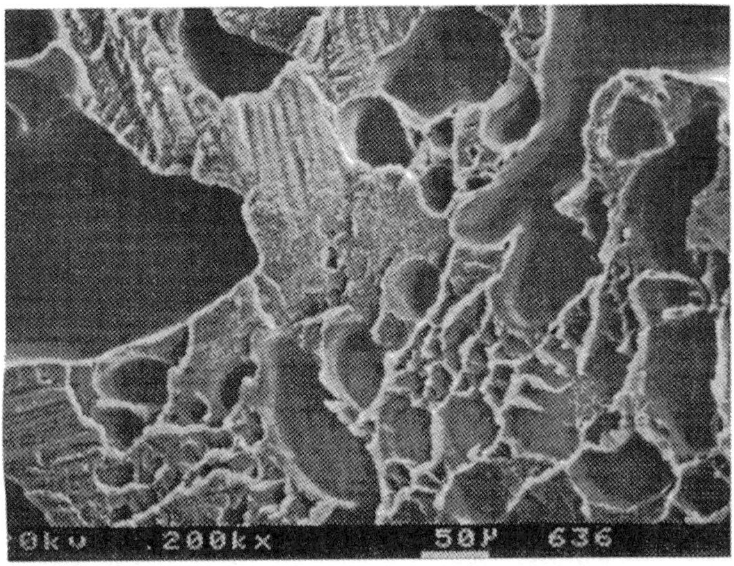

FIG. 2—*SEM micrograph of the fracture surface of a Ti-6Al-4V/FM5 bonded specimen which was cured at 350°C for 60 min under a pressure of 75 psi (0.518 MPa).*

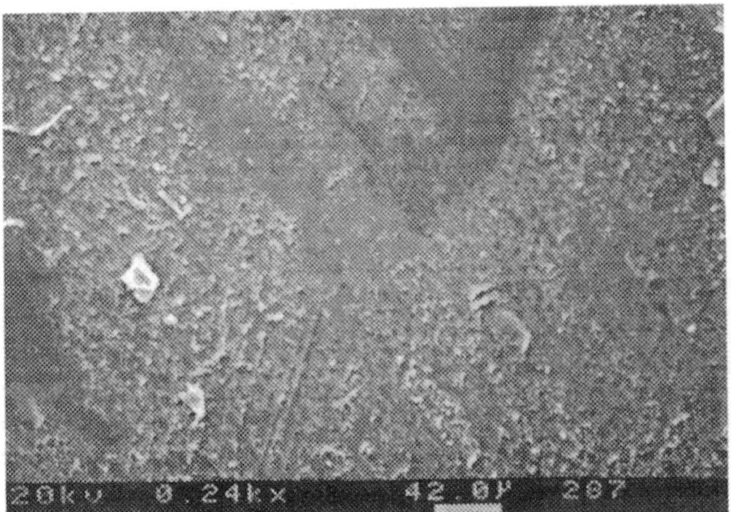

FIG. 3—*SEM micrograph of the fracture surface of a Ti-6Al-4V/FM5 bonded specimen that was cured at 250°C for 30 min followed by a 60 min cure at 350°C under a pressure of 75 psi (0.518 MPa).*

photomicrographs showing the differences in the fracture surfaces using the two curing processes. The modified cure cycle seems to have eliminated most of the air bubbles and voiding seen in the original procedure.

Specimen Conditioning

The bonded specimens were conditioned in aging ovens for periods of several months, and selected samples were periodically removed and tested at aging times that are logarithmically separated. The aging ovens are convection-type ovens manufactured by Precision Scientific Group, Chicago, Illinois, and these were maintained at three different temperatures of 150, 177, and 204°C. Samples were also aged in three different environments in each of the three ovens. Two steel chambers were placed in each one of the three ovens. A set of three chambers (one from each of the three ovens) were then interconnected by copper tubing and needle valves and connected to a vacuum pump, which maintained a constant pressure in the chambers at all times. A similar arrangement was done with the other set of three chambers. This facilitated the aging of samples in nine different environments in the three ovens. The aging environments included ambient atmospheric air, reduced air pressure of 2 psi air (13.8 KPa) corresponding to the service altitude pressure of the HSCT, and 0.2 psi air (1.38 KPa). These three environments were so chosen to mimic aging of the specimens in the three different oxygen partial pressures, and thereby study effects of chemical aging (oxidation) in the bonded joints. The neat resin samples used for taking weight measurements were also aged in the chambers within the ovens.

Wedge Testing

Wedge Data Analysis and Testing

The wedge test is a very simple and useful test to determine adhesive bond properties. The wedge test permits in obtaining a quantitative measure for the fracture energy or strain energy release rate G of an adhesive bond subjected to environmental conditioning [*5*]. In the wedge test, crack length data can be collected as a function of time and using appropriate equations, this data can be used to calculate fracture energy as a function of crack growth rate. The equation for the strain energy release rate G is [*6*]:

$$G = \frac{3E\Delta^2 h^3}{16a^4} \tag{1}$$

where

E = modulus of elasticity of adherend,
h = adherend thickness,
Δ = thickness of the wedge, and
a = debond or crack length.

The underlying assumptions for applying the above equation are, first, that there is no plastic deformation in the substrate, and secondly, that the specimen compliance is primarily due to beam bending with minimal contributions from beam shear and adhesive deformation. One disadvantage to this test is that the specimen must always be removed from the conditioning environment to make a crack length measurement. This is not easy if the samples are being conditioned in a vigorous environment. There may also be some subjectivity involved in measuring crack length in some samples where the cracks are very difficult to view especially after environmental exposure. In the current testing program five specimens were tested per environment and the values reported are averages from the five specimens. Crack measurements were taken on both sides of the test specimens.

The wedge tests conducted in this research were modified versions of ASTM D 3762 (Test Method for Adhesive-Bonded Surface Durability of Aluminum (Wedge Test)). ASTM D 3762 uses specimens cut out of plates with the dimensions of 17.8 by 21.5 by 1.5 mm. These dimensions are quite different from the 200- by 25- by 3.175-mm plates that were used for making the specimens in the present testing program. Further, the wedge specimens were all bonded one at a time in this study. The ASTM test recommends the use of a 1.5-mm teflon spacer to control the bond thickness and also for the purpose of initiating a starter crack. Due to the fact that the FM5 resin was available as a supported film, no teflon spacer was used during bonding as the glass scrim cloth helps maintain a constant bondline thickness. To ensure that a constant starter crack was obtained, wedges were very carefully inserted into the bondling by using a vice in this study. Also, based on the observations made by Cognard [*6*] and the static DCB tests conducted in this study, it was calculated that for the geometry of the specimens and wedges used in the study, (2-mm thick wedges were used in all the tests) a starter crack of at least 25 mm must be initiated to obtain satisfactory fracture energy values.

DCB Testing

Static DCB Analysis and Testing

The fracture energy in the case of the static DCB tests was determined by using Eq 2 (given below), which was derived using the compliance method with corrections for the crack length offset and adherend stiffness [*9,10,13*]. One of the assumptions for using Eq 1 for calculation of fracture energy is that specimen compliance is primarily due to beam bending; however, in the case of compliant adhesives there may be crack tip deflections and rotations. To account for these factors as well as other factors such as adherend and adhesive properties, bond thickness, and so forth, which are very system dependent, Eq 2 was used to calculate G:

$$G = \frac{9\Delta^2(EI_{eff})}{4B(a + x)^4} \tag{2}$$

where

B = specimen width,
C = compliance,
a = crack length,
x = apparent crack length offset,
EI_{eff} = effective flexural rigidity, and
Δ = specimen's opening displacement at point of load application.

The term Δ in Eq 2 refers to the specimen's opening displacement at the point of load application in the case of the DCB test as opposed to wedge thickness as defined earlier. The parameters EI_{eff} and x are determined from experimental data according to

$$EI_{eff} = \frac{2}{3m^3} \text{ and } x = \frac{b}{m} \tag{3}$$

where the experimental coefficients m and b are the slope and y-intercept, respectively, of the cube root of the measured compliance, when plotted as a linear function of the crack length. Compliance C is obtained by dividing the specimen's crack opening displacement Δ by the corresponding load P where Δ and P are experimentally measured. Typical data from a Ti-6Al-4V/FM5 static DCB test is shown in Fig. 4. Due to the linear relationship of this experimental data, a value of the crack length can be obtained for a given crack opening displacement Δ and the load P and the strain energy release rate G computed. Thus it is not necessary to manually read the crack lengths along the debond length.

Static DCB specimens were tested in an Instron machine controlled through its GPIB interface using LabVIEW software developed by National Instruments, Austin, Texas. The specimens were loaded at a crosshead speed of 1 mm/min. As a specimen was loaded, several events took place and these can be seen clearly on a typical load versus deflection plot depicted in Fig. 5. For example, when the curve deviates from linearity, the crack had begun to grow and the critical fracture energy G_{IC} has been reached. Now, due to the rate dependent failure of these adhesive systems, the loading value continued to increase. When the load reached a maximum and began to rapidly decrease, the specimen had achieved a maximum loading level G_{MAX}. Once the maximum loading was detected, the crosshead motion was stopped to allow the crack growth to continue until it approached quasi static equilibrium conditions. The term, quasi-state equilibrium, is used because the crack was not always allowed to stop completely. The criterion used to establish a reasonable arrest loading level was that the value of strain energy release rate decreased by less than 1 J/m^2 over a min. This procedure was used to identify the arrest fracture energy, G_{ARR}. The vertical lines in Fig. 5 were a result of the crack growing while the crosshead was being held constant. Once crack growth slowed sufficiently, the specimen was unloaded to make sure plastic deformation of the adherends had not occurred. If plastic deformation had occurred, the P-

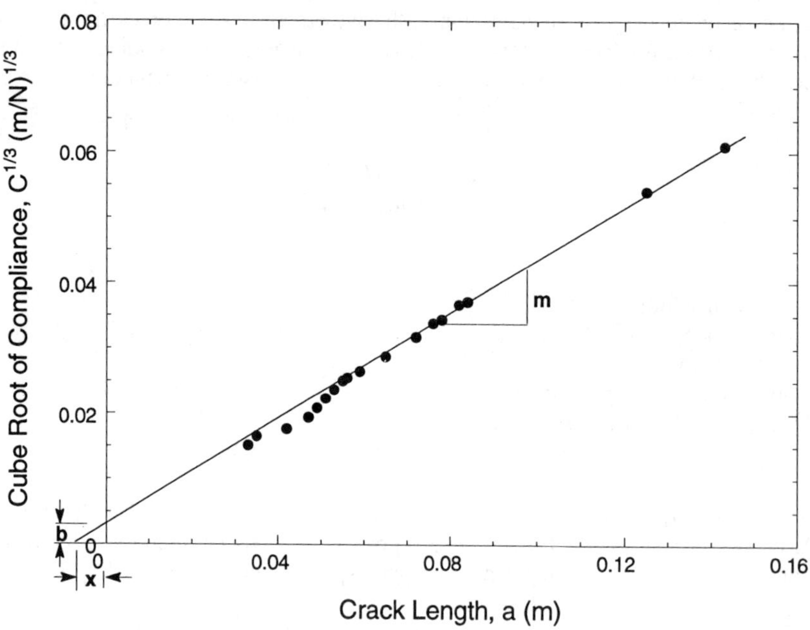

FIG. 4—*Experimental compliance versus crack length relationship for a Ti-6Al-4V/FM5 DCB joint tested in a quasi-static test.*

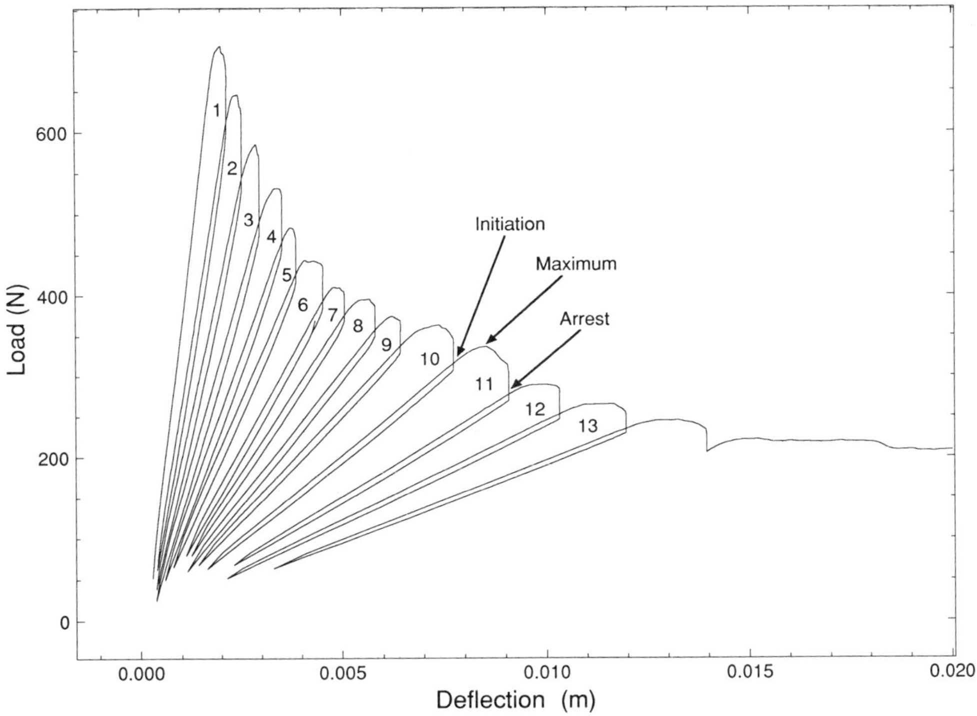

FIG. 5—*Typical load versus deflection curve for a quasi-static test on a Ti-6Al-4V/LaRC PETI-5 DCB joint.*

Δ curve will not return to the origin, but instead will intersect the deflection-axis to the right of the origin. Finally, this load-hold-unload procedure was repeated until the specimen failed. It was seen that in all the DCB bonds tested, the *P*-Δ curves did return to the origin indicating that there was no plastic deformation of the adherends.

Fatigue DCB Analysis and Testing

Fatigue DCB tests are conducted to study the cyclic debonding rate of adhesive bonds or joints as a function of strain energy release rate. In the present study, all fatigue DCB tests were conducted on a servo-hydraulic MTS 810 testing machine. Data acquisition was made possible by using LabVIEW software. All tests were conducted using displacement control at a frequency of 5 Hz and the machine was programmed to apply a sinusoidal load on the sample at a load ratio R (P_{min}/P_{max}) of 0.1. Maximum and minimum loads, number of loading cycles, and crack opening displacement data was periodically computed and written to a data file.

A graph showing the typical trends of fatigue crack growth data is depicted in Fig. 6. Many fatigue crack growth models have been offered to describe this data. Some are based on empirical observations while others are based on more rigorous theoretical foundations. The model chosen for this research is the widely used empirical equation of Paris and Erdogan [*33*], which may be expressed as

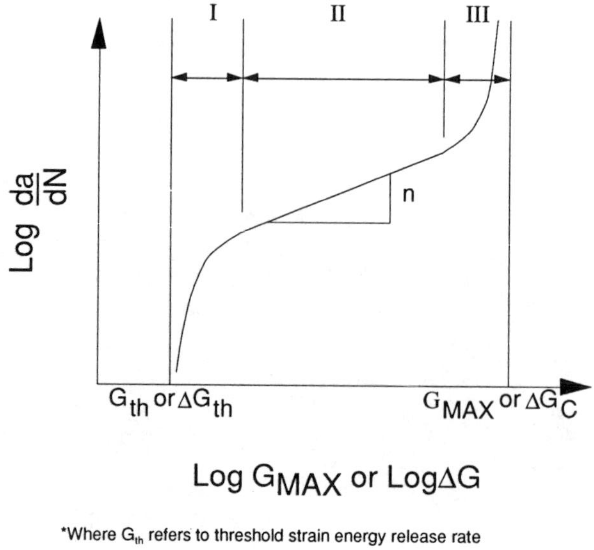

FIG. 6—*A typical fatigue crack growth curve showing the primary, secondary, and tertiary regions.*

$$\frac{da}{dN} = C(G_{MAX})^n \tag{4}$$

where

a = crack length,
N = number of cycles, and
G_{MAX} = maximum applied cyclic strain energy release rate calculated from Eq 2.

Here, C and n are considered system constants and are determined from the experimental data. Higher values of C or n indicate that an adhesive system is more susceptible to fatigue crack growth. The fatigue crack growth model of Eq 4 is intended to model the linear portion of Fig. 6 in Region II.

Solvent Sensitivity Studies and Tensile Testing

Solvent Uptake, Desorption, and Tensile Testing

Solvent uptake tests were performed on unsupported neat FM5 resin. Dog bone specimens (Type M-III) of the neat resin, 0.5-mm thick, were prepared in accordance with ASTM D 638M (Test Method for Tensile Properties of Plastics). The solvents used in the study included acetone, MEK, toluene, jet fuel, ethylene glycol, hydraulic fluid, and water. Ten dog bone specimens of the resin were immersed in each of the solvents and the samples periodically weighed to obtain equilibrium solvent mass uptake data. Data were collected on the specimens in excess of 1200 h. Five of the equilibrated samples were then tested in a miniature tensile testing machine (Minimat), according to ASTM D 638M to obtain stress-strain characteristics. In all cases, moduli, strains to failure, and yield stresses were computed, and comparisons made with unexposed control specimens. A crosshead rate of 1.5 mm/min

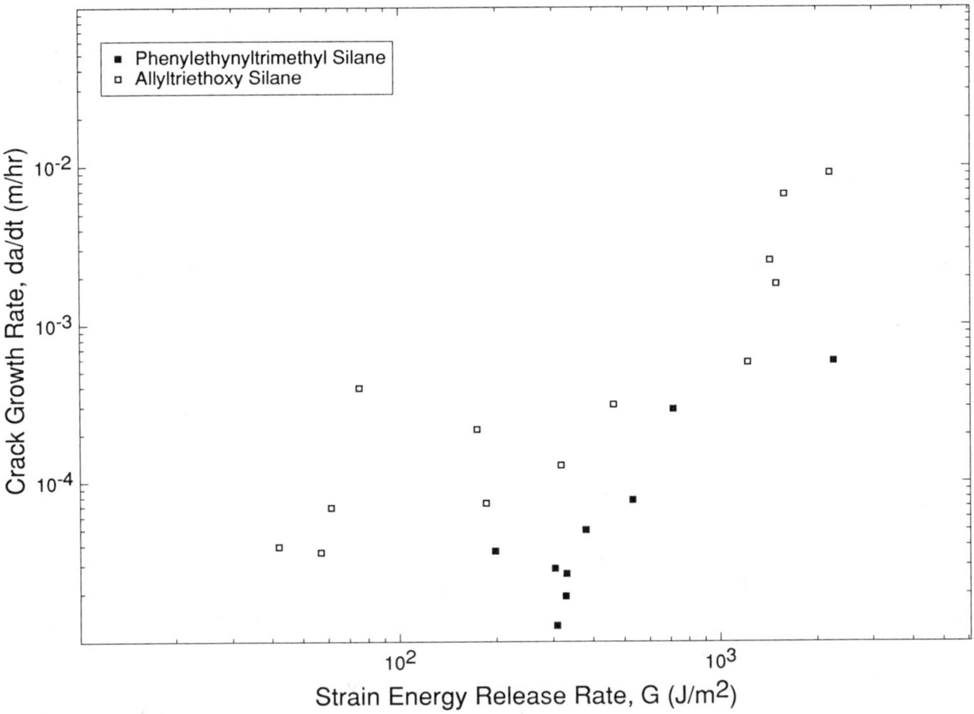

FIG. 7—*Wedge test data showing the comparison of crack growth rate versus strain energy release rate for two different silane treatments. Samples were exposed to boiling water for up to 438 h.*

was used for conducting all the tests. The other five samples equilibrated in the solvents were then desorbed in a vacuum environment at 150°C for 60 min. The temperature, 150°C, is well below the glass transition temperature of the FM5 adhesive. Following desorption these specimens were then tested on the Minimat to obtain residual properties.

Wedge Testing

To study the durability of bonded joints under solvent exposure, Ti-6Al-4V/FM5 wedge specimens were prepared and wedge tests were performed for several hours in solvent baths. In this portion of the study chromic acid anodization was used as a surface pretreatment for titanium. Based on the measured crack lengths, the strain energy release rate was computed as a function of crack growth rate. The *G* measurements help to quantify the durability of the bonded joints on exposure to the various solvents, and further help in ranking the solvent resistance of the adhesive. Five wedge specimens were exposed to each of the solvents and the results reported in a later section are average values from the five tests.

Weight Measurements

Three specimens each of the FM5 neat resin having dimensions of 60 by 10 by 0.5 mm were used to monitor weight loss/change with aging at 177 and 204°C in two different environments. The two environments were ambient atmospheric air and reduced air pressure of 2 psi air (13.8 KPa). Samples were periodically removed from the aging ovens over several months, weighed, and returned to the ovens. Samples were weighed using an electronic analytical balance from Mettler Inc. Each data point reported in the results is an average value from three specimens weighed separately.

Results and Discussion

Wedge Tests

Wedge tests were primarily used as screening tests to choose suitable and durable surface pretreatments for the Ti-6Al-4V substrates. Some of the early surface treatments used were the silane treatments. Based on some initial wedge data it was decided to evaluate adhesive bonds prepared following phenylethynyltrimethyl silane surface pretreatment. A 6 month program involving PS treated DCB samples was initiated to characterize the Ti-6Al-4V/FM5 system. The PS treatment has currently been eliminated from the testing program and all the data collected from this surface treatment have been included in this paper. Figure 7 depicts data showing a comparison of phenylethynyltrimethyl silane treatment with an allyltriethoxy silane treatment. Here crack growth rate is plotted as a function of strain energy release rate. From Fig. 7 it is evident that after 438 h of exposure to boiling water, the PS treatment has a higher G_{ARR} indicating better durability than the other silane treatment. The maximum fracture energy is in the vicinity of 2200 J/m^2. This value is important to keep in mind as similar values for maximum strain energy release rate were obtained from quasi-static DCB tests on this material system.

Additional surface pretreatments were evaluated and the results from wedge tests are summarized in Table 1. These results were obtained by immersing the samples in boiling water for 72 h. All results shown are averages of five tests and the standard deviations are at the 95% confidence level. From these tests only the chromic acid anodized (CAA) and the chromium oxide grit blast samples failed cohesively. Failure occurred at the adhesive/glass scrim cloth interface. Based on these results it was decided to incorporate the CAA treatment into the current testing program. This surface treatment will serve as a control treatment for the Ti-6Al-4V specimens. Long-term wedge data were collected on CAA treated Ti-6Al-4V/FM5 bonds, following aging of the specimens for up to 4320 h in air at atmospheric pressure at 204°C. Over the duration of the tests, the fracture energy dropped by approximately 30% of its initial value as seen in Fig. 8. Once again five specimens were conditioned and average values reported. The failure in these samples appears to be cohesive.

TABLE 1—*Comparison of arrest strain energy release rates for chronic acid anodization and different grit blast surface pretreatments. Data was obtained from wedge tests conducted on Ti-6Al-4V/FM5 bonds immersed in boiling water for periods up to 72 h.*

Surface Pretreatment	G_{ARR} (J/m^2) Std. Dev. (95% confidence)
Chromic Acid Anodization	1902+/−61 (cohesive failure)
Chromium Oxide Grit Blast	1857+/−39 (cohesive failure)
Alumina Grit Blast	367+/−44 (adhesive failure)
Titanium Dioxide Grit Blast	1167+/−151 (mixed failure)
Silicon Dioxide Grit Blast	433+/−31 (adhesive failure)
Titanium Disilicide Grit Blast	267+/−18 (adhesive failure)

NOTE—Cohesive failure is defined as a failure within the adhesive layer; adhesive failure is defined as a failure between adhesive and metal substrate; mixed failure is a combination of cohesive and adhesive failures.

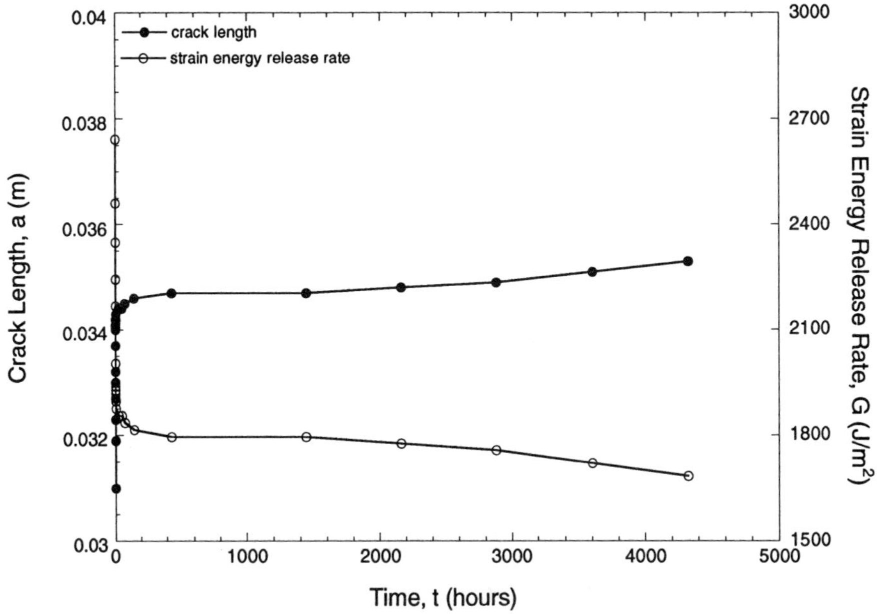

FIG. 8—*Long-term wedge test data on a Ti-6Al-4V/FM5 bond after exposure to air at 204°C for up to 4320 h. Surface pretreatment used was chromic acid anodization.*

Figure 9 is a plot of crack growth rate as a function of energy release rate for the data shown in Fig. 8. Based on the previous testing, it appears that the adhesive bonded system is quite durable.

Static DCB Tests

Static DCB tests were conducted on the Instron machine at room temperature and atmospheric pressure on Ti-6Al-4V/FM5 specimens, exposed to temperatures of 150, 177, and 204°C for periods up to several months. The effect of environment was also evaluated, the environments being ambient atmospheric air, 2 psi air (13.8 KPa), and 0.2 psi air (1.38 KPa). The two surface pretreatments used in this study included the phenylethynyltrimethyl silane and chromic acid anodization. The effect of aging temperature on the PS treated adhesive bonds is depicted in Fig. 10. The samples were aged for up to 6 months in atmospheric air and a comparison is made with respect to the energy release rate of an as-received specimen. As seen in Fig. 6, there is as much as a 20 to 25% decrease in the maximum and arrest strain energy release rate values in the aged samples with respect to the as-received sample. There also appears to be a slightly greater degradation in the fracture energy values at the highest temperature of 204°C compared to the other two temperatures. This could be partly explained by the fact that there is greater physical aging occurring in the adhesive material at the elevated temperature. Physical aging is a manifestation of a slow evolution of a polymer to thermodynamic equilibrium by time-dependent changes in volume, enthalpy, and entropy, as well as mechanical properties [*34*]. Figure 11 shows the effect of different aging environments on the bond durability for PS treated DCB specimens. The aging time and temperature were 6 months and 204°C, respectively. As seen in Fig. 11, samples aged in air

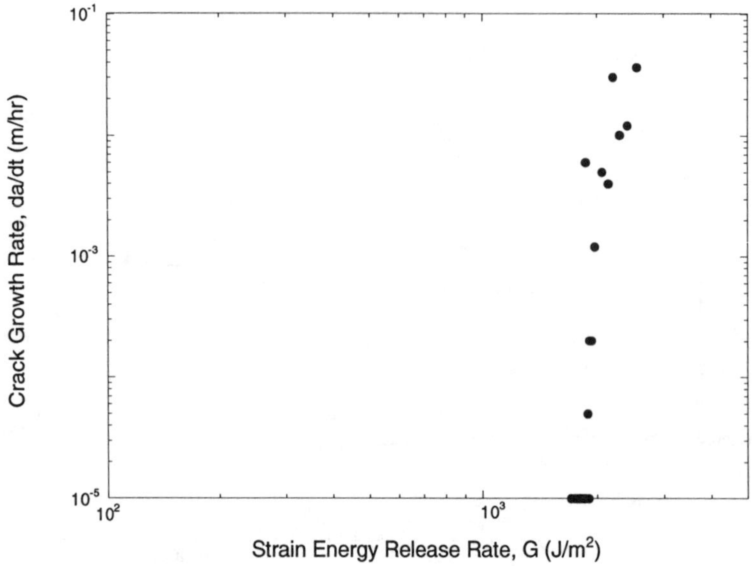

FIG. 9—*Crack growth rate versus strain energy release rate plot of wedge test data from Fig. 8.*

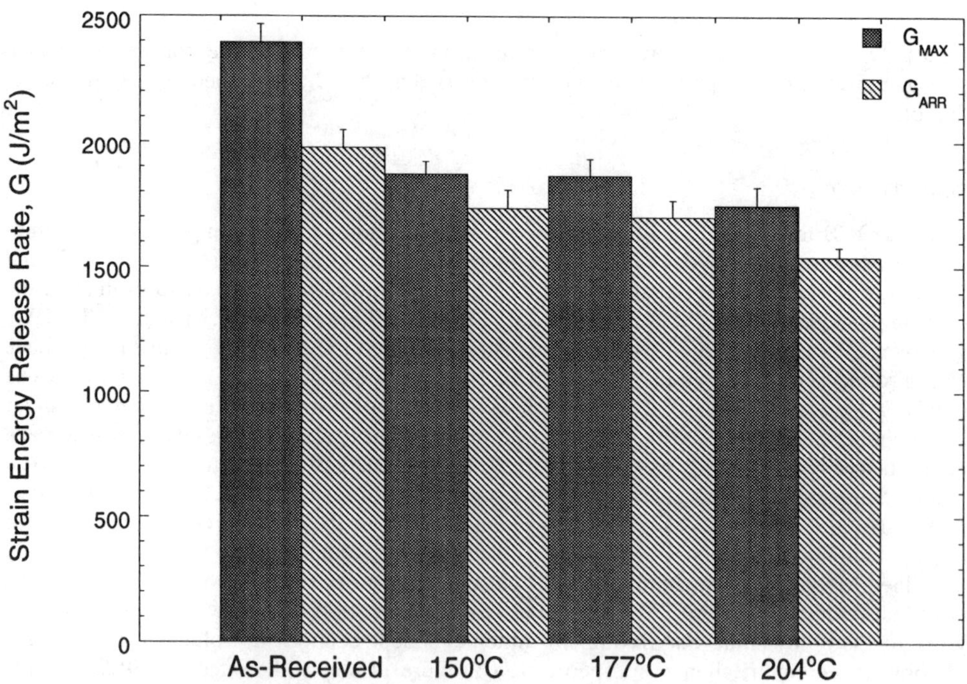

FIG. 10—*Static DCB test data on Ti-6Al-4V/FM5 bonded joints after 6 months of aging at different temperatures. Surface pretreatment was phenylethynyltrimethyl silane.*

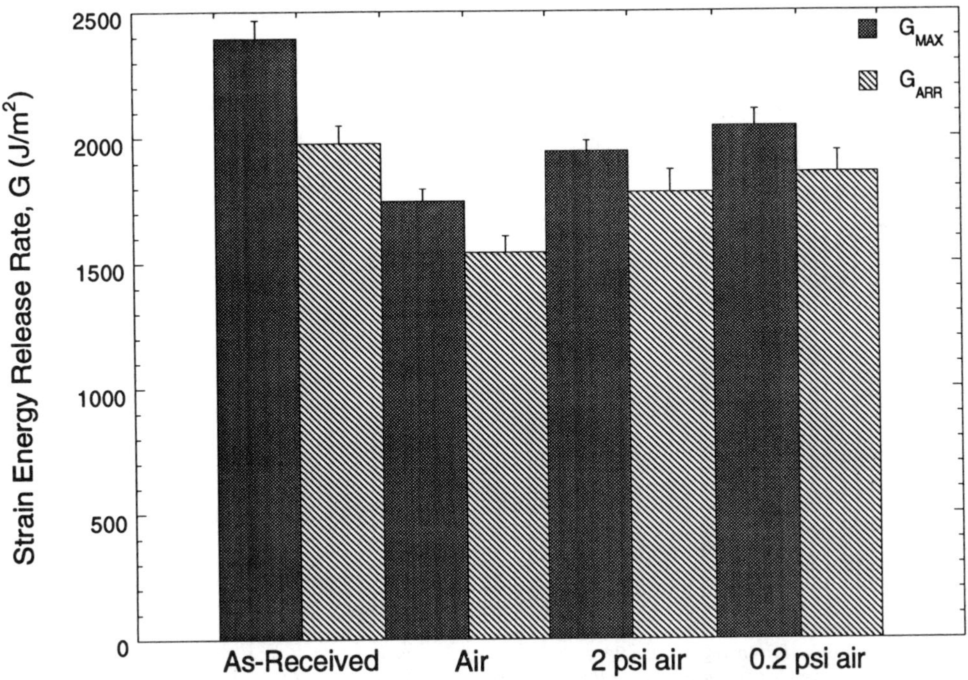

FIG. 11—*Static DCB test data on Ti-6Al-4V/FM5 bonded joints after 6 months at 204°C in different environments. Surface pretreatment was phenylethynyltrimethyl silane.*

showed a decrease of 25% in fracture energy as compared to the as-received sample, while the specimen maintained in a 2 psi air (13.8 KPa) environment showed a 15% decrease, and the 0.2 psi air (1.38 KPa) aged sample showed a 12% decrease. The error bars in Figs. 10 and 11 are standard deviations at the 95% confidence level and are obtained from a single DCB test by conducting several load and unloading cycles as shown in Fig. 5. A summary graph showing the degradation in energy release rate as function of aging time and environment (204°C in air) is shown in Fig. 12. This graph shows that there is a considerable drop in energy release rate following the first 2 weeks of aging (around 20%), but then the decrease in energy release rate is only around 8% from 2 weeks to the 6-month aging period. The above trend is very similar for both maximum as well as arrest strain energy release rates.

The effect of aging time and environment (atmospheric air at 204°C) on the degradation in the strain energy release rate on chromic acid anodized adhesive bonds is shown in Fig. 13. The data currently available for this surface treatment is up to 2 months. Trends very similar to those obtained from the silane treated static DCB tests were noted with the CAA samples. After 2 months of aging, the *G* values decreased by approximately 8% compared to the as-received specimen. The CAA samples appear to be more durable than their silane treated counterparts in that the loss in strength over the first 2 weeks was only around 6% in the CAA samples compared to 20% in the silane treated samples. It is also very evident from these tests that the greatest degradation in bond strength with both surface pretreatments occurs in atmospheric air at 204°C. There are also some consistent trends to indicate that the aging environment also plays a role in the decrease in fracture energy values: the samples

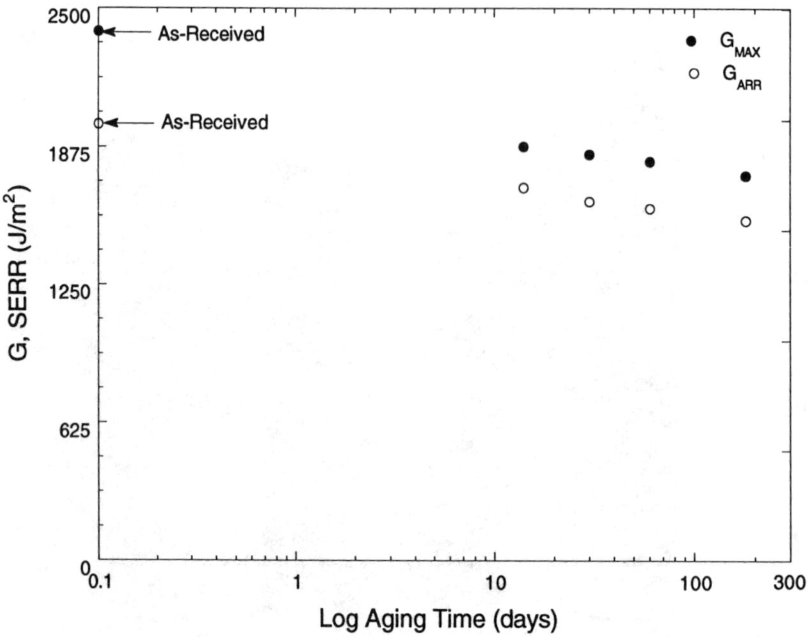

FIG. 12—*Degradation of strain energy release rate in phenyethynyltrimethyl silane treated Ti-6Al-4V/FM5 DCB specimens. Data was collected from static DCB tests and samples were aged for up to 6 months in air at 204°C.*

aged in an environment having greater oxygen concentration showing the largest drop in strength. The above trends have to be monitored over longer aging times to arrive at any definitive conclusions. It is worthwhile to note that all the samples tested so far in static DCB tests, irrespective of the surface treatment or aging conditions, have failed in a cohesive manner (that is, not at the titanium/adhesive interface). The failure tends to be very close to or at the interface between the glass scrim cloth and the FM5 adhesive.

Fatigue DCB Tests

Currently fatigue test results are available only on the silane-treated adhesive specimens. Tests were conducted at room temperature conditions and results from these tests are summarized in Figs. 14 through 16. Each figure represents data from a single specimen. Figure 14 shows a typical fatigue curve obtained on an as-received specimen, plotting crack growth rate as a function of the maximum strain energy release rate. It is clear from this curve that the energy release rate drops from slightly over 1000 J/m^2 to under 100 J/m^2 over the duration of the test, which typically runs for 2 to 3 days. The failure mode of the joint is once again cohesive and very similar to that seen in the static DCB tests. The fact that perceptible debond propagation was still occurring at energy release rates as low as 100 J/m^2 was initially viewed with concern. However, in light of similar results and low threshold energy release rate values seen by other researchers [*3,9,10*], the results obtained from this present adhesive system may be acceptable. Figures 15 and 16 show fatigue data on aged adhesive bonds, the aging times being 1 and 4 months, respectively, and the aging environment being atmospheric pressure air at 204°C. The aged samples show a slightly reduced

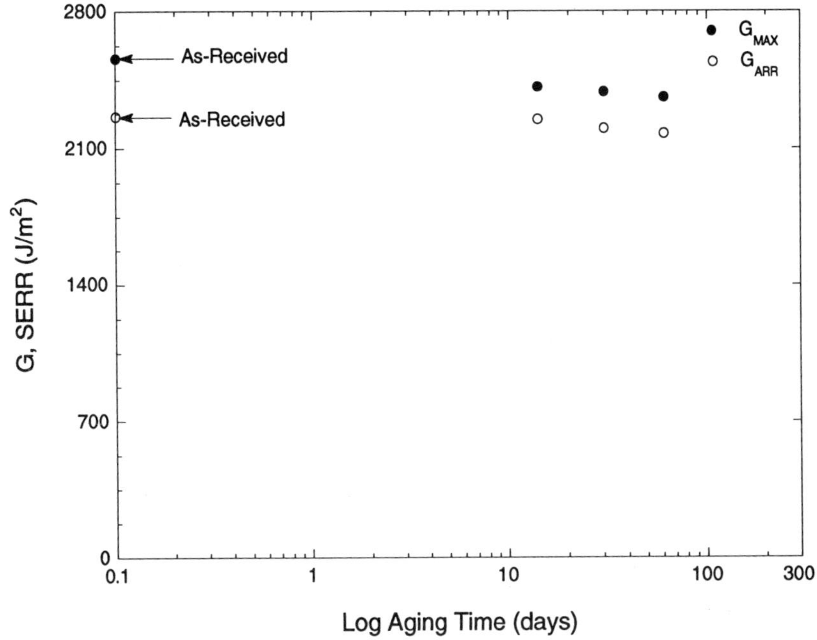

FIG. 13—*Degradation of strain energy release rate in chromic acid anodized Ti-6Al-4V/FM5 DCB specimens. Data was collected from static DCB tests and samples were aged for up to 2 months in air at 204°C.*

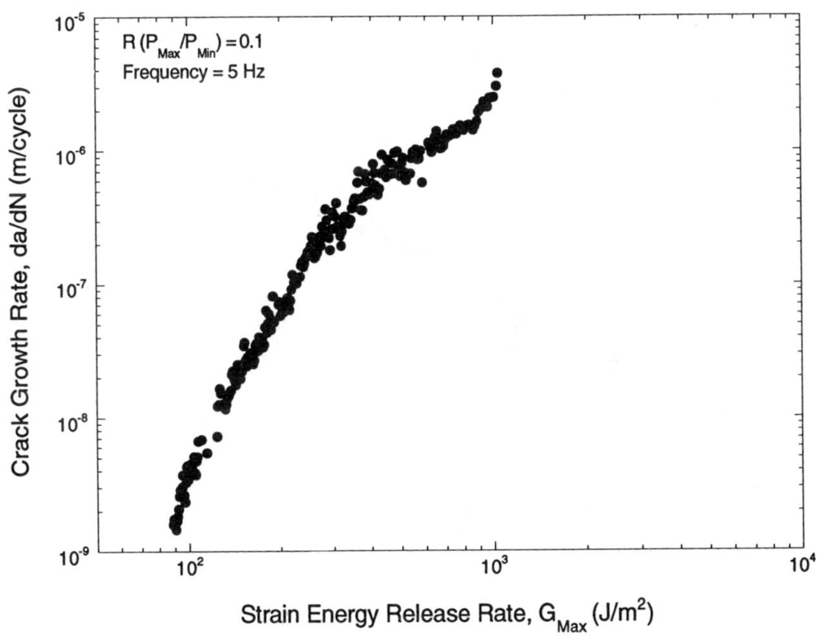

FIG. 14—*Typical fatigue DCB data on an as-received phenylethynyltrimethyl silane treatment Ti-6Al-4V/FM5 bond.*

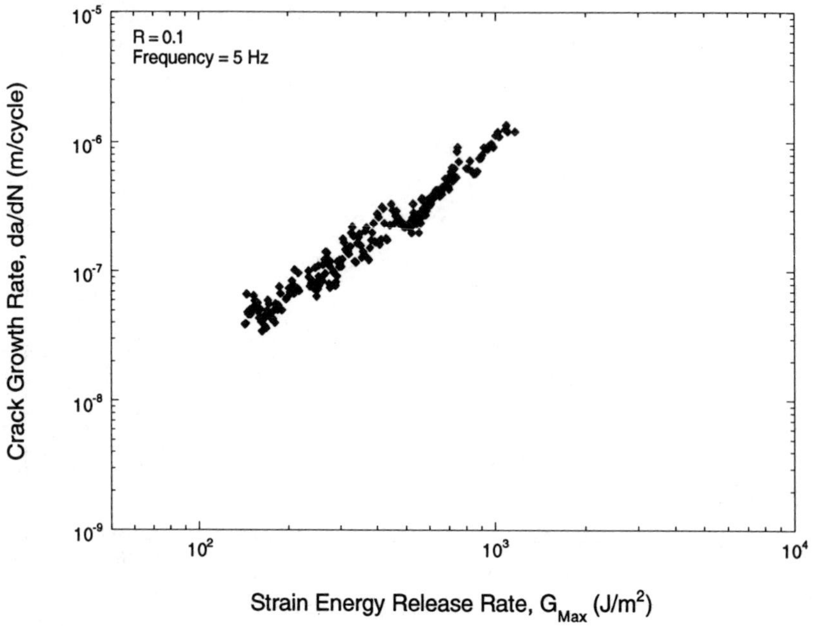

FIG. 15—*Fatigue DCB data on a phenylethynyltrimethyl silane treated Ti-6Al-4V/FM5 bond after aging in air at 204°C for 1 month.*

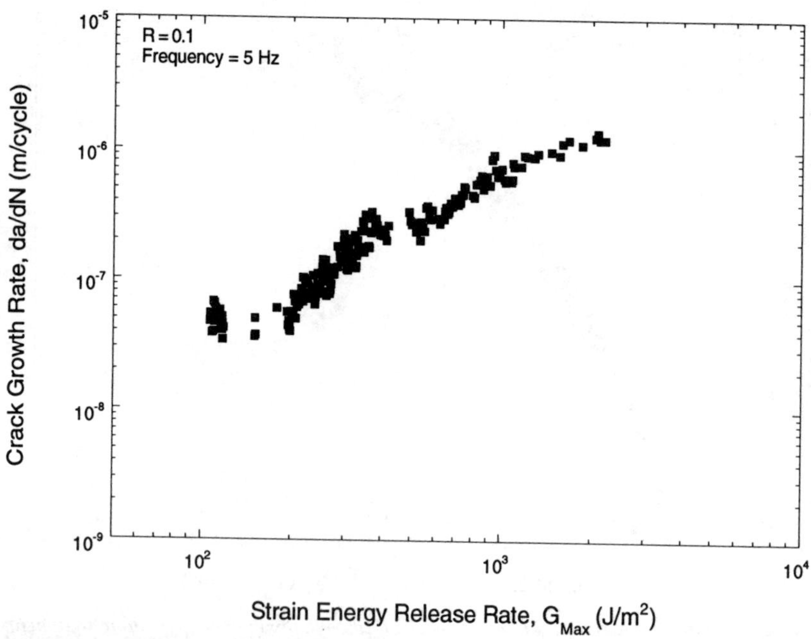

FIG. 16—*Fatigue DCB data on a phenylethynyltrimethyl silane treated Ti-6Al-4V/FM5 bond after aging in air at 204°C for 4 months.*

crack growth rate compared to the as-received specimen. Physical aging in the resin at the elevated temperature comes to mind as one of the possible causes for the above behavior; however, this behavior is not fully understood as of now and is currently being looked into in more detail.

Solvent Sensitivity and Tensile Tests

Figure 17 shows equilibrium solvent mass uptake curves for dog bone specimens of neat FM5 resin. The highest solvent uptake occurred in MEK (~6%). Jet fuel and toluene had similar uptakes (~4%). Acetone was next with around 3% uptake, followed by hydraulic fluid, ethylene glycol, and water (~1%). Following the solvent uptake tests, five of the specimens from each solvent were dried in a vacuum oven at 150°C for 60 min. The results from the desorption study are summarized in Fig. 18. As seen in this figure, all the other specimens lost most of the solvent they had absorbed, except for samples soaked in toluene that retained around 1.5% by weight of the solvent. Following the desorption tests, all specimens, both saturated with the solvents and dried, were tested in a tensile testing machine. The stress-strain curves obtained from these tensile tests are summarized in Fig. 19. Samples saturated with acetone, MEK, and toluene were plasticized, as evidenced by the lower yield stresses and the higher strains to failure. However, on drying the specimens, irrespective of the solvent to which the samples were exposed, all the stress-strain curves resembled that obtained for the as-received specimen. This finding indicates that there may be minimal residual effects of the solvents on the FM5 resin. Further, the FM5 resin system appears to

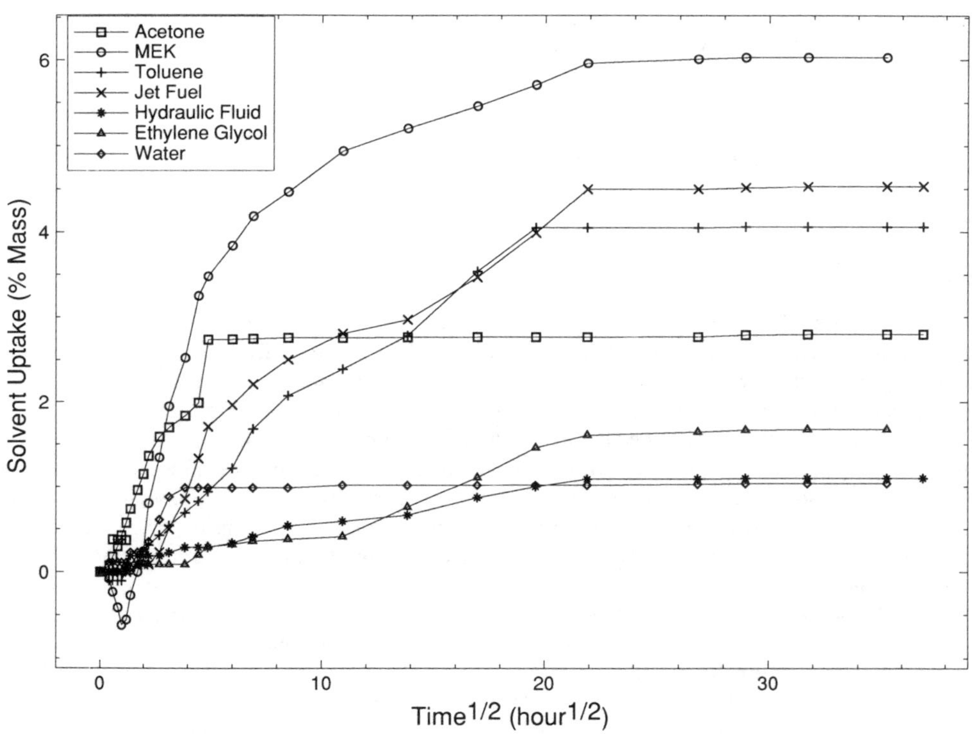

FIG. 17—*Equilibrium solvent mass uptake curves in FM5 neat resin specimens.*

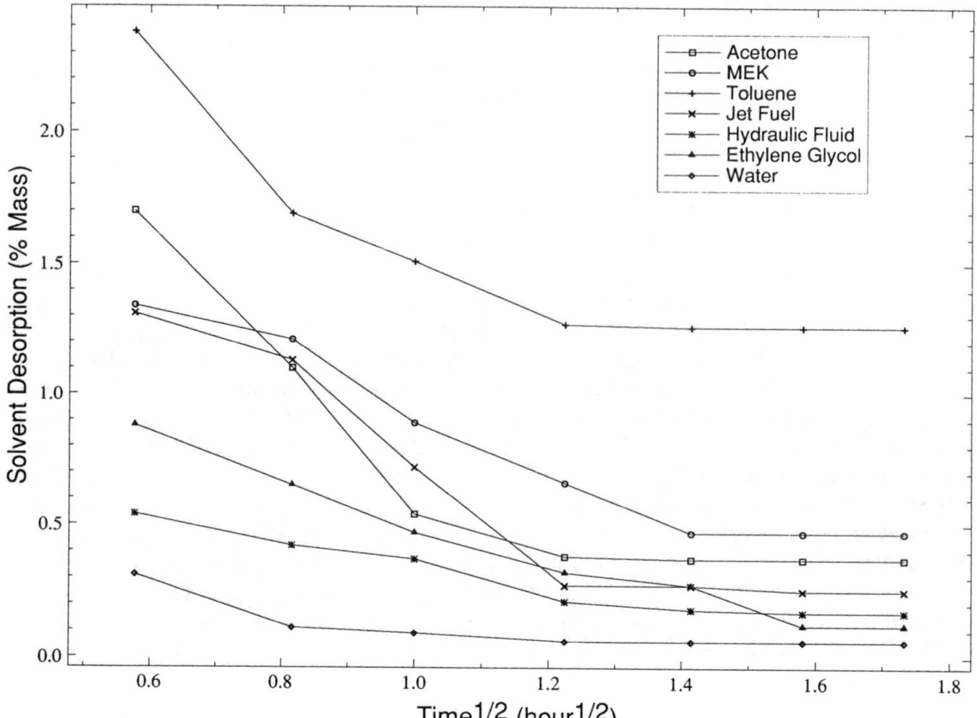

FIG. 18—*Solvent desorption curves in FM5 neat resin specimens.*

be considerably solvent resistant, compared to, for example, the polymers tested in Refs *30* and *31*.

Wedge Tests

To study the effect of solvent exposure on adhesive bond durability, Ti-6Al-4V/FM5 bonded specimens treated with CAA were immersed in solvents for periods up to 162 h. Five specimens each were immersed per solvent and crack length data collected as a function of exposure time. Results from these wedge tests are summarized in Figs. 20 and 21. Data shown is average values obtained from the 5 test specimens per solvent. As can be seen from Fig. 20 the longest crack growth was obtained from samples immersed in boiling water. The rest of the solvents had a certain ranking in terms of the crack length obtained following exposure of the bonded joint. In terms of ascending durability, the ranking order was: acetone, MEK, toluene, jet fuel, ethylene glycol, hydraulic fluid, and room temperature water. The crack growth appeared to be mixed in the boiling water, and mostly cohesive in the rest of the solvents. Figure 21 shows crack growth rate data as a function of energy release rate for all specimens immersed in the 8 solvents. The arrest strain energy release rate for the specimen immersed in the boiling water was seen to be a minimum of 1300 J/m^2, while a maximum G_{ARR} was obtained for the sample immersed in room temperature water at around 2300 J/m^2. The rest of the solvents had values of G somewhere in between. An arrest strain energy release rate is said to be obtained when the crack does not grow for a sufficiently

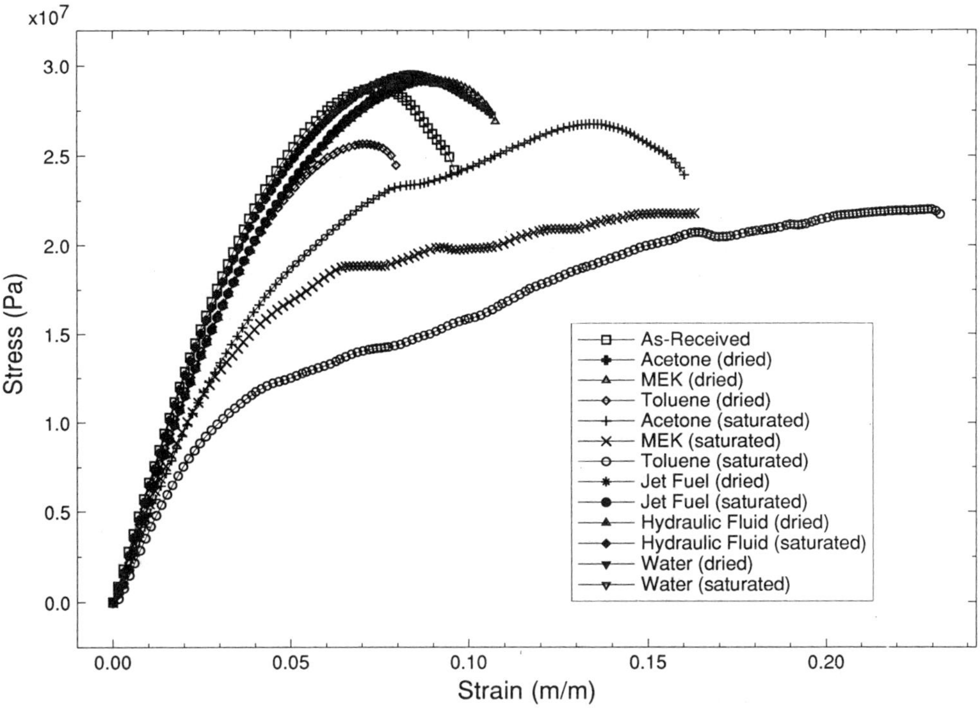

FIG. 19—*Tensile test data showing stress-strain curves of neat FM5 specimens, both in saturated and dried conditions.*

long duration of time with respect to the test time, and the system has reached a state of quasi-static equilibrium under the influence of the environment.

Weight Measurements

Three specimens of FM5 neat resin were aged at 204 and 177°C in air and reduced air pressure of 2 psi air (13.8 KPa) and periodically weighed to monitor weight loss. These results are summarized in Figs. 22 and 23. After 6 months of aging at both aging temperatures, around 2% loss in weight was seen in air and around 1.8% weight loss was seen at the reduced air pressure. Most of the weight loss was observed during the initial 2 weeks of aging, following which the weight of the specimens appeared to be quite stable. The FM5 neat resin samples contained around 0.5% by weight of solvent (NMP). Therefore, some of the initial weight loss could be due to solvent evaporating from the material as well as moisture losses. The weight loss occurring at longer aging times could be due to chemical degradation. It is seen that in the data collected so far there is a small difference in the weight loss in the two different environments, with the loss in air aged samples being slightly higher. Since the error bars obtained from the measurements are very small, it can be argued that the difference in the weight loss in the two aging environments is due to the different concentrations of oxygen in the two environments. It is not possible to positively attribute the changes to any one particular reason as of now and future trends in the data have to be monitored. Weight loss data will be collected for several more months in this testing program.

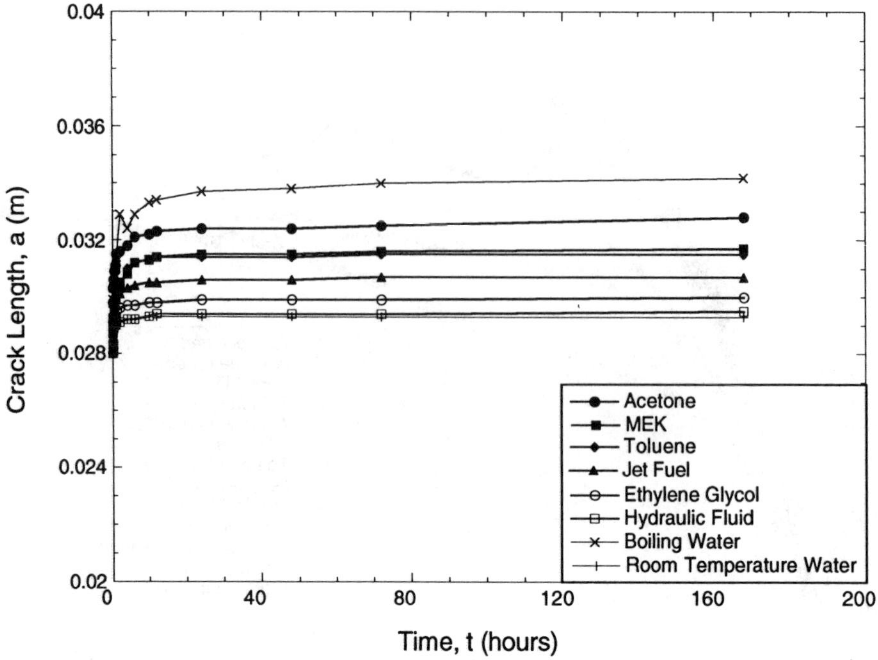

FIG. 20—*Wedge test data showing crack length as a function of time for Ti-6Al-4V/FM5 bonds immersed in different solvents for up to 162 h. Surface pretreatment was chromic acid anodization.*

Summary and Conclusions

This study was initiated to evaluate the durability of Ti-6Al-4V/FM5 adhesively bonded system and to explore the feasibility of using this material system for the high speed civil transport (HSCT) aircraft. The FM5 adhesive is a polyimide originally developed by NASA Langley Research and modified and supplied by Cytec Engineered Materials, Inc. Wedge and DCB fracture tests were used to study the properties of the bonded specimens after subjecting them to environmental aging for periods up to 6 months. The aging temperatures were 150, 177, and 204°C, and the aging environments were atmospheric air, 2 psi reduced pressure air (13.8 KPa), and 0.2 psi reduced pressure air (1.38 KPa). Two surface pretreatments for the titanium adherends were also evaluated, namely, a phenylethynyltrimethyl silane treatment, and chromic acid anodization.

Wedge tests have been used as screening tests to evaluate different surface pretreatments as well as to obtain information on the solvent sensitivity of the FM5 resin. Some long-term bench mark tests have also been initiated with wedge specimens, and one such test has been used to study the durability of the bonded joints after air aging at 204°C for 4320 h. The surface treatment used for this test is CAA, and the test shows good durability of the adhesively bonded samples.

DCB static tests were conducted on the Ti-6Al-4V/FM5 bonds after aging in the above mentioned temperatures and environments. The tests on the silane-treated samples showed a 25% drop in toughness following aging in air at 204°C for 6 months. Over a 2-month aging period, the CAA treated samples lost 8% of their original strength following similar exposure conditions. The static DCB tests indicated that the most severe aging condition was aging in atmospheric air at 204°C. It appears that there was some ordering with respect to

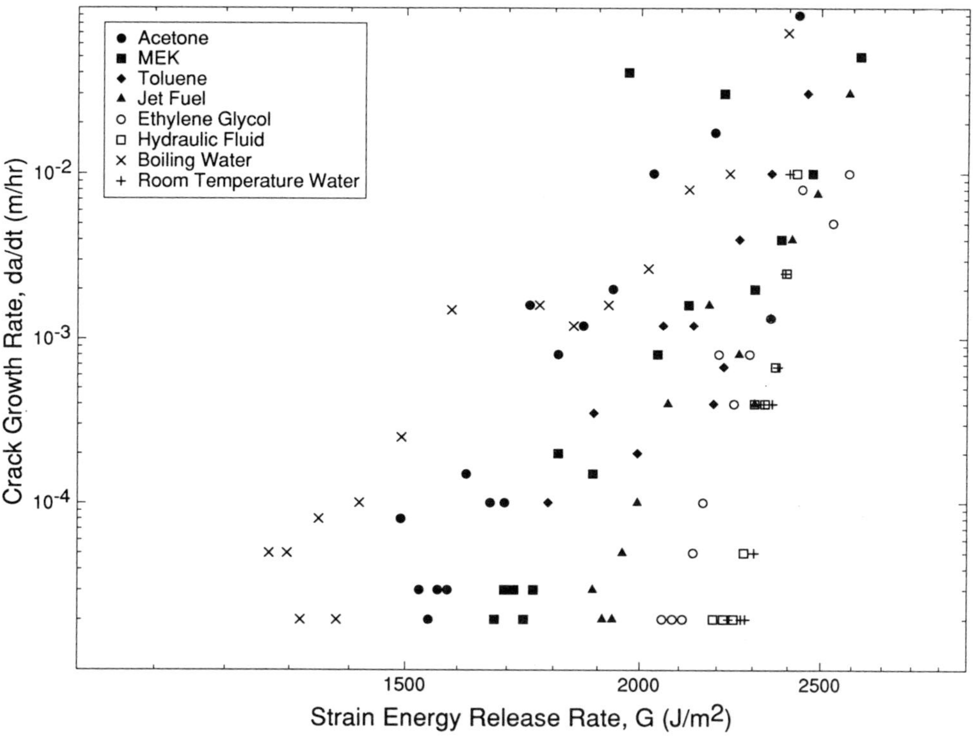

FIG. 21—*Wedge test data of crack growth rate as a function of strain energy release rate for specimens in Fig. 20.*

the degradation in the strength of the joints as a function of the exposure. While samples aged in 0.2 psi air (1.38 KPa) and 150°C showed the lowest strength reduction, those aged at 2 psi air (13.8 KPa) and 177°C showed intermediate reduction in strength, and the greatest reduction in strength occurred in samples in atmospheric air and 204°C. Some explanations for the above behavior include physical aging and chemical degradation of the adhesive material. In all the static DCB tests, cohesive failure in the region of the glass scrim cloth was obtained. DCB fatigue tests were conducted on the silane treated adhesive bonds, and the results cause some concerns about the low threshold energy release rate values obtained. Aging of the joints at 204°C in air has produced some changes in the fatigue behavior, and at this point of time there are no positive explanations for the observations made. Physical aging of the adhesive at the elevated temperature could be one of the possible reasons for the changes in the fatigue behavior.

Solvent uptake, desorption, and tensile tests were used to study the resistance of the FM5 resin to organic solvents. The results showed that although some solvents were being absorbed by the resin material on the order of 4 to 6% by weight, this solvent absorption had no bearing on residual properties measured on dried samples.

Weight loss data was also monitored on neat resin FM5 samples at 177 and 204°C in atmospheric air and 2 psi air (13.8 KPa). Weight loss of about 2% occurred at the two temperatures following air aging for 6 months. There is a slight decrease in the weight loss measured at the reduced air pressure, this may be attributed to chemical degradation; however, more data needs to be collected to substantiate this claim.

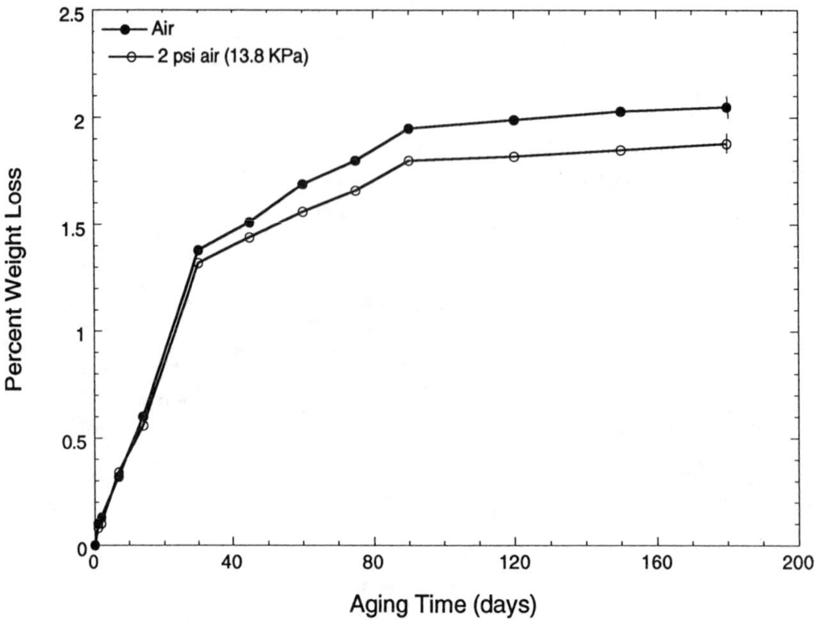

FIG. 22—*Weight loss data in neat FM5 resin at 204°C in two different environments for periods up to 6 months.*

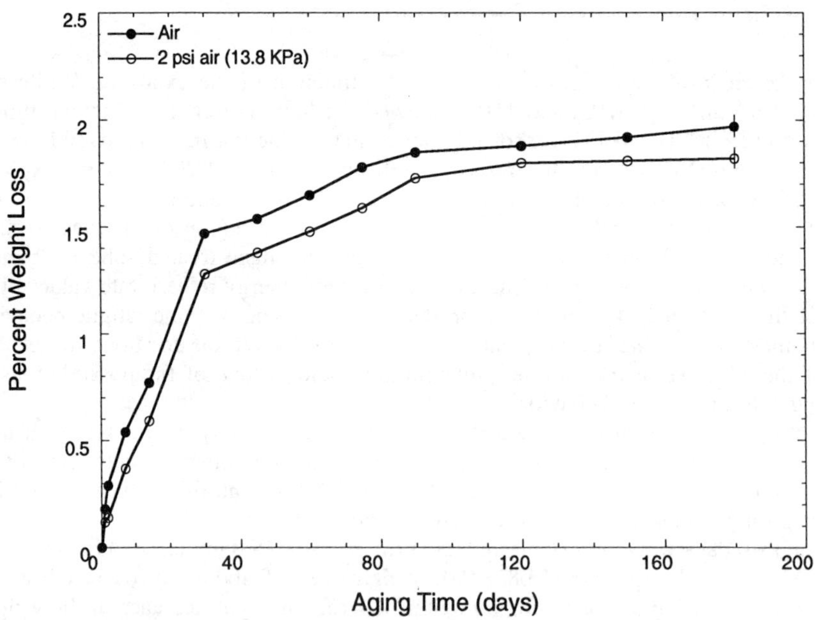

FIG. 23—*Weight loss data in neat FM5 resin at 177°C in two different environments for periods up to 6 months.*

Acknowledgments

The authors would like to acknowledge the financial support from Boeing Commercial Airplane Group, NASA Langley Research Center, and the National Science Foundation Science and Technology Center for High Performance Polymeric Adhesives and Composites (Contract DMR 9120004), Virginia Tech. The authors would also like to acknowledge Cytec Engineered Materials, Inc. for supplying the LaRC PETI-5 adhesive. The authors are grateful to Drs. Jim McGrath, Jim Wightman, Ron Kander, and Paul Hergenrother for helpful recommendations and insights. The authors would also like to acknowledge Mr. Brian Williams for his help with the sample preparation.

References

[*1*] Kinloch, A. J. and Shaw, S. J., "A Fracture Mechanics Approach to the Failure of Structural Joints," *Developments in Adhesives-2*, Chapter 3, A. J. Kinloch, Ed., Applied Science Publishers, New York, 1981, p. 83.

[*2*] Clearfield, H. M., Shaffer, D. K., Vandoren, S. L., and Ahearn, J. S., *Journal of Adhesion*, Vol. 29, 1989, p. 81.

[*3*] Johnson, W. S. and Mall, S., "Bonded Joint Strength: Static Versus Fatigue," *Proceedings of 5th International Congress on Experimental Mechanics*, Montreal, Quebec, 10–14 June 1984.

[*4*] Broek, D., *Elementary Engineering Fracture Mechanics*, 3rd Edition, Chapter 2, Martinus Nijhoff Publishers, Netherlands, 1982, pp. 38–60.

[*5*] Cognard, J., "The Mechanics of the Wedge Test," *Journal of Adhesion*, Vol. 20, 1986, pp. 1–13.

[*6*] Cognard, J., "Quantitative Measurement of the Energy of Fracture of an Adhesive Joint Using the Wedge Test," *Journal of Adhesion*, Vol. 22, 1987, pp. 97–108.

[*7*] Cognard, J., "Environmental Attack on Adhesive Joints Studied by Cleavage Fracture," *Journal of Adhesion*, Vol. 26, 1988, pp. 155–169.

[*8*] Ouezdou, M. B., Chudnovsky, A., and Moet, A., "Re-evaluation of Adhesive Fracture Energy," *Journal of Adhesion*, Vol. 25, 1988, pp. 169–183.

[*9*] Rakestraw, M. D., Vrana, M. A., Dillard, D. A., Dillard, J. G., and Ward, T. C., "A Fracture Mechanics Methodology for the Testing and Prediction of Adhesive Bond Durability," Durability and Damage Tolerance, *ASME WAM*, AD-Vol. 43, 1994, p. 65.

[*10*] Rakestraw, M. D., Taylor, M. W., Dillard, D. A., and Chang, T., "Time Dependent Crack Growth and Loading Rate Effects on Interfacial and Cohesive Fracture of Adhesive Joints," *Journal of Adhesion*, Vol. 55, 1995, pp. 123–149.

[*11*] Kinloch, A. J. and Osiyemi, S. O., "Predicting the Fatigue Life of Adhesively-Bonded Joints," *Journal of Adhesion*, Vol. 43, 1993, pp. 79–90.

[*12*] Kodokian, G. K. A. and Kinloch, A. J., "The Adhesive Fracture Energy of Bonded Thermoplastic Fibre-Composites," *Journal of Adhesion*, Vol. 29, 1989, pp. 193–218.

[*13*] Blackman, B., Dear, J. P., Kinloch, A. J., and Osiyemi, S., "The Calculation of Adhesive Fracture Energies From Double Cantilever Beam Test Specimens," *Journal of Materials Science Letters*, Vol. 10, 1991, pp. 253–256.

[*14*] Lefebvre, D. R., Dillard, D. A., and Brinson, H. F., "The Development of a Modified Double Cantilever Beam Specimen for Measuring the Fracture Energy of Rubber to Metal Bonds," *Experimental Mechanics*, Vol. 28, No. 4, 1988, pp. 38–44.

[*15*] Ouezdou, M. B. and Chudnovsky, A., "Stress and Energy Analysis of Toughness Measurement for Adhesive Bonds," *Engineering Fracture Mechanics*, Vol. 29, No. 3, 1988, pp. 253–261.

[*16*] Kanninen, M. F., "An Augmented Double Cantilever Beam Model for Studying Crack Propagation and Arrest," *International Journal of Fracture*, Vol. 9, No. 1, 1973, pp. 83–92.

[*17*] Fay, P. A. and Maddison, A., "Durability of Adhesively Bonded Steel Under Salt Spray and Hygrothermal Stress Conditions," *International Journal of Adhesion and Adhesives*, Vol. 10, 1990, pp. 179–186.

[*18*] Joseph, R., Bell, J. P., McEvily, A. J., and Liang, J. L., "Fatigue Crack Growth in Epoxy/Aluminum and Epoxy/Steel Joints," *Journal of Adhesion*, Vol. 41, 1993, pp. 163–187.

[*19*] Mall, S. and Yun, K. T., "Effect of Adhesive Ductility on Cyclic Debond Mechanism in Composite-to-Composite Bonded Joints," *Journal of Adhesion*, Vol. 23, 1987, pp. 2215–2231.

[*20*] Mall, S., Johnson, W. S., and Everett, R. A., Jr., "Cyclic Debonding of Adhesively Bonded Composites," *Adhesive Joints*, K. L. Mittal, Ed., Plenum Press, New York, 1984, pp. 639–645.

[*21*] Mostovoy, S. and Ripling, E. J., "Fracture Toughness of an Epoxy System," *Journal of Applied Polymer Science*, Vol. 10, 1966, pp. 1351–1371.

[*22*] Watts, J. F., Castle, J. E., and Hall, T. J., "The Effects of Water on the Failure Mechanism of Adhesive Joints," *Journal of Materials Science Letters*, Vol. 7, 1988, pp. 176–178.

[*23*] Rao, N. B. and Acharya, A. R., "Evaluation of Fracture Energy G/ /I/ /C Using a Double Cantilever Bean Fibre Composite Specimen," *Engineering Fracture Mechanics*, Vol. 51, No. 2, 1995, pp. 317–322.

[*24*] Hooper, S. J. and Subramanian, R., "Effects of Water and Jet Fuel Absorption on Mode I and Mode II Delamination of Graphite/Epoxy," *Fatigue and Fracture, ASTM STP 1156*, American Society for Testing and Materials, West Conshohocken, PA, 1993, pp. 318–340.

[*25*] Kambour, R. P., "A Review of Crazing and Fracture in Thermoplastics," *Journal of Polymer Science: Macromolecular Review*, Vol. 7, 1973, pp. 1–154.

[*26*] Gent, A. N., "Hypothetical Mechanism of Crazing in Glassy Plastics," *Journal of Material Science*, 1970, Vol. 5, pp. 925–932.

[*27*] Kinloch, A. J. and Young, R. J., *Fracture Behavior of Polymers*, Chapter 7, Applied Science, London, 1983, pp. 236–238.

[*28*] Stober, E. J., Seferis, J. C., and Keenan, J. D., "Characterization and Exposure of Polyether etherketone (PEEK) to Fluid Environments," *Polymer*, 1984, Vol. 25, pp. 1845–1852.

[*29*] Hay, J. N. and Kemmish, D. J., "Environmental Stress Crack Resistance and Absorption of Low-Molecular Weight Penetrants by Poly (aryl ether ether ketone)," *Polymer*, 1988, Vol. 29, pp. 613–620.

[*30*] Dillard, D. A., McDaniels, P. R., and Hinkley, J. A., *Journal of Materials Science Letters*, Vol. 12, 1993, p. 1258.

[*31*] Parvatareddy, H., Heithoff, C. A., Clifton, A. P., Dillard, D. A., and Kander, R. G., "Environmental Stress Cracking of High Performance Polymeric Composites," *Composite Materials: Testing and Design*, Twelfth Volume, *ASTM STP 1274*, Ravi B. Deo and Charles R. Saff, Eds., American Society for Testing and Materials, West Conshohocken, PA, 1995, pp. 56–68.

[*32*] Smith, J. G. and Hergenrother, P. M., *Polymer Pre-prints*, Vol. 35, 1994, p. 353.

[*33*] Paris, P. C. and Erdogan, F., *ASME Trans., Journal of Basic Engineering*, Vol. 4, 1963, p. 528.

[*34*] Sullivan, J. L., Blais, E. J., and Houston, D., "Physical Aging in the Creep Behavior of Thermosetting and Thermoplastic Composites," *Composites Science and Technology*, Vol. 47, 1993, pp. 389–403.

Kenneth J. Bowles,[1] *Gary D. Roberts,*[1] *and John E. Kamvouris*[2]

Long-Term Isothermal Aging Effects on Carbon Fabric-Reinforced PMR-15 Composites: Compression Strength

REFERENCE: Bowles, K. J., Roberts, G. D., and Kamvouris, J. E., **"Long-Term Isothermal Aging Effects on Carbon Fabric-Reinforced PMR-15 Composites: Compression Strength,"** *High Temperature and Environmental Effects on Polymeric Composites: 2nd Volume, ASTM STP 1302*, Thomas S. Gates and Abdul-Hamid Zureick, Eds., American Society for Testing and Materials, 1997, pp. 175–190.

ABSTRACT: A study was conducted to determine the effects of long-term isothermal thermo-oxidative aging on the compressive properties of T-650-35 fabric reinforced PMR-15 composites. The temperatures that were studied were 204, 260, 288, 316, and 343°C. Specimens of different geometries were evaluated. Cut edge-to-surface ratios of 0.03 to 0.89 were fabricated and aged. Aging times extended to a period in excess of 15 000 h for the lower temperature runs.

The unaged and aged specimens were tested in compression in accordance with Test Method for Compressive Properties of Rigid Plastics (ASTM D 695). Both thin and thick (prism) specimens were tested. Three specimens were tested at each time/temperature/geometry condition. The failure modes appeared to be initiated by fiber kinking with longitudinal, interlaminar splitting.

In general, it appears that the thermo-oxidative degradation of the compression strength of the composite material may occur by both thermal (time-dependent) and oxidative (weight-loss) mechanisms. Both mechanisms appear to be specimen-thickness dependent.

KEY WORDS: composites, polymer matrix, carbon fiber, durability, compression properties, thermal aging

Today, planning is in progress for advanced 21st century aircraft with greater and faster payload delivery and higher fuel efficiency. Of vital importance in the development of the propulsion systems for these aircraft is the use of cost-effective materials and manufacturing procedures. The current philosophy in the development of these programs is to use the light-weight polymer composites in the higher temperature sections of the new engines. In some cases, the structures may be load bearing. Because of these plans, it is imperative to develop reliable, long-term-durability models so the feasibility of using polyimide-matrix composites as efficient propulsion system materials over the required lifetime of the aircraft can be demonstrated.

Numerous investigations, regarding the thermo-oxidative stability (TOS) of PMR-15 and other high-temperature polyimides, have been conducted and reported [*1–7*]. A significant

[1] NASA Lewis Research Center, Cleveland, OH 44135.

[2] National Research Council-NASA research associate at Lewis Research Center, Cleveland, OH 44135.

amount of this attention has been directed toward edge and geometry effects [*1–6*], reinforcement fiber influences [*3,4*], and modeling of elevated temperature weight loss behavior [*7*]. This was done as part of an on-going effort to provide input for use in the prediction of the long-term elevated temperature behavior of structures fabricated from the advanced polymer matrix composite materials. Some of these composite materials are currently being used in aircraft engine secondary structural applications.

In the analysis of numerous TOS results, it has been determined that the specimens lose weight from the different surfaces at different rates. The surfaces of the specimens that were studied were found to undergo two different types of thermo-oxidative transformations during extended periods of elevated temperature aging. The cut edge surfaces experienced significant amounts of crack development and inward growth from the surface. The molded surfaces that are oriented parallel to the ply surfaces suffered from surface cracking, but the cracking did not penetrate as far inward as the cut surface cracks. This was due to the cross-ply weave of the fabric which, in effect, positioned crack arrestors at each side of the woven tows. It was concluded that the latter surface degradation should bring about changes in the surface-dependent mechanical properties of the aged composites. If the surface degradation controls the changes in mechanical properties during aging, then the properties retention can be related to aging time at a particular temperature for a particular composite system and specimen size. If this is not the case, then other mechanisms, along with the surface oxidation, are contributing to the degradation of these materials at elevated temperatures.

The results of a study at the Lewis Research Center [*9*] showed that aged, unidirectional PMR-15 matrix composites displayed a reduction in the matrix-dependent transverse flexural strength. The magnitude of the decrease increased with aging time. Also in this study, and in another study [*10*], it was observed that mechanical properties that are not sensitive to surface properties, such as in-plane shear strength, do not exhibit significant property degradation due to exposure to the hostile environment and aging times up to 1000 h at 316°C.

The purpose of the study described herein was to confirm the role of the oxidizing surface layer in mechanical properties degradation during elevated temperature aging. It was also initiated to point out possible causes, evolving from the formation and growth of this layer over extended periods of times approaching 20 000 h for the degradation of properties. The compression properties were chosen for study because they play a significant role in impact, delamination, and other types of damage failures caused by out-of-plane forces.

Materials

The materials used in this study were PMR-15 polyimide composites reinforced with T-650-35, 24 by 23, 8-harness satin carbon fiber fabric. The material was supplied by ICI Fiberite. The composites were fabricated by autoclave techniques at 316°C. The molded surfaces exhibited a roughened appearance from the bleeder cloth that was used. The cure was followed by a free-standing post cure at 316°C in an air-circulating oven for 16 h. All processing was done at the General Electric Aircraft Engine Plant in Evendale, Ohio.

The specimen designations and nominal dimensions are presented in Table 1 and Fig. 1. The dimensions were selected to provide specimens with different molded surface (resin rich areas adjacent to the vacuum bag containment materials) to cut surface area ratios. The percentage of cut surface areas varied from 2.7 to 89% of the total surface area. The numeral designations of the specimens in Table 1 reflect the nominal percent cut edge areas. The material in this study was found to have void contents less than 2 vol% and to be free from delaminations. The fiber volume fractions ranged from 0.55 to 0.59. A typical photomicrograph of the material prior to aging is shown in Fig. 2. Some resin-rich areas are present between the tows.

TABLE 1—*T-650-35 graphite/PMR-15 composite dimensional data.*

Specimen	Length, cm	Width, cm	Thickness, cm	Edge, %
T-3	8.94	10.83	0.13	2.7
T-5	8.94	10.83	0.25	5.4
T-12	8.94	10.83	0.75	11.9
T-50	4.61	1.80	1.30	50.3

Test Procedures

The composite specimens were aged in air-circulating ovens. The ovens maintained an air flow rate of 100 cm^3/min. The temperatures that were studied were 204, 260, 288, 316, and 343°C. The specimens were removed from the ovens at regular intervals and placed in a desiccator where they cooled to room temperature. The specimens were not removed from the desiccator until they were ready to be weighed. The weights were recorded and the specimens were returned to the ovens. At scheduled times, selected specimens were permanently removed from the ovens for various tests, including compression tests.

The glass transition temperature was measured using dynamic mechanical analysis. Measurements were made with rectangular pieces, using a Rheometrics RMS-800 rheological spectrometer. The test pieces were stressed in torsion across the specimen widths. The test pieces measured 60 mm in length, 10 mm in width, and ranged in thickness from 1.3 to 7.5 mm. They were mounted in grips that were separated to give a gage length of 45 mm. The two components of the complex shear modulus were measured as a function of temperature. A frequency of 1 Hz with an amplitude of 0.1% strain and a heating rate of 5°C/min were used. All specimens were dried in an air-circulating oven at 125°C for 24 h before they were tested. This was a sufficient amount of time to reduce the moisture to very low values [*11*]. The glass transition temperatures were determined by measuring the intersection of the two tangents to the G′ curve where it changes slope. This is illustrated in Fig. 3.

The compression tests were done at the Cincinnati Testing Laboratories in Forest Park, Ohio. There were two types of compression tests that were performed in accordance with Test Method for Compressive Properties of Rigid Plastics (ASTM D 695M). One was the

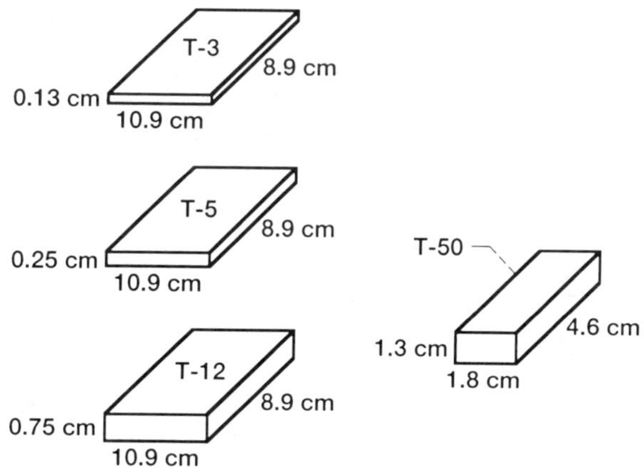

FIG. 1—*Test specimens.*

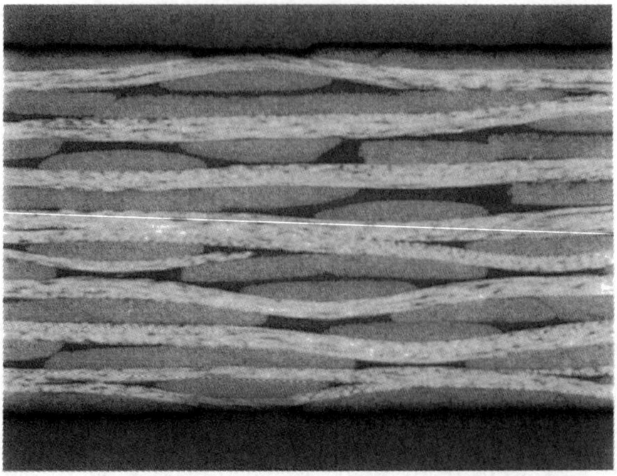

FIG. 2—*Unaged T650-35/PMR-15 fabric reinforced composite.*

test for specimens of thickness less than 3 mm, and the second was specified for specimens with thickness greater than 3 mm. The composites that were tested in accordance with the thin specimen specifications were the T-3, T-5, and T-12 plates. The T-50 composite was tested as described in ASTM D 695M for thick specimens. The cross-head speed was 1.2 mm/min. Strain measurements were made using an extensometer. No end tabs were used. All specimens were conditioned at 125°C for 16 h prior to testing. The tests were run at 23.3°C and a relative humidity of 50%.

The fiber contents of the composites were measured by acid digestion as described in Test Method for Fiber Content of Resin-Matrix Composites by Matrix Digestion (ASTM D 3171). The void volume was calculated as the difference between the specific volume of a composite specimen, measured using the immersion technique as specified in Test Methods for Density

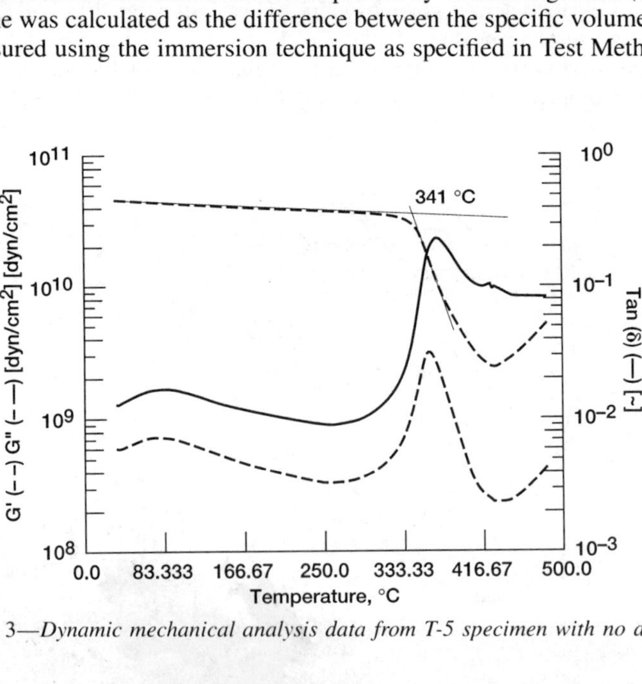

FIG. 3—*Dynamic mechanical analysis data from T-5 specimen with no aging.*

and Specific Gravity (Relative Density) of Plastics by Displacement (ASTM D 792), and the theoretical specific volume calculated from the acid digestion results. The differences were divided by the measured specific volume and multiplied by 100 to give the void volume in percentage. The densities of the polymer and the fiber were taken to be 1.32 and 1.78g/cm^3, respectively.

Results

Dynamic Mechanical Analysis

The T_{Gs} of the specimens that were aged at different temperatures for different durations were measured and the data are presented in Table 2. The types of specimens that were tested were the T-3, T-5, and T-12 plates, which contained over 90% of molded exposed surfaces.

The T_G measurements in Table 2 present values only for aged T-5 specimens. They indicate that at temperatures below 260°C, the T_G is dependent on the temperature of aging and not the period of aging. This is probably due to the large differences between the aging temperatures and the 316°C 16-h post-cure treatment temperature. Most chemical changes that would occur over an extended time period at the lower temperatures occurred quickly at the higher post-cure temperature. At temperatures above 260°C, there are differences due to both aging times and temperatures.

Compression Strength

Effects of Aging on Compression Strength

Compression strength values for the T-5 specimens are compared as a function of both temperature and time in Figs. 4 and 5. The data group differently for the two independent variables. The compression strength data for the specimens that were aged at the higher temperatures appear to be represented by a single curve when they are plotted against percent weight loss. By regression analysis, the data for the T-5 specimens that were aged at 260, 288, 316, and 334°C were fitted to an exponential curve with a slope of -10.3×10^{-2} and an intercept of 100.9 (in ksi units). The coefficient of variation R^2 value is 0.970. This indicates good correlation between the strength and weight loss. The intercept in MPa units is 695.7. The slope and R^2 remain the same. The resulting equation is

$$Ln\ S_c = 4.614 - 10.259 \times 10^{-2} w \tag{1}$$

TABLE 2—*Glass transition temperatures of T-5 specimens at different times and temperatures.*

	Temperature, °C				
Time, h	RT	204	260	288	316
0	330	...	...	...	...
0	342	...	...	...	...
0	341	...	...	...	...
500	...	...	...	...	358
1000	...	...	...	349	364
2000	...	328	330	352	391
5000	...	323	...	346	...
10 000	...	322	350	...	...

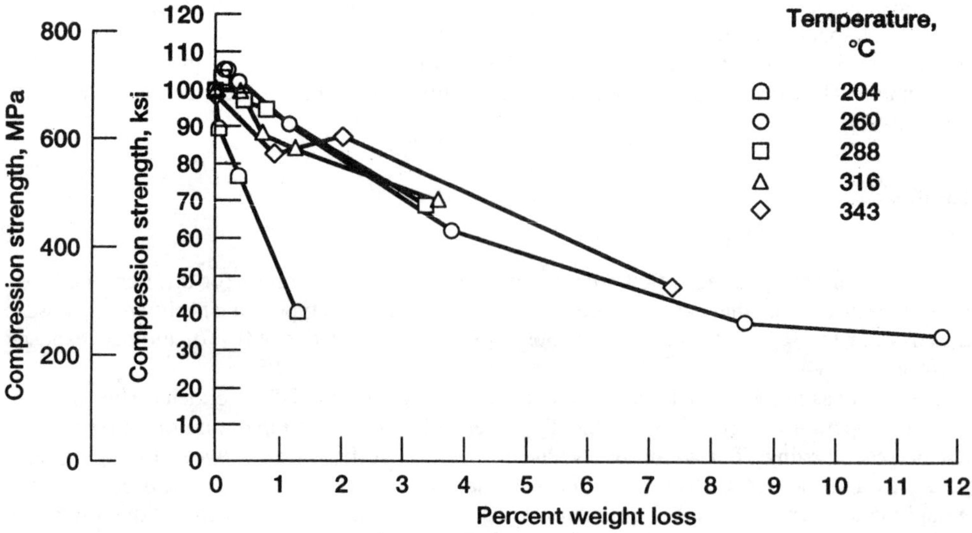

FIG. 4—*Compression strength of T650-35/PMR-15 T-5 composite specimens as a function of percent weight loss at various temperatures.*

where S_c is the compression strength in MPa and w is the weight loss in percent. When the strengths are plotted against aging time, the materials aged at 204 and 260°C produce data that form a common curve with an exponential slope of -6.480×10^{-5}, an intercept of 111.57 in ksi units and an R^2 value of 0.926. The intercept is 769.26 in MPa units. The resulting mathematical relationship is

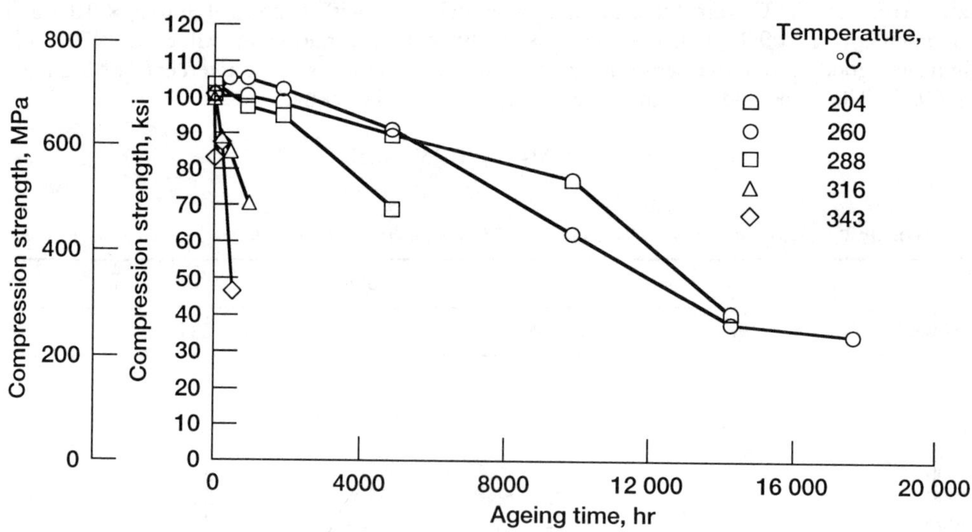

FIG. 5—*Compression strength of T650-35/PMR-15 T-5 composite specimens as a function of aging time at various temperatures.*

$$Ln\ S^c = 4.710 - 6.480 \times 10^{-5}t \tag{2}$$

for the same strength units as in Eq 1 and t is aging time in hours. At these two temperatures, the weight loss does not control the degradation of the compression strength since the weight loss at 204°C is about one-tenth of that at 260°C as shown in Fig. 4. It appears that there are significant differences in relationships between the data sets depending on the exposure temperatures. These data indicate that weight loss behavior cannot be used to evaluate structural integrity of PMR-15 composite materials over temperature levels below 260°C. The effect of the formation of microcracks may be a factor that has a great influence on the mechanical property degradation. However, since the concentration of microcracks in the material aged at 204°C is less than that of the material aged at 260°C, and the compression strengths are very close, it is not the single factor involved.

Figures 6, 7, and 8 show how the effect of increasing specimen thickness in decreasing compression strengths for aged specimens. The data in Fig. 6 are from T-12 specimens that are thicker than the T-5 material. It is evident that the time to reduce the compression strength by 50% for the T-5 material aging at 260°C is about 12 000 h (Fig. 5). The time for strength reduction to the same level for the thicker T-12 specimens appears to be greater than 20 000 h. This is a significant increase.

The contrast between specimen thicknesses and also between test methods is shown graphically in Figs. 7 and 8. The data at 0 time (unaged specimens) shown in Fig. 7 for the T-3, T-5, and T-12 composites do not show statistically different strength values due to the thickness differences. As the aging time at 316°C increases past 1%, the strength values start separating. The thinnest (T-3) specimens lose their strength the fastest. After about 2% weight loss, the slopes of the three curves appear to be parallel, which indicates the same rate of loss of strength. The differences in the aged T-3, T-5, and T-12 compression strengths are probably due to the damage incurred during aging and not the thickness. These specimens

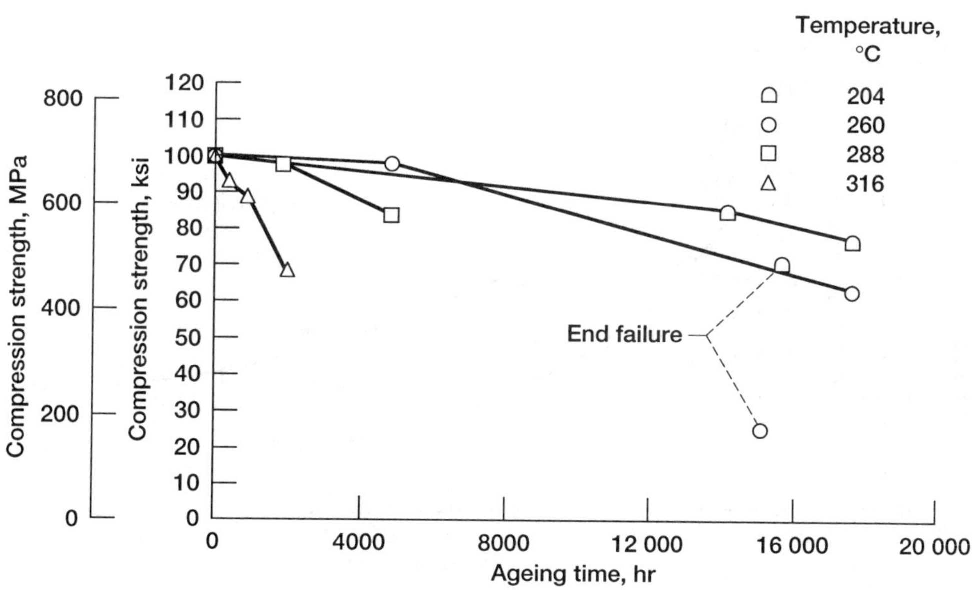

FIG. 6—*Compression strength of T650-35/PMR-15 T-12 composite specimens as a function of aging time at various temperatures.*

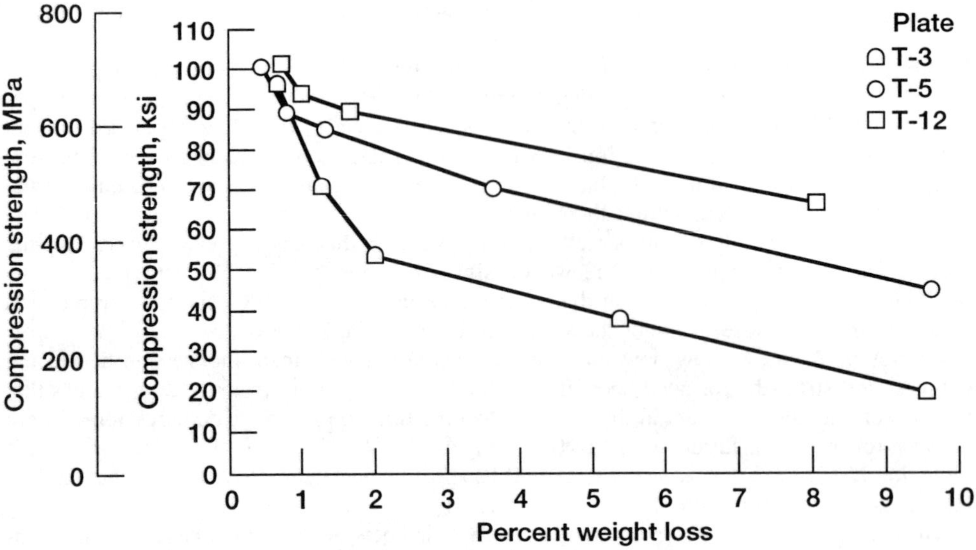

FIG. 7—*Compression strength of T650-35 fabric/PMR-15 composites of different thicknesses after aging in air 316°C.*

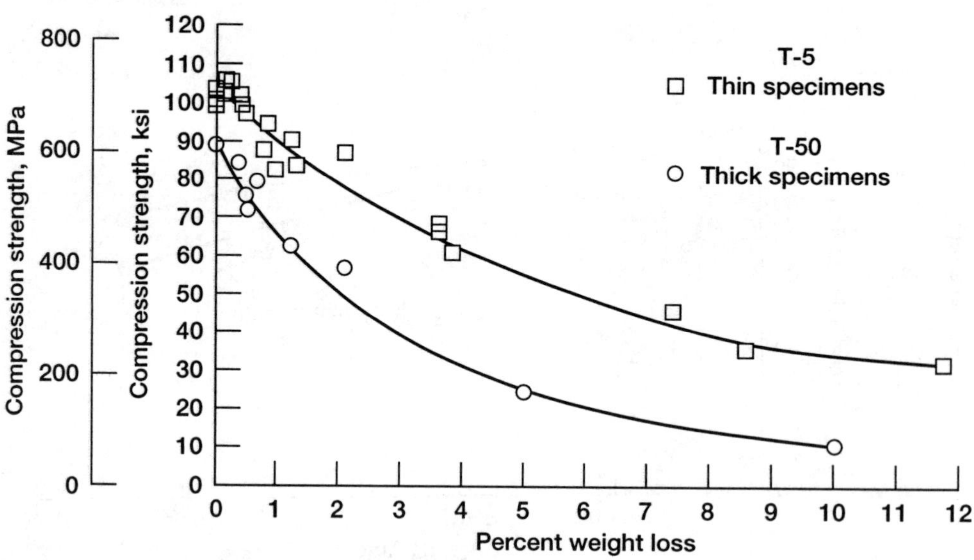

FIG. 8—*Compression strength of T650-35/PMR-15 composites as a function of percent weight loss.*

have 4, 8, and 20 plies, respectively, through the thickness. From measurements using optical microscopy methods to measure surface layer thicknesses, the changes in residual strengths appear to match up mathematically with the differences in effective thickness. For example, if one calculates the measured compression strength for each size specimen by assuming the outer oxidized layer is not strong enough to sustain the full test failure load, and the failure load is carried only by the cross-sectional area minus the area of the two oxidized layers, the calculated surface layer thicknesses are consistent between the three different samples. This will be studied further.

In Fig. 8, T-5 and T-50 compression strength data are plotted against percent weight loss. The data were measured from specimens aged at all temperatures except 204°C. When the specimen thickness is increased to that of the T-50 specimen (1.3 cm), then the measured unaged specimen test data are about 90% of the values for the thinner specimens. Again the two curves are similar, but this time neither specimen seems to lose strength at a faster rate than the other after the initial 1% of weight loss has taken place. It is apparent that the unaged material data shown in Fig. 7 are not affected by thickness differences as they are in Fig. 8. The differences observed in Fig. 7 are therefore thermal and oxidative effects.

Compression Modulus

Figures 9 and 10 show the measured moduli of some of the specimens that were aged isothermally in this study. Initially, the moduli increase slightly during the first 4000 h. The time for maximum modulus increase decreases with increasing aging temperature. The increase is attributed to an increase in cross-linking at the aging temperature. Again, the composites that were aged at the lower temperatures show almost identical relationships with aging time. Only the specimens that were aged at 316 and 343°C lose their moduli at a faster rate than the other specimens. One interesting observation is that the specimens that were

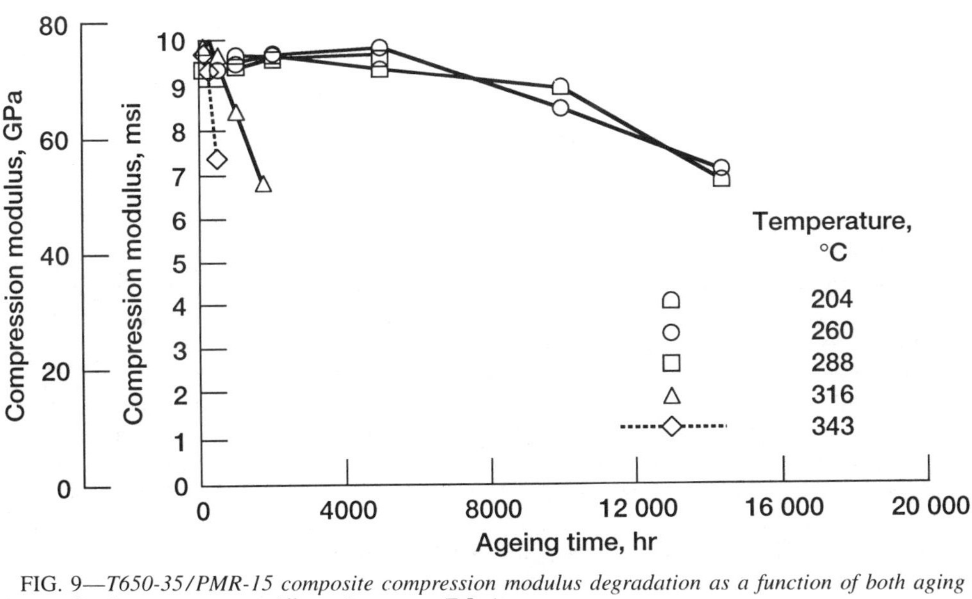

FIG. 9—*T650-35/PMR-15 composite compression modulus degradation as a function of both aging time and aging temperature. All specimens are T-5 size.*

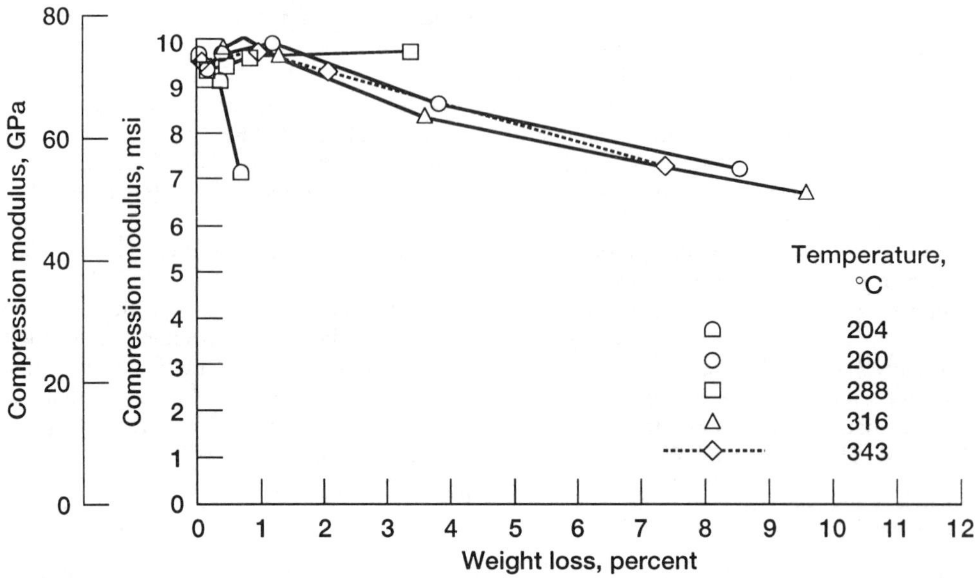

FIG. 10—*T650-35/PMR-15 composite compression modulus as a function of both aging temperature and weight loss. All specimens are T-5 size.*

aged at 204°C show no increase in moduli similar to the other two lower temperature specimens. In contrast, when the percent weight loss is the independent variable as in Fig. 10, only the material aged at 204°C falls outside of the band of data produced by the tests of the material aged at the higher temperatures.

Figure 11 presents data measured using T-12 specimens. It is obvious that the thicker specimens retain their modulus longer than the thinner T-5 specimens. The moduli are definitely thickness-dependent for the aged material.

The moduli were calculated from the formula, $E_C = (P_{1500} - P_{500}/.001bd$. E_C = modulus of elasticity; P_{1500} = load at 1500 microstrain; P_{500} = load at 500 microstrain; b = specimen width; and d = specimen thickness. The microstrains were measured with the extensometer.

Microscopy

Figures 12 through 15 show comparisons between two specimens aged at 288 and 316°C and two that were aged at 204 and 260°C. Figures 12 and 13 show the composites that were aged at 316°C for 500 and 1000 h with weight losses of 3.6 and 11.3%, respectively. These pictures show distinct surface layers of damage with the 1000 h of aging producing more microcracking than the shorter duration of aging. The lower temperature aging for 10 000 h produced weight losses of 0.40 and 3.9%, respectively. In contrast, Figs. 14 and 15 show that damage from the thermal exposures is not confined to the surfaces. In fact, no distinct surface layer is observed in either figure. The entire cross-sections of both composites are filled with evenly distributed intraply cracks, some of which develop into delaminations as shown in Fig. 15. The concentrations of the cracks in the two figures are not the same. The materials aged at 204°C contain less cracks. Figures 13 through 15 also show backscattering from oxygen atoms in the specimens as produced by scanning electron microscope backscattering electrons (SEM-BSE). The extent of damage becomes clearer due to these pictures.

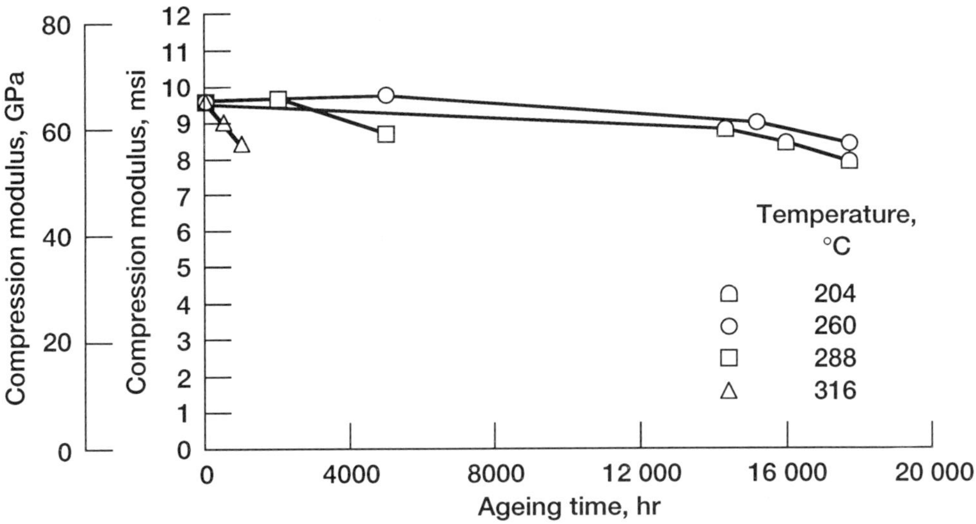

FIG. 11—*Compression modulus of T650-35/PMR-15 T-12 specimens as a function of aging time at various temperatures.*

The light areas are sites where the oxygen atoms are highly concentrated. The oxygen is only concentrated around the cracks at the surfaces in Figs. 12 and 13. It is evident that an increase in aging time, as observed when comparing these two figures, causes not only an increase in oxygen concentration around the surface cracks, but it also causes a significant increase in surface damage by cracking. The SEM-BSE pictures in Figs. 14 and 15 show the damage produced in the composites aged at the two lowest temperatures. It is evident that the oxygen pickup is not concentrated exclusively at the surfaces of the material. Oxygen has penetrated to areas within the inner composite material. It is significant that the oxygen is not concentrated around the internal cracks that are distributed throughout these specimens. Oxygen penetration from the outside probably does not occur through a network of these cracks. It does appear that the areas of oxygen concentration follow the tows that run parallel

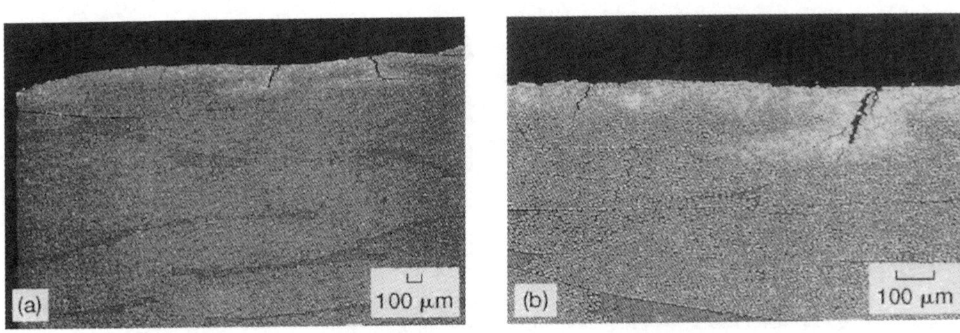

FIG. 12—*SEM-BSE picture of oxygen distribution in T650-35/PMR-15 composite. White areas are areas of high oxygen concentration. (a) Magnification is X20. (b) Magnification is X100. Composite was aged at 316°C for 500 h.*

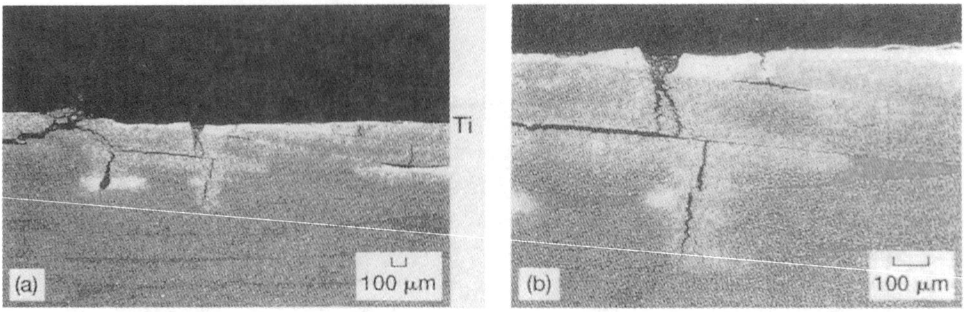

FIG. 13—*SEM-BSE picture of oxygen distribution in T650-35/PMR-15 composite. White areas are areas of high oxygen concentration. (a) Magnification is X20. (b) Magnification is X100. Composite was aged at 316°C for 1000 h.*

to the plane of the polished surface. Since no obvious horizontal cracks are evident in these figures (Figs. 14 and 15), it might be that oxygen diffusion is a significant factor in the formation of damage. The damage that has occurred does involve the formation of internal cracks, delaminations, and the reaction of the matrix with oxygen.

Failure Modes

Figures 16 to 18 reveal failure modes for specimens that were aged in this study. The larger T-50 specimens failed by a combination of fiber kinking and delamination. The delaminations appear to be formed by expansion due to the formation of a wedge at one end of the thick specimen (Fig. 16). The movement of the wedge inward toward the middle of the length of the specimen forces the outer surfaces of the specimen to separate from the specimen. The thinner T-12 specimens failed as shown in Fig. 17. The failure is similar to that for the T-50 specimen. However, it occurs in the gage section and not at the end.

As the aging time increases at a particular temperature, the failure appears to occur by "brooming" at one end of the specimen (Fig. 18). This suggests that the transverse strength has deteriorated significantly. The end failures occurred in the T-12 specimens that were exposed at the lower temperatures for times of 15 830 h but the specimens aged for 17 720

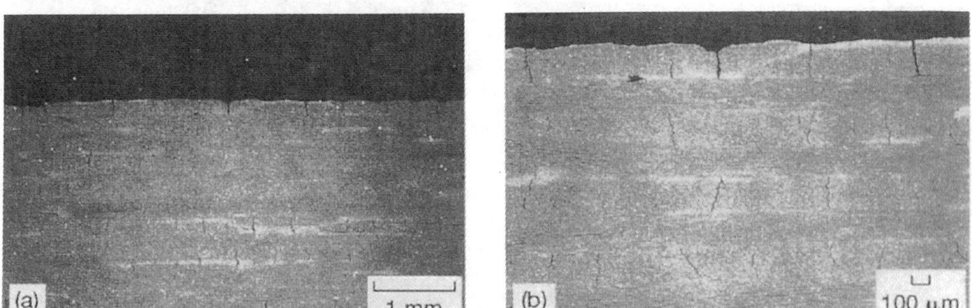

FIG. 14—*SEM-BSE picture of oxygen distribution in T650-35 fabric/PMR-15 composite. White areas are areas of high oxygen concentration. (a) Magnification is X20. (b) Magnification is X100. Composite was aged at 204°C for 10 000 h.*

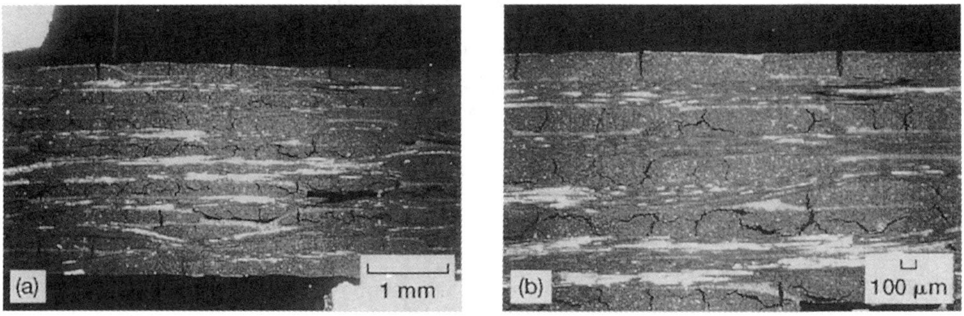

FIG. 15—*SEM-BSE pictures of oxygen distribution in T650-35 fabric/PMR-15 composite. White areas are areas of high oxygen concentration. (a) Magnification is X20. (b) Magnification is X100. Composite was aged at 260°C for 10 000 h.*

h failed in the gage sections. The T-3 specimen failed at the ends after 1510 h at 316°C. These data are included in the strength plot shown in Fig. 6 for the T-12 specimens. The strengths are significantly lower than the values for gage area failures.

Discussion and Conclusions

The results of this study indicate that thermo-oxidative stability at elevated temperature by itself is not a valid measure of long-term durability of polymer-matrix composites. This was evidenced in the data from the aging tests conducted at 204 and 260°C. Property measurements must be made to properly assess the potential of polymer-matrix composites. Results similar to this were published in Ref *12* that addressed the use of epoxy-matrix composites as long-term materials of construction for aircraft engine components. Mathematical models that are developed to predict long-term mechanical property durability must include terms that describe thickness, weight loss, and microcrack density effects. These

FIG. 16—*T-50 specimen after compression test. Wedge formed at the top.*

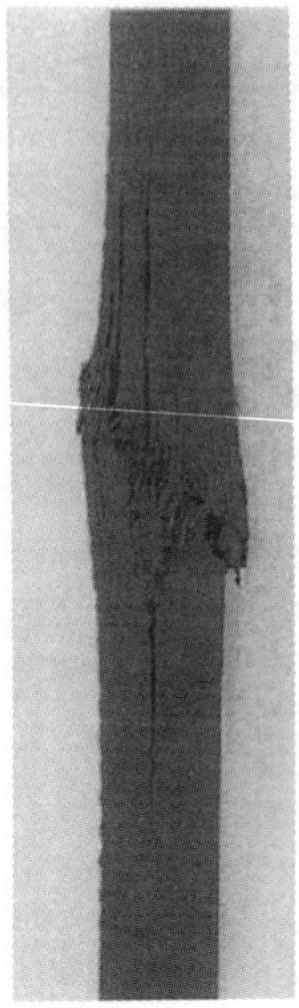

FIG. 17—*Compression failure in the gage section of T-12 specimen.*

results also indicate that attempts to predict long-term durability of mechanical properties of the polyimide composite from accelerated testing or extrapolation of short-term aging test data may be unreliable even when extrapolating downward in temperature.

Two more important conclusions can be drawn from the results of the work described herein. The first is that the compression properties retention is dependent not only on the

FIG. 18—*Compression end failure of a T-50 specimen. Failed surface has "broomed" out at the upper surface of the specimen.*

weight loss but also on the aging time over the temperature range that was studied. This is observed only for specimens of the same geometric size. The second conclusion is that specimen thickness has a significant affect at all temperatures. Any modeling that is to be developed to describe the durability of these composites should include both of these effects. The data indicate a strong relationship between these two independent variables when regression analyses are performed. The most effective mathematical relationship is an exponential one.

A cursory examination of Figs. 4 and 5 may lead one to believe that there may be at least two different mechanisms involved in the thermo-oxidative degradation of the PMR-15 composite laminates over the temperature range that was studied. At lower temperatures, the weight loss data indicate that the oxidative-degradation rate is very low while the compression strength degrades significantly more than at identical weight loss levels at higher temperatures. This is contrary to the initial premise that the surface degradation may control mechanical properties changes. It might be concluded that the degradation which diminishes the compression strength at the lower temperatures affects the bulk of the composites much more than the surfaces. At the higher temperatures, the loss in compression strength is shown to be directly related to the percent weight loss for the same thickness of specimens. If the weight loss is assumed to be concentrated in the oxidizing surface layers, then the initial assumption holds true and the weight loss can be related to the surface layer thickness at these temperatures. Earlier work reported in Ref *3* indicates a change in oxidative degradation activation energy over the temperature range from 260 to 343°C. This suggests that there is a change in mechanisms over the temperature range that was studied.

Pictures showing oxygen distribution throughout the specimens aged at the lower temperatures do not indicate a uniform concentration of oxygen throughout the internal volume. The oxygen is observed to concentrate along the lateral surfaces of some fiber tows and not around the internal cracks that have developed during the aging. When one compares Figs. 14 and 15, the damage in the specimen aged at 204°C is much less than the damage in the specimen aged at 260°C. However, the two specimens have the same compression strengths and moduli. Because of this, it appears that the internal cracking is not the dominant factor in the reduction of properties.

The variation of compression strength with specimen thickness at the lower temperatures suggests that a chemical change does not occur evenly throughout the entire volume of the composite material. The change in material probably advances inward from the surfaces by diffusion along the fiber tows as the aging time progresses. The initiation and growth of intraply cracks might be due to a density change in the matrix material caused by chemical changes that result in matrix shrinkage. This was reported in Ref *8*. The difference in crack density within the specimens aged at 204 and 260°C suggest that any chemical change that may be taking place proceed at different rates and are further from completion in the former specimens. During the analysis of thermal cycling test specimens, it has been observed that microcracks develop in polymer matrix composites from the outer surfaces inward and then they multiply. This type of crack initiation could also contribute to the low-temperature property degradation observed in this study.

If a strength loss of 50% is arbitrarily set, the data indicate that composite thickness over 0.66 cm may allow structures to sustain design loads for times over 30 000 h. The upper thickness limit has not been established in this study, but the potential for thick structure durability appears to be there. The data also suggest that materials having a thickness less than 0.66 cm in all probability cannot survive temperatures of 316°C and possibly 288°C for the 30 000 h.

References

[*1*] Bowles, K. J. and Meyers, A., Proceedings of 31st SAMPE International Symposium and Exhibition, J. L. Bauer and R. Dunaetz, Eds., SAMPE, Covina CA, 1986, pp. 1285–1299.

[*2*] Bowles, K. J. and Nowak, G., *Journal of Composite Materials*, Vol. 22, 1988, p. 966.

[*3*] Bowles, K. J., *SAMPE Quarterly*, Vol. 21, 1990, p. 6.

[*4*] Magendie, F. J., "Thermal Stability of Ceramic and Carbon Fiber Reinforced Bismaleimide Matrix Composites," Master's Thesis, Dept. of Chemical Engineering, University of Washington, Seattle, 1990.

[*5*] Nan, D. J. and Seferis, J. C., *SAMPE Quarterly*, Vol. 24, 1992, p. 10.

[*6*] Bowles, K. J. and Kamvouris, J. E., *Journal of Advanced Materials*, Vol. 26, 1994, p. 2.

[*7*] McManus, H., "Coupled Materials and Mechanics Analyses of Durability Tests for High Temperature Polymer Matrix Composites," Second Symposium on High Temperature and Environmental Effects on Polymer Composites, American Society for Testing and Materials, Norfolk, VA, 13 Nov. 1995.

[*8*] Bowles, K. J., Jayne, D., and Leonhardt, T. A., *SAMPE Quarterly*, Vol. 24, 1993, p. 2.

[*9*] Bowles, K. J., *SAMPE Quarterly*, Vol. 24, No. 2, 1993, p. 49.

[*10*] Madhukar, M. S., Bowles, K. J., and Papodopoulos, D. S., Proceedings of the First International Conference on Composites Engineering, David Hui, Ed., New Orleans, LA, 1994.

[*11*] Roberts, G. D., Malarik, D. C., and Robaidek, J. O., Proceedings of the Eighth International Conference on Composite Materials, S. W. Tsai and G. S. Springer, Eds., SAMPE, Covina, CA, pp. 12-H-1 to 12-H-10.

[*12*] Tsotsis, T. K., *Journal of Composite Materials*, Vol. 29, 1995, p. 410.

Lawrence C. Bank[1] *and Moshe Puterman*[2]

Microscopic Study of Surface Degradation of Glass Fiber-Reinforced Polymer Rods Embedded in Concrete Castings Subjected to Environmental Conditioning

REFERENCE: Bank, L. C. and Puterman, M., **"Microscopic Study of Surface Degradation of Glass Fiber-Reinforced Polymer Rods Embedded in Concrete Castings Subjected to Environmental Conditioning,"** *High Temperature and Environmental Effects on Polymeric Composites: 2nd Volume, ASTM STP 1302*, Thomas S. Gates and Abdul-Hamid Zureick, Eds., American Society for Testing and Materials, 1997, pp. 191–205.

ABSTRACT: The surface degradation of glass fiber-reinforced polymer (GFRP) pultruded rods when embedded in concrete castings and subjected to environmental conditioning is discussed in this paper. Investigation of the degradation of the GFRP rods was performed using optical microscopy and scanning electron microscopy (SEM). Unidirectionally reinforced pultruded rods (6.3- and 12.7-mm diameters) containing E-glass fibers in polyester and vinylester matrices were conditioned at standard laboratory conditions (21°C, 65% relative humidity) or submerged in aqueous solutions (tap water) at 80°C for durations of 14 and 84 days. Observations of the surfaces and cross-sections of the rods by optical microscopy and SEM revealed a variety of degradation phenomena. Embedded hygrothermally conditioned rods were found to have developed surface blisters of different sizes and depths. SEM studies of the surface revealed degradation of the polymer matrix material and exposure and degradation of the fibers close to the surface of the rods. The rods with the vinylester resin matrix showed less extensive degradation than those with the polyester resin matrix; however, the degradation characteristics of the two types of rods appear to be similar.

KEY WORDS: glass fibers, polyester matrix, vinylester matrix, environmental degradation, accelerated tests, concrete environment

Fiber-reinforced polymer composites are currently being evaluated as promising new materials for the construction industry. One particular area of intense interest is the use of composite materials as reinforcements for concrete structural members. The composite reinforcements are intended to take the place of conventional steel reinforcements that have historically been plagued by corrosion-related problems. Recent research and development activity in the field can be found in the proceedings of international symposia [*1*,*2*] devoted to this topic.

[1] Professor, Department of Civil and Environmental Engineering, University of Wisconsin-Madison, Madison, WI 53706.

[2] Senior research associate, National Building Research Institute, Technion, Haifa 32000, Israel.

One of the major issues to be resolved with regard to the use of composite materials in the construction industry and, in particular, in conjunction with highly alkaline portland cement-based concrete materials, is that of long-term durability. To date, the overwhelming majority of exploratory research in use of composite materials as either conventional or pretentioned reinforcements for concrete has concentrated on short-term properties of the composite materials and structural performance of concrete members reinforced with composite materials. A recent review of the development of accelerated test methods for predicting long-term performance of polymer composites for use in the construction industry subjected to environmental effects uncovered very few studies on the durability of composite materials reinforcements for concrete structures [*3*]. As expected, the bulk of the literature on environmental effects on composites is related to aerospace applications. In this literature, the interaction of composite materials with Portland cement-based concretes has not been considered. Consequently, the results of extensive testing described in this "aerospace" literature may not be directly applicable to the materials described in the current paper. Where durability studies have been performed on composites for construction, the approach tends to be that of subjecting the structural members to an exposure condition and performing structural tests on these members at periodic intervals as a means of determining degradation. As discussed in Ref *3*, this approach can only be used to obtain empirical data and no fundamental understanding of the degradation mechanism is obtained. The alternative proposed approach is to use a combination of physicochemical, microscopic, and mechanical investigations to develop a fundamental understanding of the degradation mechanisms and thereby to develop predictive models based on "accelerated" testing. Preliminary results of research into the use of microscopic studies are presented in this paper, while those related to physicochemical methods are presented in a companion paper [*4*] in this symposium.

Background

All major non-metallic fiber types (glass, carbon, aramid, polyvinylalcohol) are being considered as possible reinforcements for the composites to be used for reinforcing concrete. Polyester, vinylester, and, to a lesser degree, epoxy thermosetting resins are being considered for the matrix materials. Economic cost-effectiveness will play a central role in the introduction of composite materials into the construction industry. Consequently, there is a strong incentive to use glass fibers and lower cost polyester and vinylester resins. However, it remains to be determined if composites using these lower cost constituent materials will be durable in the construction environment, that is, that they will be able to provide sufficient and predictable performance over the typical design life of a concrete structure (50 to 100 years). Based on extensive studies [*5,6*] over the last 30 years into the use of fibers in glass fiber-reinforced cement (GRC) there is substantial evidence to suggest that glass fibers will not be durable when used in conjunction with concrete. Even more disturbing is that even after 30 years of research in this field it appears that a reliable accelerated test method has still not been developed to predict the long-term durability of GRC composites [*6*]. Less is known with regard to the durability of carbon and aramid fibers when used with concrete, however, and it would appear that based on experience with these fibers (especially the carbon-based fibers) their resistance to concrete environments will be better than that of glass fibers. While most attention in the construction materials industry is usually focused on the issue of the durability of the fibers, it is equally important to develop an understanding of the durability of the resin systems to be used in the composite bars for concrete reinforcements [*7*]. The superior performance of vinylester relative to polyester resins in typical industrial corrosive environments has been well documented [*3*] in the reinforced plastics industry and specialty vinylester resins are currently being developed for use in composite reinforcements. Since most construction applications will not require the superior elevated temperature characteristics of higher-cost epoxy resins use of these resins may not warranted. In reinforcing members, the role of the polymer matrix will be critical in both protecting and transferring high sustained loads to the reinforcing fibers. Degradation in the matrix material that reduces its effectiveness to transfer loads to the fibers over the long term is of major concern with respect to both the ultimate capacity and the serviceability of the concrete structure reinforced by the composite.

Objective

The objective of this research project was to perform a preliminary study of the degradation of glass fiber, polyester and vinylester matrix, composite rods when embedded in concrete and subjected to environmental conditioning. In order to examine the surface characteristics of the composite rods by microscopy, smooth, pultruded rods were used in this study. Results of a parallel study in which commercially produced "deformed" composite reinforcing rods were used will be reported elsewhere. A secondary objective was to perform short-term conditioning under aggressive conditioning and simultaneous long-term conditioning under standard laboratory conditions in order to develop data for development of accelerated test methods for these composites.

Materials

Commercially pultruded glass-fiber composite rods of two diameters were used in the investigation. Rod diameters, constituent materials, manufacturers, and volume fractions (obtained according to ASTM D 2584, Test Method for Ignition Loss in Cured Reinforced Resins) are detailed in Table 1.

Rods were saw cut to 500-mm lengths. An "embedded rod test specimen" was fabricated by casting a concrete cylinder of 100-mm diameter axisymmetrically around the lower 250 mm portion of the rod. PVC pipe was used for the molds and normal portland cement concrete (67.7 kg aggregate, 59.7 kg sand, 23.1 kg cement, 14.0 kg water, 44 MPa, 28-day compressive strength) was hand-poured into the molds. The specimens were submerged in water for 7 days for concrete curing. Following this, the molds were removed and the concrete was allowed to cure for an additional 21 days in standard laboratory conditions (21°C, 65% relative humidity). The specimens were then conditioned as described in the next section. Following the conditioning, the concrete part of the specimen was cut into three sections using a diamond saw. A 25-mm "slice" was cut from the bottom end, a 160-mm cylinder from the central section of the concrete and a 65-mm section that included the protruding 250-mm unembedded rod from the top end. The 160-mm cylindrical section was split open using the procedure recommended (for a different purpose) in ASTM C 496, Test Method for Splitting Tensile Strength of Cylindrical Concrete Specimens, to enable extraction of the rod from the concrete and subsequent observation of the rod surface and cross-section by microscopy. This part of the study is reported in this paper. The top section with the protruding rod was used for pull-out type tests while the bottom "slice" was used to study the interface between the rod and concrete in the cross-sectional plane.

For the optical and scanning electron microscopy studies, samples were cut from portions of the rods in the embedded section. Two types of samples were used. For the surface studies, rod pieces measuring approximately 10-mm long were cut from the rods. The circular cross-section was cut lengthwise and the resulting half-cylinder bonded to a mounting plate. No other treatment was used on these specimens and they were viewed "as-is." For the cross-section studies, 10-mm long pieces were cut from the rods, and these were potted in epoxy

TABLE 1—*Specifications of composite rods.*

Rod type identification number	R1	R2	R5	R6
Number of specimens	12	12	12	12
Resin type	polyester Aropol 2036C Ashland Chemical	vinylester DFV-1004-D 922 HV, 7030 Ashland Chemical	polyester Aropol 2036C Ashland Chemical	vinylester DFV-1004-D 922 HV, 7030 Ashland Chemical
E-glass fiber manufacturer and yield	Owens Corning 4 400 tex (113 yd/lb)	Vetrotex/Certainteed 4 400 tex (113 yd/lb)	Owens Corning 4 400 tex (113 yd/lb)	Vetrotex/Certainteed 4 400 tex (113 yd/lb)
Rod manufacturer	Creative Pultrusions Inc	DFI Pultruded Composites Inc	Creative Pultrusions Inc	DFI Pultruded Composites Inc
Nominal diameter	12.7 mm (1/2 in.)	12.7 mm (1/2 in.)	6.3 mm (1/4 in.)	6.3 mm (1/4 in.)
$\nu_{\text{glass fiber}}$	60%	57%	not measured	not measured
ν_{resin}	40%	43%	not measured	not measured

NOTES—(1) Volume fraction test conducted on 12.7-mm (1/2 in.) rods only as per ASTM D 2584. Calculations based on densities, $\rho_{\text{glass}} = 2.54$, $\rho_{\text{polyester}} = 1.151$, $\rho_{\text{vinylester}} = 1.043$ g/cm^3. (2) Only continuous unidirectional glass fiber rovings were used to reinforce the rods. Rods contained approximately 5 pph resin of inorganic clay filler. (3) 1 tex = 1 g/km.

molds and then polished to a fines of 0.5 to prepare the surface for SEM. All SEM samples were sprayed with the required gold coating.

Conditioning

The specimens were conditioned in five different environments as shown in Table 2. The results reported in this paper are those obtained to-date from the short-term environments, E1, E2, E3. After completion of the short-term testing, specimens placed in Environment E4 were divided into two groups (E4 and E5) to obtain data on long-term conditioning in water.

TABLE 2—*Conditioning environments and number of specimens.*

Name	Description	Number of Replicates for each Rod Type
E1	14 days submerged at 80°C	3
E2	84 days submerged at 80°C	3
E3	84 days at standard laboratory conditions (21°C, 65% RH)	3
E4	24 months (2 years) at standard laboratory conditions	1
E5	7 months at standard laboratory conditions (E4) and 17 months submerged in water at room temperature (21°C)	2

This was after analysis of the E3 data showed no observable degradation of the rods when placed in standard conditions for 84 days.

The temperature of 80°C was chosen for this work based on data reported in the GRC literature where the attempted development of accelerated test methods [*8*] have used this as the "upper bound" temperature for degradation of glass fibers in cementitious matrices. The choice of 80°C over 60°C (the other "often-used" elevated temperatures in the GRC literature) was based on the need to obtain a "as short as possible" accelerated conditioning time without changing the basic degradation mechanism in the composite, since a long-term objective of this research is the development of accelerated test methods for polymer composites for the construction industry.

Optical and Scanning Electron Microscopy (SEM) Observations

Photographs of the concrete cylinders after being split open and showing the embedded composite rods are shown in Figs. 1 and 2, respectively. The photographs show the specimens from the three environments (E1, E2, E3) reported in this paper. (Actual rod diameters shown in Fig. 1 are 6.3 mm while those shown in Fig. 2 are 12.7 mm.) A close up of the R5-E2 (6.3-mm diameter polyester rod) specimen shown in Fig. 3 reveals a phenomenon observed in all of the 80°C conditioned rods. This phenomenon is the development of locally degraded regions, referred to as "white-spots" on the rod, adjacent to cavities that were formed in the concrete cylinder during fabrication and curing of the concrete. In some cases the spot was even in the form of a visible blister (Fig. 3). These phenomena, both the whiteness of the spots and the blistering formation, were, in the case of the polyester rods, much more severe than in the case of the vinylester rods, and more severe after the 84 days in water at 80°C than after the 14-day submersion. Evidence of spotting can be seen in the R5-E1 specimen in Fig. 1 and also in vinylester rod shown in Fig. 2. A close-up photograph of the spots on the R2-E2 (12.7-mm diameter vinylester rod) specimen is shown in Fig. 4.

It appeared that the cavities in the concrete were filled with an alkaline concrete pore-water solution (pH = 12 to 13) that degraded and damaged the polymeric resin, causing a separation between the resin and the embedded fibers, the "white color" being the optical result of the disintegrating resin and the separation of the fibers. It is no surprise that the degradation phenomena seem to be much more severe in the case of the polyester than in the case of the vinylester, as the latter is more stable chemically, and hence, less susceptible to alkaline hydrolytic degradation reactions. The electron-microscopy investigations were directed mainly, yet not solely, to these areas of damage concentration.

Figure 5 shows a portion of a polished cross-sectional area of an "unaccelerated" (E3) polyester (R1) rod near the surface of the rod. Glass-fibers can easily be detected at the very edge of the rod along the circumference. In many instances, these fibers, close to the rod surface, were hardly covered, if at all, by the polymeric resin. Except for minor chipping of the fiber as a result of the mechanical cutting, and some minor separation of the resin from the fiber near the very edge of the rod, the internal structure of the composite seems intact. At some other locations along the circumference, as shown in Fig. 6, fibers were not that close to the surface and were covered by a thicker layer of resin. In some cases (not shown here), some small cracks and minor cavities could be seen within the resin, in-between the fibers. The "unaccelerated" vinylester rods (for example, R2-E3) were similar in structure in that in regions around the circumference the fibers were barely covered at the rod surface. However, they did not show evidence of degradation.

After 84 days of submersion in water at 80°C (E2), a severe phenomenon of damage was detected along the surfaces of the rods, particularly in the case of the polyester resin, as can be seen in Figs. 7*a* and 7*b*. The polyester resin close to the surface is cracked and degraded;

FIG. 1—*6.3-mm glass/polyester rods (R5) in concrete cylinders following conditioning in three environments (E1, E2, E3) and splitting.*

separation of fibers from the resin occurs, and the fibers themselves seem to have broken under the mechanical stresses of the cutting and polishing. This type of damage characterizes most of the surface of the rod.

The situation is less critical and less severe in the case of the vinylester composites, as the degradation is concentrated mainly in the areas next to the cavities in the concrete, and as seen from the SEM investigation, only where fibers are close enough to the surface. Figures 8*a* and 8*b* show a small area where a fiber very close to the surface was completely detached from the rod after the thin layer of resin surrounding it degraded. A similar example can be seen in Fig. 9 where fibers on the edge of the rod, that are hardly protected by resin, become completely exposed, with some signs of penetration of cracking and degradation advancing further into the rod. Fibers in this region close to the surface of the rod also appear to show degradation phenomena although instances of fiber degradation do not appear to be

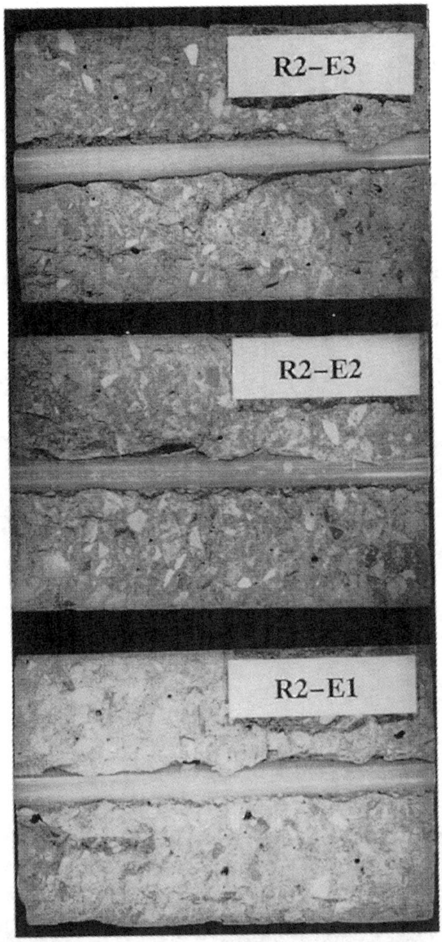

FIG. 2—*12.7 mm glass/vinylester rods (R2) in concrete cylinders following conditioning in three environments (E1, E2, E3) and splitting.*

nearly as widespread as those of the resin degradation in the surface region. Figure 10 shows an enlarged view (X4000) of a single fiber in the surface region. Fiber degradation in this region is similar to that reported by Sen at al. [*9*] where micrographs of degradation of glass fibers close to the interface with the concrete and submerged in water were presented.

The cracking of the polymeric resin on the rod-surface, exposing the fibers and allowing the alkaline solution to penetrate into the rod, seems to be a typical and common phenomenon as can be seen in Figs. 11*a* and 11*b* that show micrographs of the surface of a polyester rod embedded in concrete after 14 days in water at 80°C. The cracks and exposed fibers are clearly visible. Micrographs of higher magnification (not shown here) reveal a dense network of fine cracks along the entire surface area of the polymer.

Discussion

While degradation of glass fibers in cementitious composites has traditionally been a source of concern to the construction industry (since they often represent the only reinforce-

FIG. 3—*Close-up of the white spots and blister on the surface of a glass/polyester rod (R5) after 84 days submersion at 80°C (E2). The "spots" are seen to be adjacent to the cavities in the concrete.*

FIG. 4—*Close-up of the white spots on the surface of a glass/vinylester rod (R2) after 84 days submersion at 80°C (E2).*

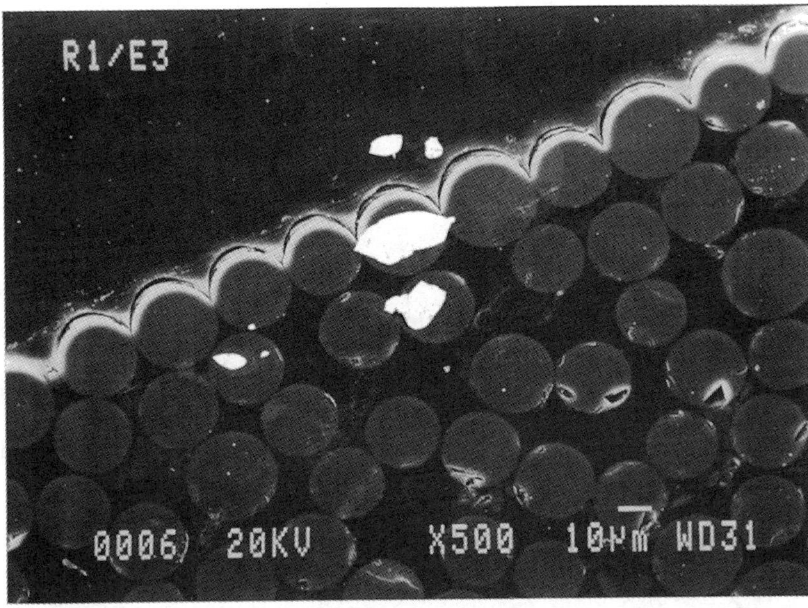

FIG. 5—*SEM micrograph of a polished surface of a glass/polyester rod (R1) embedded in concrete for 84 days at standard laboratory conditions (E3) showing fibers at the very edge of the rod surface.*

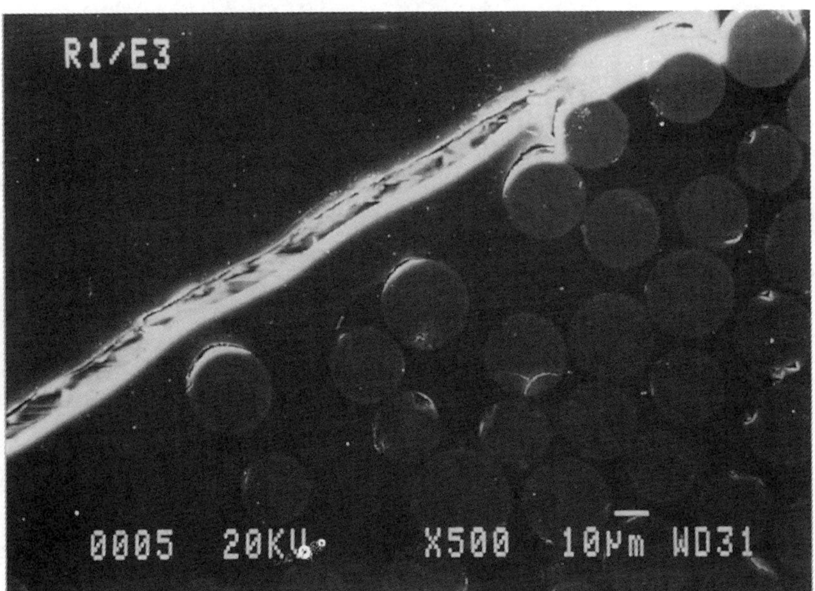

FIG. 6—*SEM micrograph of a polished surface of a glass/polyester rod (R1) embedded in concrete for 84 days at standard laboratory conditions (E3) showing fibers at the rod surface cover by a resin layer.*

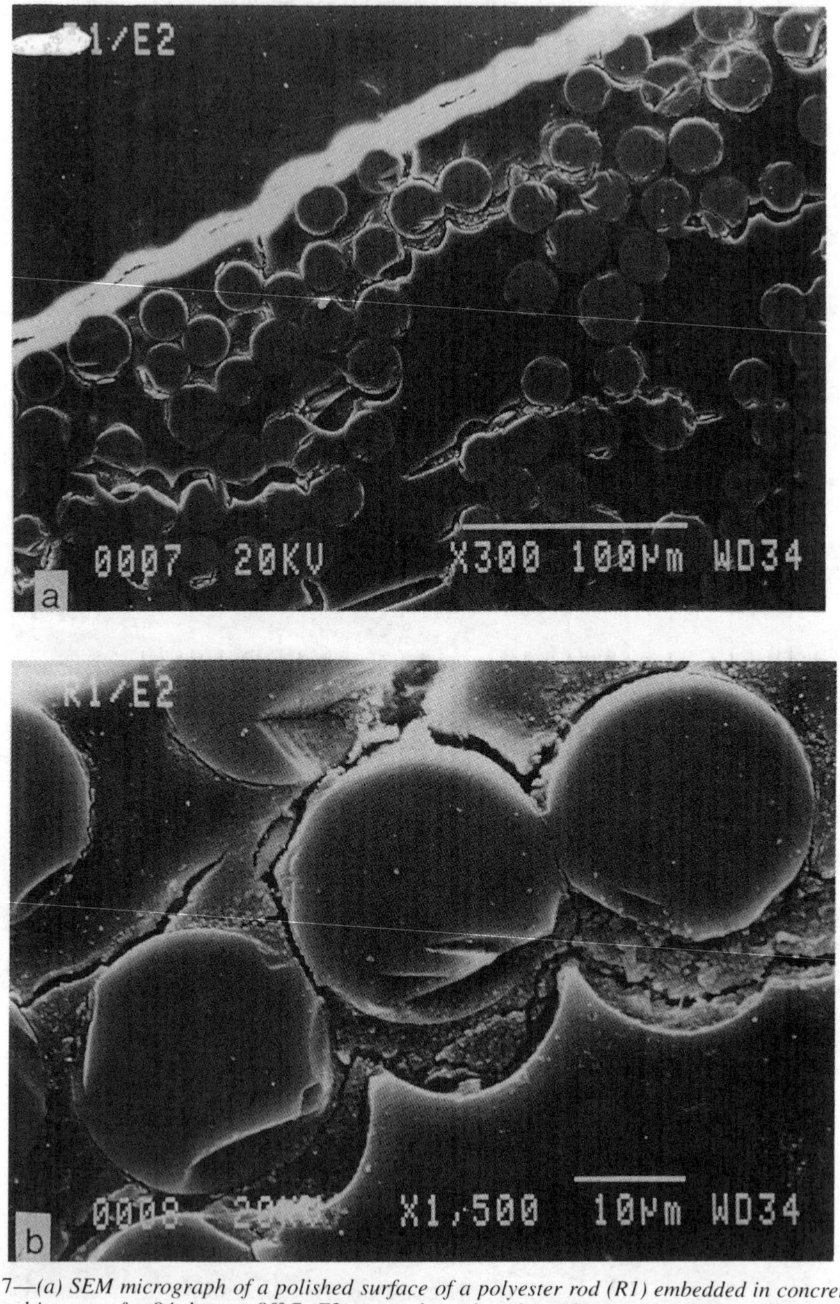

FIG. 7—*(a) SEM micrograph of a polished surface of a polyester rod (R1) embedded in concrete and submerged in water for 84 days at 80°C (E2), near the rod surface. (b) An enlarged detail of the above.*

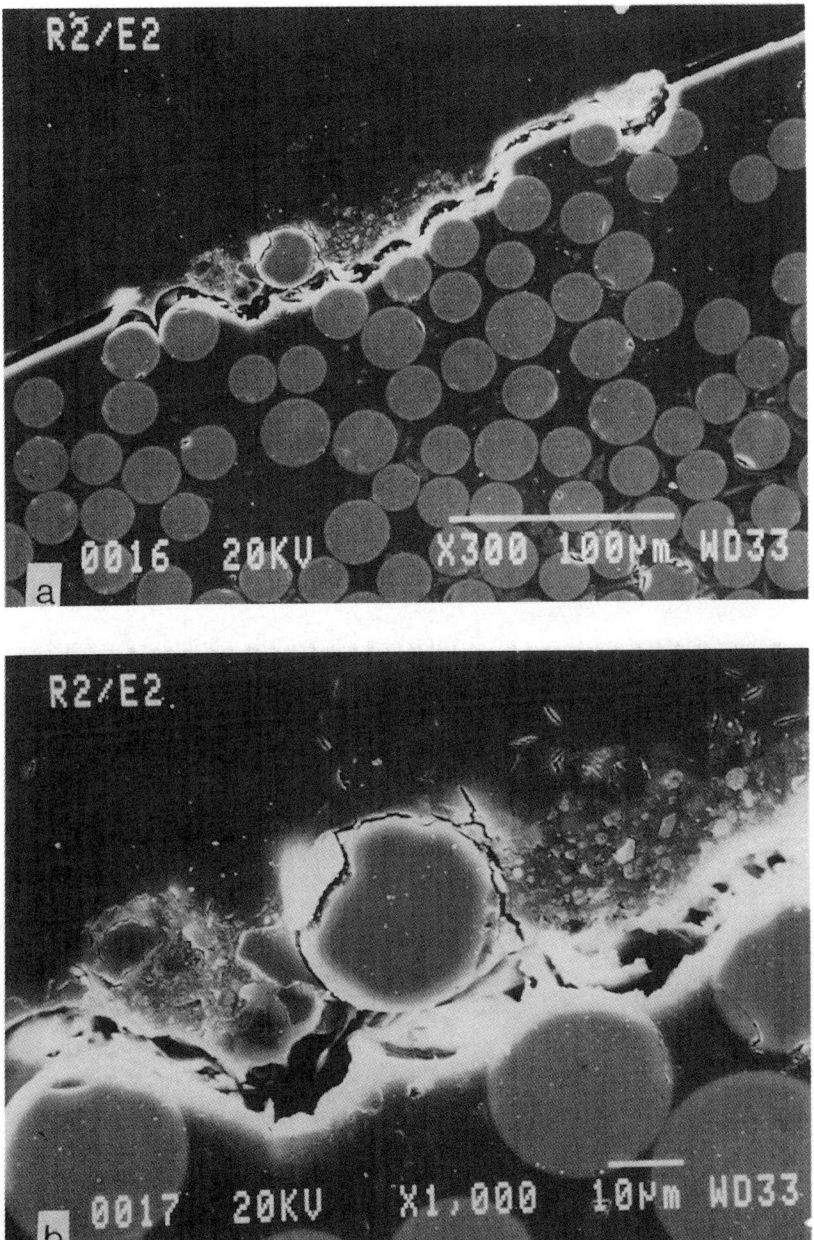

FIG. 8—*(a) SEM micrograph of a polished surface of a vinylester rod (R2) embedded in concrete and submerged in water for 84 days at 80°C, near the rod surface. (b) An enlarged detail of the above.*

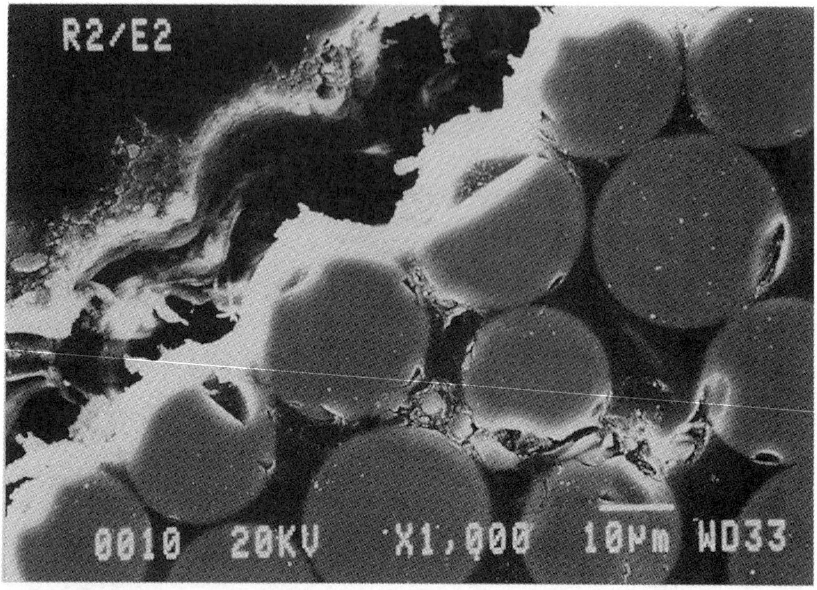

FIG. 9—*SEM micrograph of a polished surface of a vinylester rod (R2) embedded in concrete and submerged in water for 84 days at 80°C, near the rod surface showing penetration of cracking and resin degradation into the rod core.*

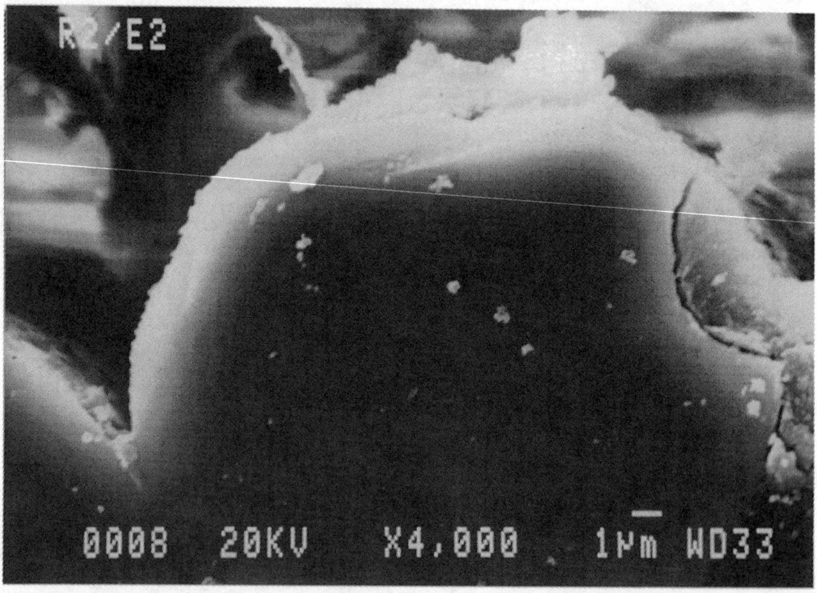

FIG. 10—*SEM micrograph of enlarged view of a single fiber in the surface region of a vinylester rod (R2) embedded in concrete and submerged in water for 84 days at 80°C, showing fiber degradation.*

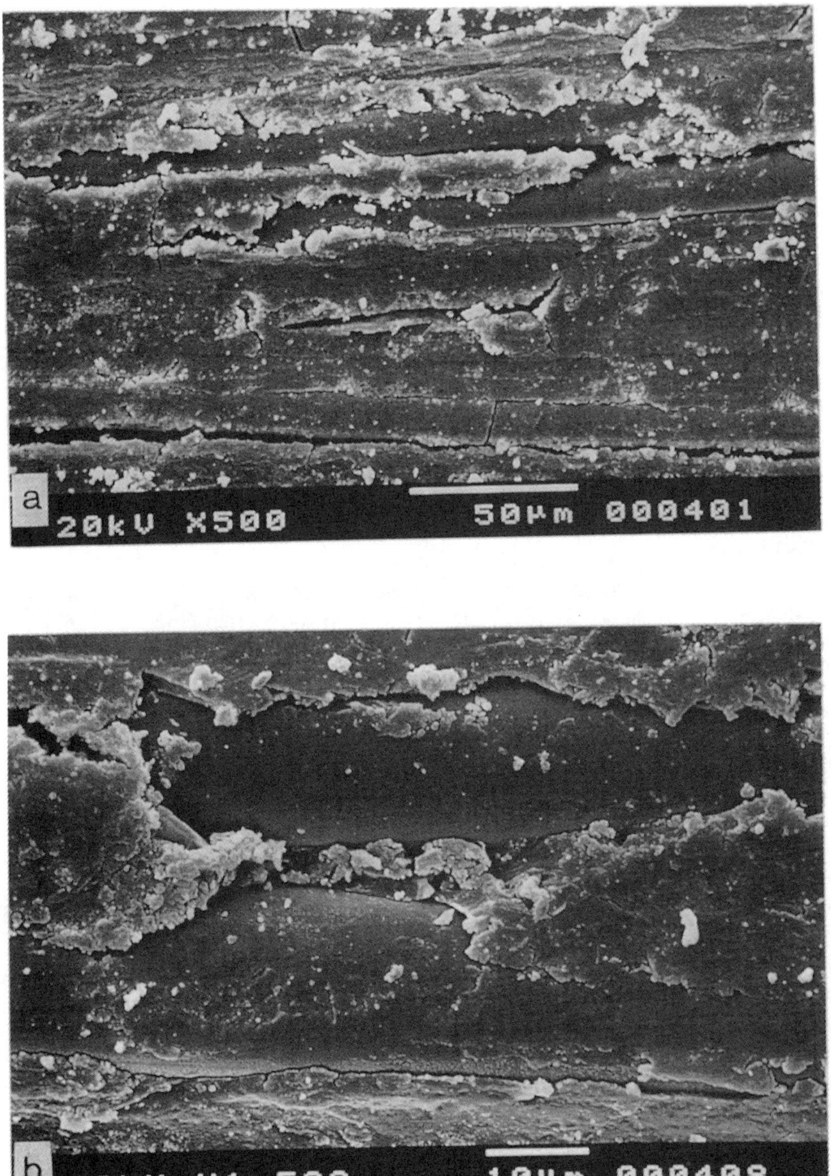

FIG. 11—*SEM micrograph of the external surface of a polyester rod embedded in concrete and submerged in water for 14 days at 80°C. (a) Cracked and degraded surface. (b) Exposed fibers.*

ment in GRC products), it appears that the degradation of the polymer resin systems that are used in composite rods for concrete reinforcing applications may also be a source of concern. In reinforcing bar applications, the polymer matrix in the composite rod is intended to perform the traditional role of the matrix in transferring load to the fibers and protection the fibers from damage. In the case of reinforcing bar applications where a "composite-action" must exist on a macro-scale between the concrete and the reinforcing bar, effective transfer of stresses between the composite bar and the concrete must occur. This is often referred to as "bond" in reinforced concrete theory. If, as observed in the present study, the surface layer of the bar in contact with the concrete degrades such that cracking and disintegration of the resin occur in this region, the transfer of stresses from the concrete to the load-bearing fibers in the composite rod may be significantly impaired. This will, in turn, decrease the composite-action between the bar and the surrounding concrete and the bar will not function as a reinforcing member. It appears that this phenomenon may be of more, or at least of equal, concern to that of the degradation of the fibers themselves. In fact, it would appear that based on the current study, very few fibers themselves are significantly degraded and that degradation of this insignificant number of fibers (in comparison to all the fibers in the rod) should not decrease the load carrying capacity in a meaningful way. However, if the polymer matrix degrades in the surface region (to a depth of 5 to 10 fibers), the interior fibers may never be mobilized to carry the load and ensuing failure of the reinforced concrete element may occur.

Conclusions

Based on observations to-date, it can be stated that there is clearly significant degradation of both glass/polyester and glass/vinylester pultruded rods when embedded in concrete and submerged in water at 80°C. This degradation occurs at the surface of the rods in contact with the concrete and particularly at locations where cavities exist in the concrete adjacent to the rod surface. Degradation of the polymer resins in the surface region (to a depth of 5 to 10 fiber diameters) is widespread. As expected, the degradation of polyester resin is more pronounced than that of the vinylester resin. Together with the degradation of the polymeric matrix, the reinforcing glass fibers in the vicinity of the surface are also effected by the alkaline concrete solution and degrade as suggested by other authors. These fibers (in the surface region) prove, in the micrographs, to be brittle and subjected to damage by mechanical stresses much more than those further away from the surface.

Based on the research reported in this paper, it appears that composite material rods consisting of unidirectional fibers in a polymer matrix may not be appropriate for concrete-reinforcing applications. Regardless of the external "deformation" that may be added to the surface to improve bond, the internal unidirectional core will most likely be subjected to the types of matrix and fiber degradation observed in this study. In order to prevent, or at least to contain this degradation, it is recommended that multidirectional fiber architectures, more chemically resistant resins and fibers, and resin-rich (but tough) surface coatings be investigated for composite reinforcing bar applications.

Acknowledgments

The research described in this report was conducted at the Israel National Building Research Institute (NBRI), Technion—Israel Institute of Technology, Haifa, Israel, while the first author was on Sabbatical leave as a Lady Davis Visiting Fellow. The support of the Lady Davis Foundation Trust is gratefully acknowledged. Financial support for the research was provided by the NBRI, The Israel Ministry for Construction and Housing, and the U.S.

Federal Highway Administration, Department of Transportation under Contract DTFH61-93-C-00012. Assistance during the course of the research provided by Dr. A. Katz, Mr. M. Lanzman, and Dr. Y. Frostig is acknowledged.

References

[1] Nanni, A. and Dolan, C. W., Eds., *Fiber Reinforced Plastic Reinforcement for Concrete Structures, International Symposium*, SP 138, America Concrete Institute, 1993.

[2] Taerwe, L., Ed., *Non-Metallic (FRP) Reinforcements for Concrete Structures*, Proceedings of the RILEM Symposium, EP&N Spon, London, 1995.

[3] Bank, L. C., Gentry, T. R., and Barkatt, A., "Accelerated Test Methods to Determine the Long-Term Behavior of FRP Composite Structures: Environmental Effects," *Journal of Reinforced Plastics and Composites*, Vol. 14, No. 6, 1995, pp. 559–587.

[4] Prian, L., Pollard, R., Shan, R., Mastropietro, C. W., Gentry, T. R., Bank, L. C., and Barkatt, A., "Use of Thermogravimetric Analysis to Develop Accelerated Test Methods to Investigate Long-Term Environmental Effects of Fiber-Reinforced Plastics," *High Temperature and Environmental Effects on Polymeric Composites: 2nd Volume*, (this publication), *ASTM STP 1302*, Thomas S. Gates and Abdul-Hamid Zureick, Eds., American Society for Testing and Materials, West Conshohocken, PA, 1997.

[5] Bentur, A. and Mindess, S., *Fiber Reinforced Cementitious Composites*, Chapter 8, Elsevier Science Publishers, 1990.

[6] Majumdar, A. J. and Walton, P. L., "Durability of Fibre Cement Composites," *Durability of Concrete*, SP-126, V. M. Malhotra, Ed., American Concrete Institute, 1991, pp. 745–771.

[7] Bank, L. C., "Properties of FRP Reinforcements for Concrete," *Fiber-Reinforced Plastic (FRP) for Concrete Structures: Properties and Applications*, A. Nanni, Ed., Elsevier Science Publishers, New York, 1993, pp. 59–86.

[8] Litherland, K. L., Oakley, D. R., and Proctor, B. A., "The Use of Accelerated Ageing Procedures to Predict Long Term Strength of GRC Composites," *Cement and Concrete Research*, Vol. 11, 1981, pp. 455–466.

[9] Sen, R., Mariscal, D., and Shahawy, M., "Investigation of S-2 Glass/Epoxy Strands in Concrete," *Fiber Reinforced Plastic Reinforcement for Concrete Structures, International Symposium*, SP-138, A. Nanni and C. W. Dolan, Ed., America Concrete Institute, 1993, pp. 15–33.

Luca Prian,[1] Ritsuko Pollard,[1] Rongmin Shan,[2] Christopher W. Mastropietro,[1] T. Russell Gentry,[3] Lawrence C. Bank,[4] and Aaron Barkatt[1]

Use of Thermogravimetric Analysis to Develop Accelerated Test Methods to Investigate Long-Term Environmental Effects on Fiber-Reinforced Plastics

REFERENCE: Prian, L., Pollard, R., Shan, R., Mastropietro, C. W., Gentry, T. R., Bank, L. C., and Barkatt, A., **"Use of Thermogravimetric Analysis to Develop Accelerated Test Methods to Investigate Long-Term Environmental Effects on Fiber-Reinforced Plastics,"** *High Temperature and Environmental Effects on Polymeric Composites: 2nd Volume, ASTM STP 1302*, Thomas S. Gates and Abdul-Hamid Zureick, Eds., American Society for Testing and Materials, 1997, pp. 206–222.

ABSTRACT: The development of accelerated test methods to characterize long-term environmental effects on fiber-reinforced plastics (FRPs) requires the use of physicochemical methods, as well as macromechanical measurements, in order to investigate the degradation processes and predict their course over long periods of time. Thermochemical and mechanical measurements were performed on a large number of FRPs exposed to neutral, basic, and acidic media between 23 and 80°C over periods of 7 to 224 days. The resin matrices used in the present study included vinylester, polyester, and epoxy, and the fiber materials were silicate glass, aramid, and carbon. TGA was used to study the effects of aqueous media on FRPs. In particular, the relative weight loss upon heating the previously exposed material from 150 to 300°C was found to be indicative of the extent of matrix depolymerization. Indications were obtained for correlation between this weight loss and the extent of degradation of various measures of mechanical strength. The measured weight change of the tested materials during exposure was found to reflect the extent of water absorption and could be related to the extent of the weight loss between 150 and 300°C. In basic environments, weight loss, rather than gain, took place as a result of fiber dissolution. Among various FRPs, those with aramid fibers were found to be especially prone to degradation due to extensive water uptake. Basic environments caused a large decrease in strength due to fiber dissolution. Acetic acid buffers caused significant matrix depolymerization. Raising the temperature to 80°C appeared to be the most effective method of enhancing the degradation effects, as reflected in both the thermochemical and mechanical data. Furthermore, the trends observed at 80°C (for example, with respect to solvent effects) were identical with the trends observed at lower, service-like temperatures. Thus, elevating the temperature is a promising acceleration factor. In agreement with the enhanced matrix depolymerization observed in the TGA measurements, higher temperatures also caused a decrease in T_g, as measured by DSC.

KEY WORDS: fiber-reinforced plastics, thermogravimetric analysis, accelerated tests, environmental degradation, depolymerization

[1] The Catholic University of America, Department of Chemistry, Washington, DC 20064.
[2] The Catholic University of America, Department of Civil Engineering, Washington, DC 20064.
[3] Assistant professor, Architecture Program, Georgia Institute of Technology, Atlanta, GA 30332-0155.
[4] Professor, Department of Civil and Environmental Engineering, University of Wisconsin-Madison, Madison, WI 53706.

Establishing the long-term durability of fiber-reinforced plastic materials (FRPs) over periods as long as several decades is a very important factor in evaluating the potential of these materials for use in civilian structural applications. As the kinetics of degradation of FRPs is not well understood and may be quite complex, monitoring of the macromechanical properties alone is not sufficient for predicting the progress of degradation with time. Measurements of physicochemical properties can yield important mechanistic information for the development of predictive models. Accordingly, the best route to satisfactory characterization of the long-term environmental stability of FRPs involves the acquisition of both mechanical and physicochemical data in corrosion tests. Accelerated tests are very useful in acquiring data to serve as a basis for model development, as well as for model validation within relatively short periods of experimentation. In order to ensure that the mechanisms controlling the degradation under test conditions are the same as those predominating under service conditions, tests under the latter conditions are also required.

A previous paper [*1*] described initial findings related to the development of accelerated tests for FRP composites. The tests described in Ref *1* were performed on a single composite material, consisting of pultruded vinylester/E-glass rod (VGR), over periods ranging between 3 and 28 days. The principal physicochemical measurement employed was thermogravimetric analysis (TGA). Based on the initial findings, it was concluded that TGA, combined with measurement of the initial weight change resulting under given conditions of environmental exposure is a valuable method of providing information about FRP degradation.

Weight gain following exposure to an aqueous medium or humid atmosphere has been known to be a measure of the extent of water uptake by a polymeric matrix [*2,3*]. On the other hand, weight loss of glass fiber-reinforced plastics (GFRPs) can be observed if the aqueous environment is basic, reflecting the susceptibility of the glass fibers to dissolution in alkaline environments [*4,5*]. This phenomenon is important in view of the high pH of water in contact with concrete environments [*4,5*].

The initial weight change, however, does not provide direct information about the degradation of the composite matrix. Valuable information concerning such degradation can be obtained by monitoring weight loss as a function of temperature during subsequent TGA analysis. TGA was used to monitor loss of compounds with low molecular weight during the aging of composites [*6*]. Studies of the use of TGA to monitor the degradation of vinylester/E-glass composites upon exposure to aqueous media recently showed that the weight loss which takes place upon heating from room temperature to 150°C following the exposure mostly represents loss of water absorbed during the contact with the aqueous medium. Heating above 300°C causes rapid and massive loss of organic material, and this loss is insensitive to the conditions of previous environmental exposure. The extent of weight loss between 150 and 300°C is the best indicator of the effects of the exposure on the material [*1*]. This weight loss is generally larger the larger the amount of water absorbed during the exposure. Analysis of the vapors evolved upon heating between 150 and 300°C by means of infrared and nmr spectroscopy as well as molecular weight determination showed that these vapors consist of monomeric units of the plastic matrix. Thus, the weight loss between 150 and 300°C reflects the extent of hydrolytic depolymerization during the preceding exposure. This conclusion was supported by the observation that depolymerization induced by gamma irradiation, even under dry conditions, also enhances the weight loss upon subsequent heating between 150 and 300°C. Returning to depolymerization caused by exposure to aqueous environments, it was found that increased concentrations of hydrolyzing reagents (ammonia, acetic acid), longer exposure periods, and higher temperatures all cause the weight loss observed between 150 and 300°C to become larger. The same factors result in considerable decrease in the tensile strength of the composite material [*1*].

Another thermochemical method that has been used in the characterization of plastics and plastic-based composites is differential scanning calorimetry (DSC) [*6–8*]. This method has been used to follow structural changes in the plastic matrix upon aging of composites [*6*].

Experimental

Following the initial findings obtained with vinylester/E-glass rods and described in Ref *1*, studies of the effects of various aqueous environments were extended to five common types of FRPs with exposure periods ranging up to 228 days. The materials investigated included (a) vinylester/E-glass rods (helically overwrapped, identical with the material used in Ref *1*, VGR) or plain tendons (VGT); (b) vinylester/E-glass plate (VGP); (c) polyester/E-glass rod (PGR); (d) vinylester/aramid tendon (VAT); (e) vinylester/carbon tendon (VCT); (f) epoxy/carbon tendon (ECT). The length of each sample was 25 mm and its diameter was 6.4 mm, except in the case of VGR, where samples with diameters of both 6.4 and 12.7 mm were used in order to assess the effect of changing the aspect ratio and increasing the average distance of the fibers from the center of the rod. The large-diameter VGR rods were designated VGR-LD. The VGP samples had a slab-like, rather than cylindrical, shape, with dimensions of 6.4 by 12.7 by 25.4 mm. The detailed descriptions of the various materials are given in Table 1. The manufacturers listed in Table 1 are Creative Pultrusions (CP), DFI Pultrusions, and Morrison Molded Fiberglass (MMFG), respectively. The environments to which the materials were exposed included de-ionized water, aqueous solutions of 0.3, 3, and 30% ammonia (NH_3), and two 0.6% (0.100 M) acetic acid buffers formulated to simulate acidic groundwater [*9*]. The more acidic of these two acetic acid buffers was a solution of 0.6% acetic acid (HOAc) adjusted with NaOH to pH 2.88 ± 0.05. The less acidic buffer was prepared by adjusting the pH of the 0.6% HOAc solution with NaOH to pH 4.93 ± 0.05. In addition, ambient air was used as a baseline environment.

The characterization techniques used on the samples before and after exposure to aqueous environments included both mechanical and physicochemical methods. The former included measurements of tensile strength, tensile modulus, flexural strength, flexural modulus, short-beam shear (SBS) strength, and diametral compression (DC) strength. The test equipment consisted of an MTS mechanical tester with suitable accessories.

The main physicochemical techniques included: (a) TGA measurements using an ATI/Cahn TG-131 analyzer at a heating rate of 2°C/min over the temperature range 23 to

TABLE 1—*FRP test materials.*

Material	Diameter, mm	Fiber Volume Fraction, %	Manufacturer	Comments
VGR	6.35	50 low-filler	CP	Helical overwrap
VGR-LD	12.7	50 low-filler	CP	Helical overwrap
VGT	6.35	60	DFI	
VGP	6.35 thickness	30 high-filler	CP	Used in second set of tests
VGP	6.35 thickness	30 high-filler	MMFG	Used in third set of tests
PGR	6.35	60	CP	
VAT	6.35	60	DFI	
VCT	6.35	60	DFI	
ECT	6.35	65	DFI	

NOTE—*Resin* (*first letter*): V = vinylester; P = polyester; E = epoxy. *Fiber* (*second letter*): G = glass; A = aramid; C = carbon. *Shape* (*third letter*): R = rod; T = tendon; P = plate.

335°C, using nitrogen as a reaction gas and helium as a purge gas. (Air was used in the second set of measurements, described below.); (b) measurements of weight change during environmental exposure, conducted by drying the sample for 16 h in a desiccator after completion of the environmental exposure and then weighing the sample and comparing this weight with the original weight of the same sample before exposure. (Prior to the exposure, samples were stored in ambient air for at least several weeks. The drying in a desiccator for 16 h after exposure was intended to remove excess moisture quickly and return the samples to a state approximating their original condition.); and (c) DSC measurements using a Perkin-Elmer DSC-7 calorimeter at a heating rate of 10°C/min. The results of the TGA measurements were usually expressed in terms of the relative weight retention (WR), defined as the percentage ratio of the weight at 300°C to the weight at 150°C. The results of the weight measurements on the exposed samples (WE) were expressed in terms of the percentage ratio between the weight of the exposed, re-dried sample and its original weight before the exposure. (Since the true dry weight of the specimens prior to the aqueous exposure was not accurately determined, and given the various shapes, aspect ratios, and resin/fiber ratios for the various materials used here, WE is mainly useful for comparing the conditions of exposure on the same material rather than for making comparisons among different shapes.) The data recorded in the DSC measurements included the glass transition temperature T_g and the change in heat capacity associated with it, ΔC_p.

Results

The first series of tests consisted of allowing samples of VGR, PGR, VCT, and VAT, each of them with a diameter of 6.4 mm, to stand for 7 days in each of three environments, consisting of ambient air, 3% ammonia at room temperature, and 3% ammonia at 50°C, respectively. Following drying for 16 h in a desiccator to remove the excess molecular water absorbed during the exposure, the final sample weight of each sample was determined, followed by TGA measurements. The results of the TGA measurements, expressed as residual weight during heating relative to the weight of each exposed sample after drying and prior to heating, are shown in Figs. 1 through 4 for the cases of VGT, PGR, VCT, and VAT, respectively. The results clearly show that exposure to an ammonia solution causes a decrease in WR, the weight retention between 150 and 300°C (that is, an increase in weight loss between these two temperatures) during the subsequent TGA measurement. In addition, exposure at an elevated temperature of 50°C has a much larger effect than exposure at 23°C, and VAT exhibits larger weight losses upon heating than the other composites. In general, the results indicated that the weight loss observed with the TGA can be considered in terms of three temperature ranges. Between 20 and 100 to 150°C, the weight changes are rather small and mostly reflect the loss of absorbed bulk water [*1*]. (A small weight gain observed below 80°C in the case of VCT can be attributed to moisture re-absorption into the dried sample.) Above 300°C, the TGA loss is massive and is not sensitive to previous environmental exposure. The range best reflecting the effects of environmental degradation is between 150 and 300°C [*1*].

A second set of measurements focused on weight changes of FRP specimens in the course of environmental exposure (obtained by comparing the weight after exposure and air-drying with the weight prior to exposure, see above) as well as on weight losses between 150 and 300°C during the subsequent TGA run. The results of the measurements of the weight change of a VGR rod with a 6.4-mm diameter exposed to an aqueous 0.3% ammonia solution at 80°C for periods of up to 60 days are shown in Fig. 5, and the corresponding weight losses between 150 and 300°C are shown in Fig. 6. The results indicate that the weight of the samples progressively decreases with increasing exposure time to the ammonia solution while

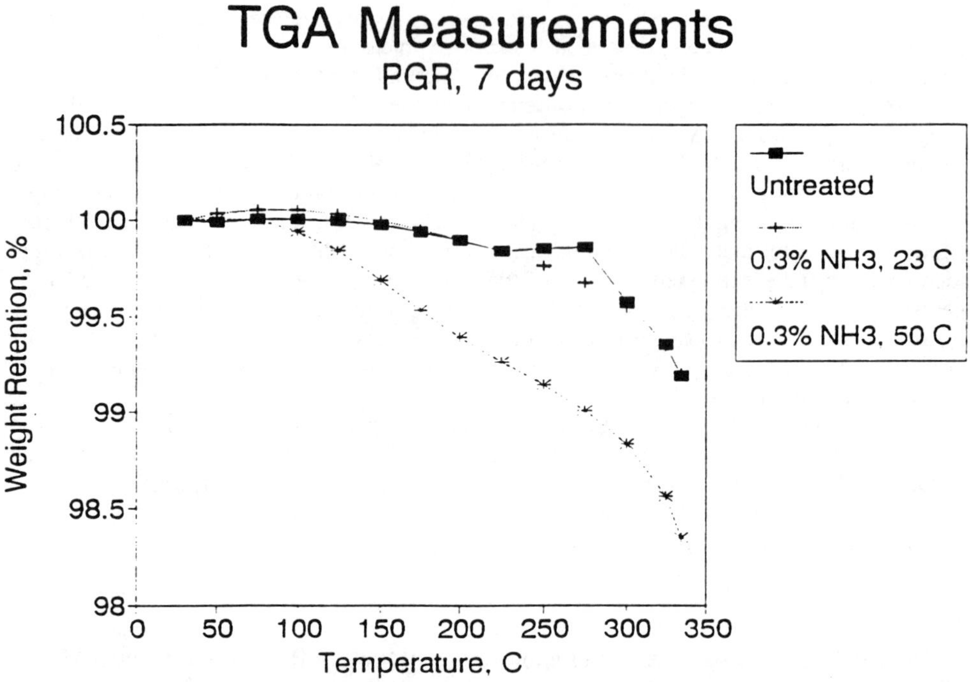

FIG. 1—*TGA measurements on PGR after a 7-day exposure to ambient air, 0.3% ammonia at room temperature, and 0.3% ammonia at 50°C.*

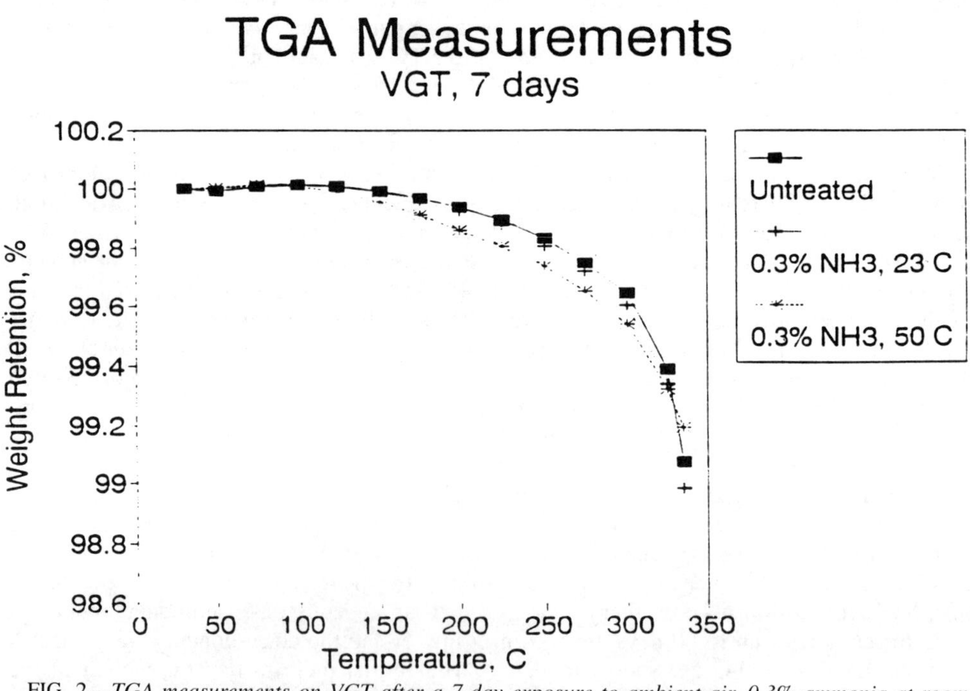

FIG. 2—*TGA measurements on VGT after a 7-day exposure to ambient air, 0.3% ammonia at room temperature, and 0.3% ammonia at 50°C.*

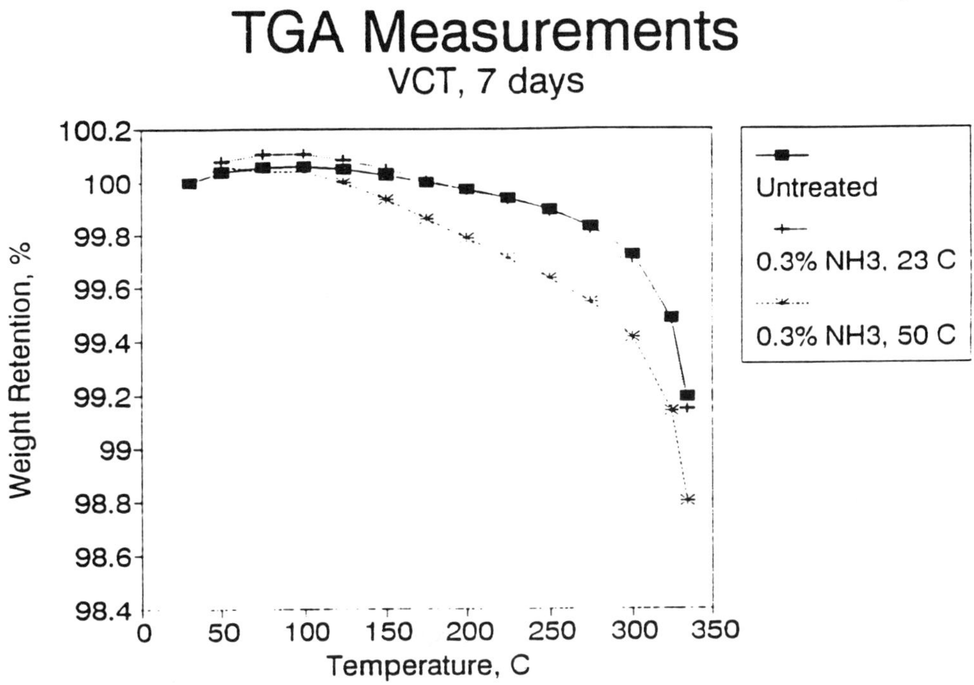

FIG. 3—*TGA measurements on VCT after a 7-day exposure to ambient air, 0.3% ammonia at room temperature, and 0.3% ammonia at 50°C.*

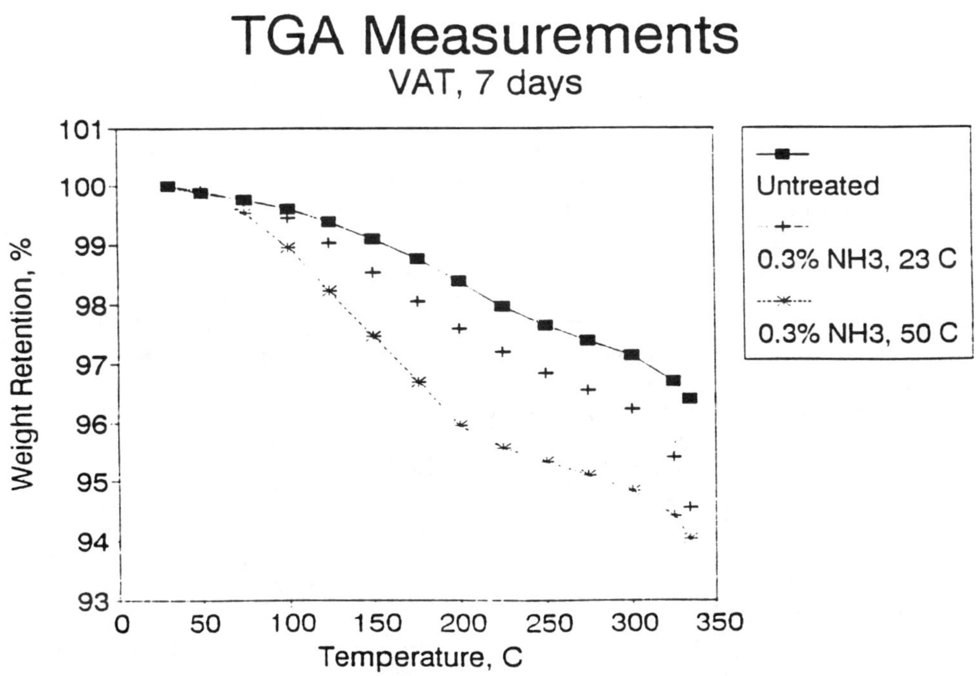

FIG. 4—*TGA measurements on VAT after a 7-day exposure to ambient air, 0.3% ammonia at room temperature, and 0.3% ammonia at 50°C.*

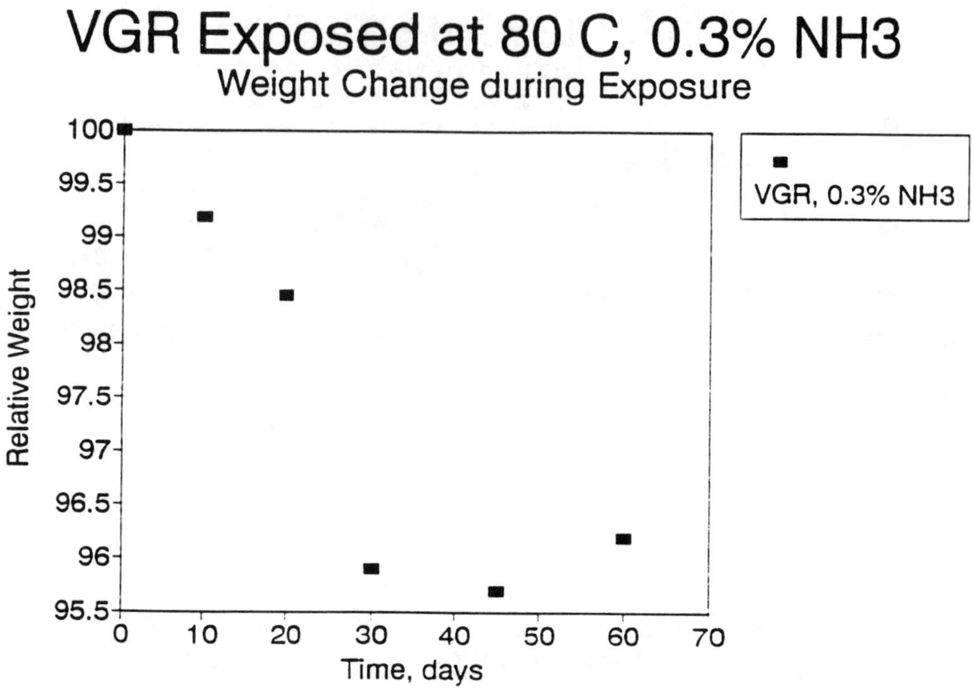

FIG. 5—*Changes in weight of VGR samples as a result of exposure to 0.3% ammonia at 80°C as a function of exposure time.*

the weight loss between 150 and 300°C increases. However, the increase in weight loss during the exposure, which is representative of the extent of fiber dissolution, as a function of time, is more gradual than the corresponding increase in weight loss between 150 and 300°C, which reflects the extent of matrix depolymerization [1].

Several TGA measurements were run on VGR (vinylester/glass rod) and VGP (vinylester/glass plate) materials previously exposed to various basic and acidic media at 80°C. The results obtained for the weight change during the exposure and for the relative weight retention between 150 and 300°C in various tests are shown in Figs. 7 and 8, respectively. The results of the weight change measurements show that significant weight loss during the exposure took place only when the composite was exposed to basic media, in particular at the higher ammonia concentration (3%), and not to an acidic solution. This agrees with the assumption that this weight loss represents the extent of dissolution of the glass fibers. The weight loss during the exposure was higher the higher the basicity of the solution. The weight loss during the exposure in the case of the rod material (VGR) was higher than in the case of the plate material (VGP), resulting in a smaller WE in the case of VGR. The results of the TGA run showed that the weight loss upon heating from 150 to 300°C was highest in the case of the buffered acetic acid solution, indicative of this solution being the most effective in causing hydrolytic depolymerization of the matrix. Such hydrolytic depolymerization has also been directly observed using spectroscopic techniques, such as Raman spectroscopy [1] and infrared spectroscopy [13]. Furthermore, WR, the weight retention between 150 and 300°C, was lower in the case of the rod material (VGR) than in the case of the plate material (VGP). In other words, the weight loss between 150 and 300°C

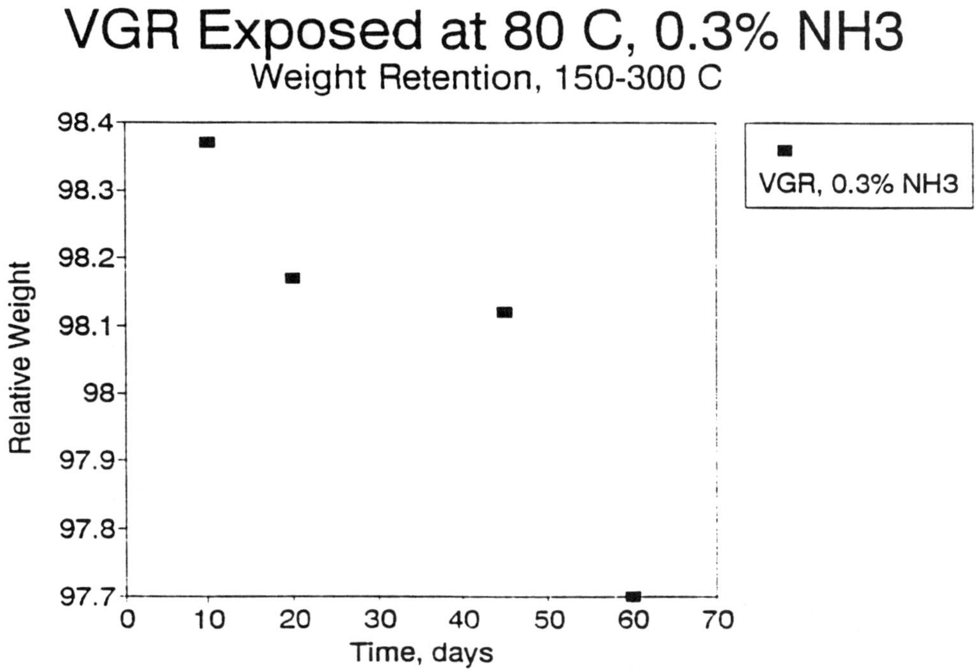

FIG. 6—*Relative weight retention between 150 and 300°C of VGR samples as a result of exposure to 0.3% ammonia at 80°C as a function of exposure time.*

was larger in the case of VGR than in the case of VGP. As mentioned above, the weight loss during the exposure (WE) was also larger in the case of VGR than in the case of VGP. This is probably due to the fact that in the case of the VGP material the composite is surrounded by a relatively thick layer of pure resin. (See the discussion of the DSC results below.)

The results of exposure to an aqueous solution of 0.3% ammonia on five different types of composites are shown in Figs. 9 and 10. In the case of VGR, tests were run both on samples with the usual diameter of 6.4 mm and on large-diameter (12.7-mm) rods, designated VGR-LD. The results show that the larger the diameter of the rods, the smaller the weight loss during the exposure. In the case of a thicker rod, the average distance of the fibers from the surface of the rod increases, the fibers become more protected by the matrix, and the pH rise in the center of the rod is expected to be more gradual. On the other hand, the weight loss upon heating from 150 to 300°C is larger in the case of the thicker rod. As expected, VAT (vinylester/aramid tendon), VCT (vinylester/carbon tendon), and ECT (epoxy/carbon tendon) do not exhibit weight loss during exposure, as the fibers with which they are reinforced are not made out of alkali-soluble silicate glass. Instead, these composites exhibit some weight gain during the exposure, especially in the case of VAT, reflecting the absorption of aqueous solution during the exposure. In contrast, upon heating from 150 to 300°C, VAT exhibits the largest relative weight loss by far. These results are an example of a large extent of water absorption being associated with extensive hydrolytic depolymerization. As mentioned above, the exposure studied in this set of tests consisted of immersion in a 0.3% aqueous solution of ammonia. However, very similar trends were observed when the exposure conditions consisted of immersion in de-ionized water at room temperature. In

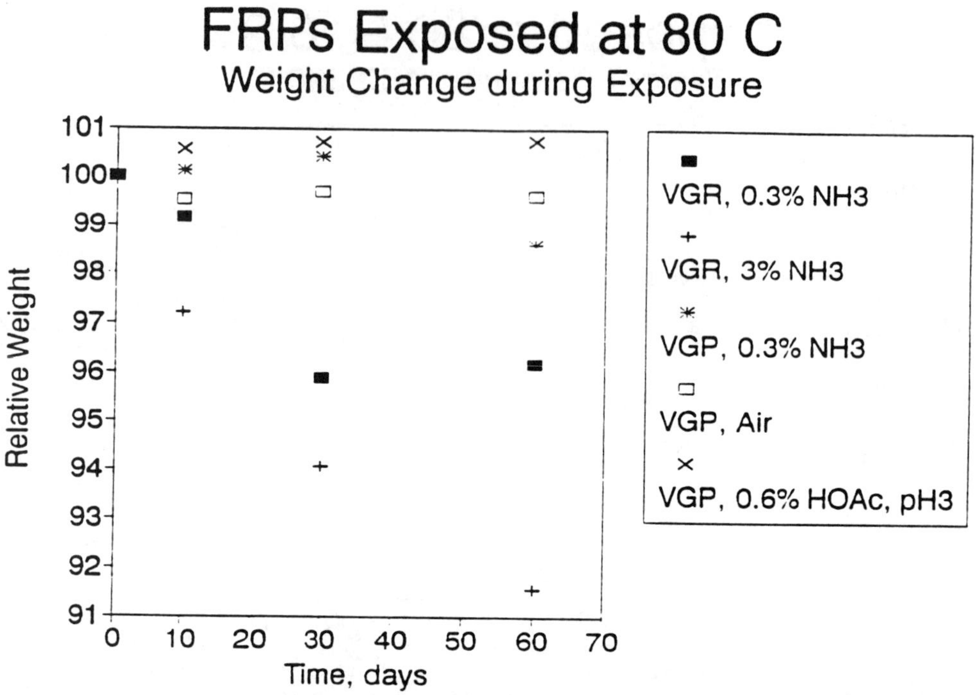

FIG. 7—*Changes in weight of VGR and VGP samples as a result of exposure to basic and acidic media at 80°C as a function of time.*

addition, comparison between results obtained with VGP and PGR exposed to de-ionized water at room temperature and at 50°C, respectively, showed that the increased temperature caused, on one hand, an increase in the amount of water absorbed by the composites and, on the other, an increased weight loss upon subsequent heating from 150 to 300°C, indicative of more extensive hydrolytic depolymerization.

A third series of tests was added to the first two with a view to examining the effects of a broader range of conditions and of longer periods of exposure to aqueous environments on the thermochemical and mechanical properties of FRPs. The results of exposure to de-ionized water for 28 days on the properties of two FRPs, viz. PGR (polyester/glass rod) and VGP (vinylester/glass plate), are shown in Table 2. The thermochemical properties listed are the weight after exposure relative to the original weight (WE) and the weight retention (WR) between 150 and 300°C. The reported mechanical properties include tensile strength (TS), tensile modulus (TM), flexural strength (FS), flexural modulus (FM), short beam shear strength (SBS), and diametral compression strength (DC). It can be seen that the effect of temperature is very significant. Raising the temperature of the exposure from 23 to 50°C, and, in particular, from 50 to 80°C, results in large changes in the values of the mechanical properties, especially those which are matrix-dependent (SBS and DC). These changes are more noticeable in the case of PGR than in the case of VGP. Among the mechanical properties, measurements related to strength (TS, FS, SBS, DC) are much more affected by exposure to aqueous environments than those representing moduli (TM, FM).

The thermochemical and mechanical properties of PGR (Table 3) and VGP (Table 4) were also studied as a function of the composition of the solvent (the aqueous phase) at a typical

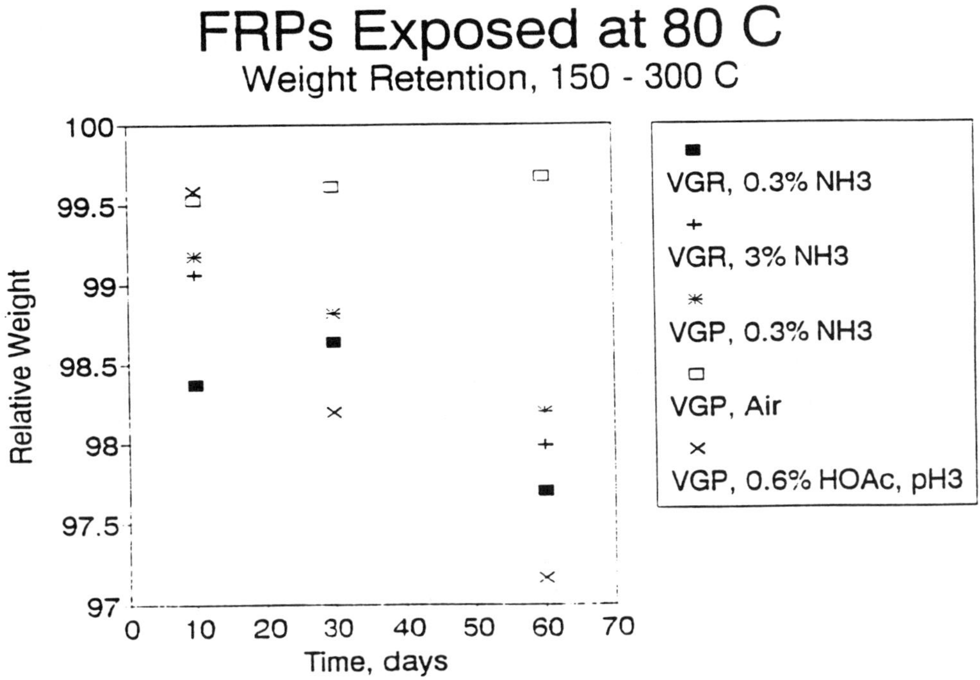

FIG. 8—*Relative weight retention between 150 and 300°C of VGR and VGP samples as a result of exposure to basic and acidic media at 80°C as a function of time.*

service temperature (23°C). The results of these measurements indicate that larger extent of water absorption and larger weight loss between 150 and 300°C correlate well with greater degradation of mechanical properties. Ammonia solutions have the largest effects, and 3% ammonia has larger effects than 0.3% ammonia solution. The results tabulated in Table 5 show that this is also true when the tests are accelerated by conducting them at a temperature of 80°C. In the latter case, the differences among various aqueous media are less pronounced as the degradation observed in all the solutions becomes magnified due to the use of a higher temperature.

The results of measurements carried out on various types of FRP materials in de-ionized water at a typical service temperature of 23°C over periods as long as 224 days are shown in Figs. 11 and 12. The data indicate that, in general, larger water absorption correlates with a larger loss of weight upon heating from 150 to 300°C, in agreement with the trend reported in Ref *1*. Among the various FRP materials tested, the one which shows the largest sensitivity to exposure to aqueous environments is VAT. The effects on VGP are much smaller and the effects on the other three components, at the test temperature of 23°C, are very small. Tests on samples previously exposed under accelerated conditions, that is, at a temperature of 80°C, exhibit the same trends and also show that the effects on the mechanical properties, like those on the thermochemical properties, are larger in the case of VAT than in the cases of the other FRP materials, as shown in Table 6.

The data summarized above indicate that, in general, raising the temperature of the exposure to 80°C has larger effects on the properties of various FRPs than changing the composition of the aqueous medium. DSC measurements were carried out on VGP stored at

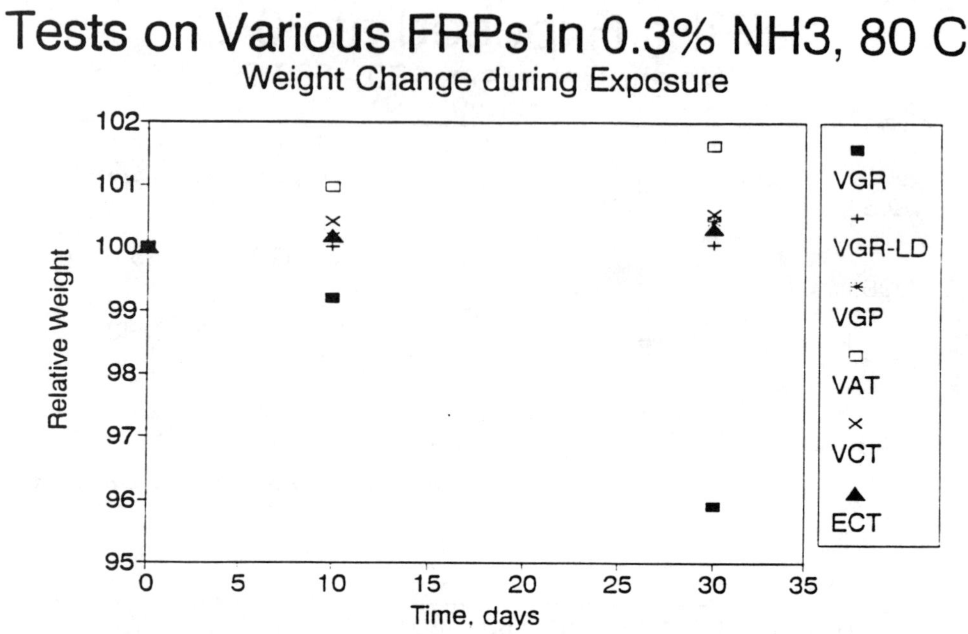

FIG. 9—*Changes in weight of various types of FRPs as a result of exposure to 0.3% ammonia at 80°C as a function of time.*

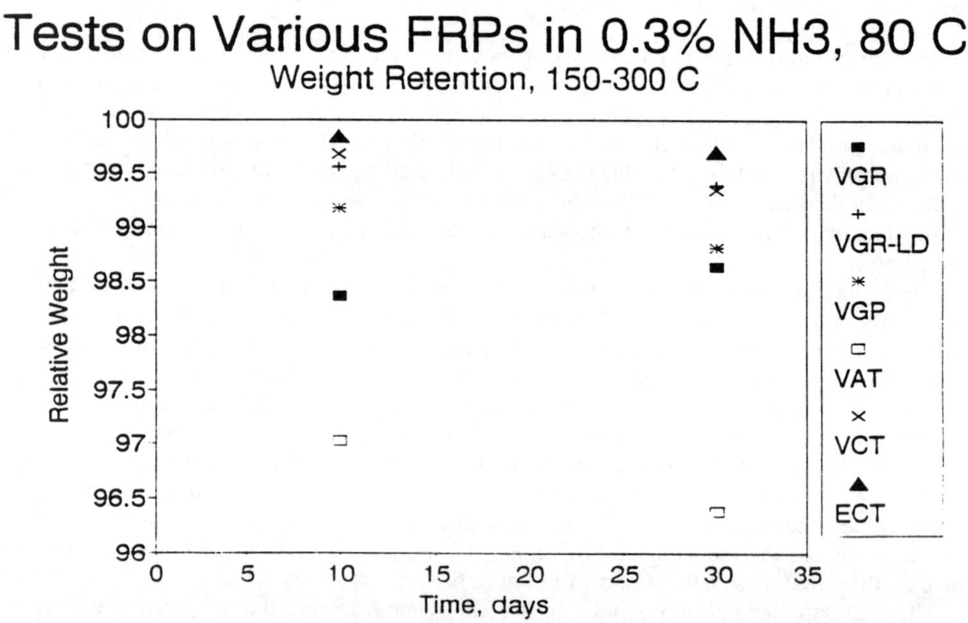

FIG. 10—*Relative weight retention between 150 and 300°C of various types of FRPs as a result of exposure to 0.3% ammonia at 80°C as a function of time.*

TABLE 2—*Temperature effect on the change in the properties of FRP materials upon exposure to deionized water for 28 days. Values of properties (except WR) relative to unexposed samples, %.*

	PGR			VGP		
Property	23°C	50°C	80°C	23°C	50°C	80°C
WE	99.99	100.32	100.88	100.14	100.25	99.80
WR	99.46	99.26	98.03	99.43	99.02	99.88
TM	100.2	102.0	100.7	98.2	100.0	101.4
TS	96.9	89.6	24.1	98.3	92.8	69.4
FM	99.6	100.6	102.3	97.9	90.5	91.6
FS	96.4	97.0	42.7	98.2	88.6	68.2
SBS	99.1	103.3	59.2	96.2	90.9	76.0
DC	94.1	90.4	40.8	n/d	n/d	n/d

NOTE—n/d = not determined.

23°C and on VGP held for 28 days in air at 80°C. The VGP was sectioned into three regions before the DSC measurements. These regions included the outer surface, the intermediate region, and the core. Each measurement was repeated five times, each time with a different sample of the same material. The results are shown in Table 7. They indicate that the surface region has a higher T_g than the inside of the material, and that the material shows a decrease in T_g (by about 3.5 K at the surface and 1.5 K in the inner regions) as a result of exposure to a temperature of 80°C for 28 days. ΔC_p was higher for the surface, lower for the core, and lowest for the intermediate region. ΔC_p did not exhibit a significant change as a result of thermal exposure.

Discussion

The results of the three series of experiments described above give rise to several important conclusions concerning the characterization of the environmental degradation of FRP materials. Thermogravimetric analysis has been shown to be an important indicator of the extent

TABLE 3—*Solvent effect on the change in the properties of PGR upon exposure to various solvents at 23°C for 28 days. Values of properties (except WR) relative to unexposed samples, %.*

	Solvent			
Property	De-ionized Water	0.3% NH_3	3% NH_3	0.6% HOAc pH 4.93
WE	100.01	100.03	100.39	99.97
WR	99.46	99.42	99.34	99.44
TM	100.2	n/d	104.3	99.3
TS	96.9	n/d	82.8	92.7
FM	99.6	n/d	99.1	100.9
FS	96.4	n/d	85.4	97.8
SBS	100.3	n/d	98.5	101.2
DC	94.1	n/d	73.1	95.2

NOTE—n/d = not determined. WE = weight after exposure/weight before exposure; WR = weight at 300°C/weight at 150°C; TM = tensile modulus; TS = tensile strength; FM = flexural modulus; FS = flexural strength; SBS = short beam shear strength; DC = diametral compression strength.

TABLE 4—*Solvent effect on the changes in the properties of VGP upon exposure to various solvents at 23°C for 28 days. Values of properties (except WR) relative to unexposed samples, %.*

	Solvent			
Property	De-ionized Water	0.3% NH_3	3% NH_3	0.6% HOAc pH 4.93
WE	100.14	100.10	100.15	100.13
WR	99.43	99.31	98.58	99.47
TM	98.2	100.2	97.0	101.8
TS	98.3	88.6	88.6	100.0
FM	97.9	86.9	82.7	96.3
FS	98.2	87.4	75.7	96.4
SBS	96.2	88.0	78.6	94.1

NOTE—WE = weight after exposure/weight before exposure; WR = weight at 300°C/weight at 150°C; TM = tensile modulus; TS = tensile strength; FM = flexural modulus; FS = flexural strength; SBS = short beam shear strength.

of degradation. The weight loss at elevated temperatures (specifically, between 150 and 300°C) primarily reflects the extent of depolymerization of the matrix during the preceding exposure to an aqueous environment. It is generally higher the larger the extent of water absorption during the exposure. Most significantly, the TGA weight loss generally increases with increasing degradation of the various mechanical strength parameters of an FRP material. (Under the same conditions, tensile and flexural moduli are not greatly affected by environmental exposure, most probably because they reflect the structure of the entire specimen rather than that of the regions most exposed to attack [*1*].)

The extent of environmental degradation depends on the temperature and composition of the composite as well as on the nature of the surrounding medium. In basic media, such as ammonia solutions, weight loss rather than weight gain may be observed during the exposure period in the cases of composites based on silicate glass fibers because the absorption of water is offset by dissolution of fibers. This effect is larger the higher the basicity of the solution. Conversely, the largest TGA weight loss is observed in buffered acetic acid solutions, that can be expected to cause the most extensive matrix depolymerization. Matrix

TABLE 5—*Solvent effect on the change in the properties of VGP upon exposure to various solvents at 80°C for 7 and 28 days. Values of properties (except WR) relative to unexposed samples, %.*

	7 days			28 days		
Property	Deionized Water	3% NH_3	0.6% HOAc pH 4.93	Deionized Water	3% NH_3	0.6% HOAc pH 2.88
WE	100.83	99.90	99.93	99.80	101.88	99.91
WR	98.50	98.74	99.17	98.88	96.32	98.72
TM	97.8	93.0	89.4	101.4	92.2	95.0
TS	74.6	73.1	75.9	69.4	60.9	59.5
FM	76.4	77.5	83.8	91.6	79.6	91.1
FS	80.7	74.2	78.1	68.2	65.4	68.4
SBS	80.4	66.8	75.7	76.0	62.0	73.6

NOTE—WE = weight after exposure/weight before exposure; WR = weight at 300°C/weight at 150°C; TM = tensile modulus; TS = tensile strength; FM = flexural modulus; FS = flexural strength; SBS = short beam shear strength.

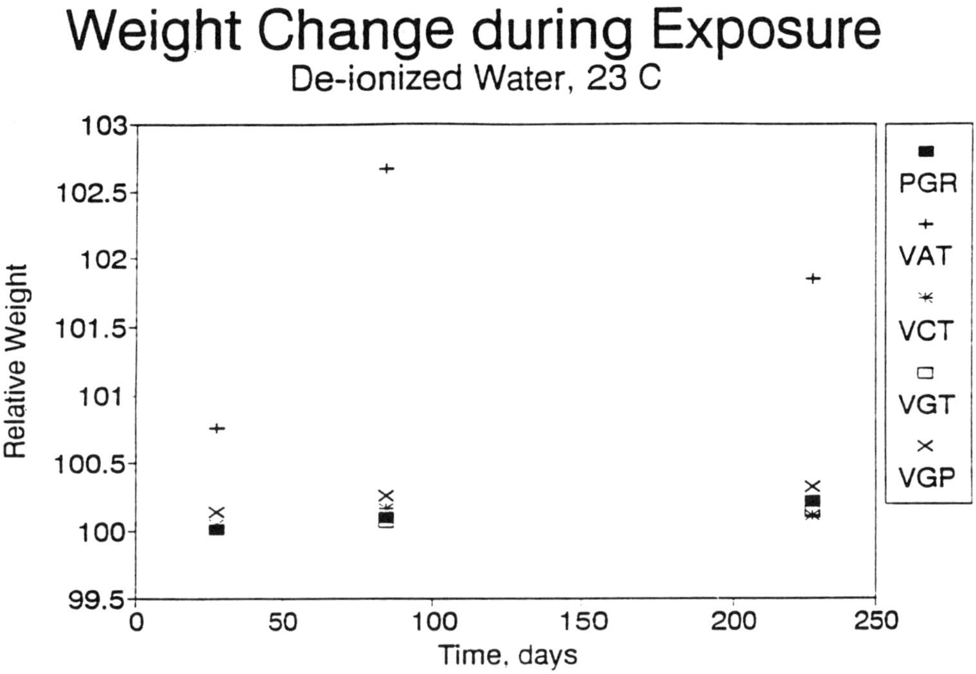

FIG. 11—*Changes in weight of various types of FRPs as a result of exposure to de-ionized water at 23°C as a function of time over prolonged periods.*

depolymerization proceeds most rapidly during the initial period of exposure, reflecting the Fickian nature of water absorption [*2,10*], while fiber dissolution appears to be more gradual, probably due to the linear kinetics of silicate glass dissolution in water prevailing almost since the beginning of the exposure [*11*]. The Fickian nature of the attack on the matrix is also reflected in the decrease in the extent of degradation with increasing diameter of the specimen. The effect of temperature is very noticeable. Raising the temperature from 23 to 50°C gives rise to a pronounced increase in the extent of change of both the thermochemical and mechanical properties, and further raising the temperature has an even larger effect. It is encouraging to note that the trends observed upon short-term exposure at 80°C with respect to the dependence of property degradation on the composition of the material and of the attacking medium parallel the trends observed over longer periods of time at room temperature. Thus, the use of elevated temperature at least up to 80°C appears to be an effective method of test acceleration.

Among various FRPs, VAT is the most susceptible to environmental attack. This can be attributed to the tendency of the aramid fibers to absorb and transmit water [*12*]. This results in rapid hydrolytic attack on the fibers as well as on the matrix. In general, the results of tests on specimens of different materials should be qualified by noting that the characteristics of a given specimen are influenced not only by the nature of the fiber and matrix materials, but also by the manufacturing technique, which determines factors such as porosity [*1*], volume fraction of fibers, and the presence of fillers.

The fact that at elevated temperatures FRP materials interact much more strongly with their environment is also reflected in the DSC results. In the case of VGP, it is observed that the outside surface has a higher T_g than the bulk, probably because it cools more rapidly

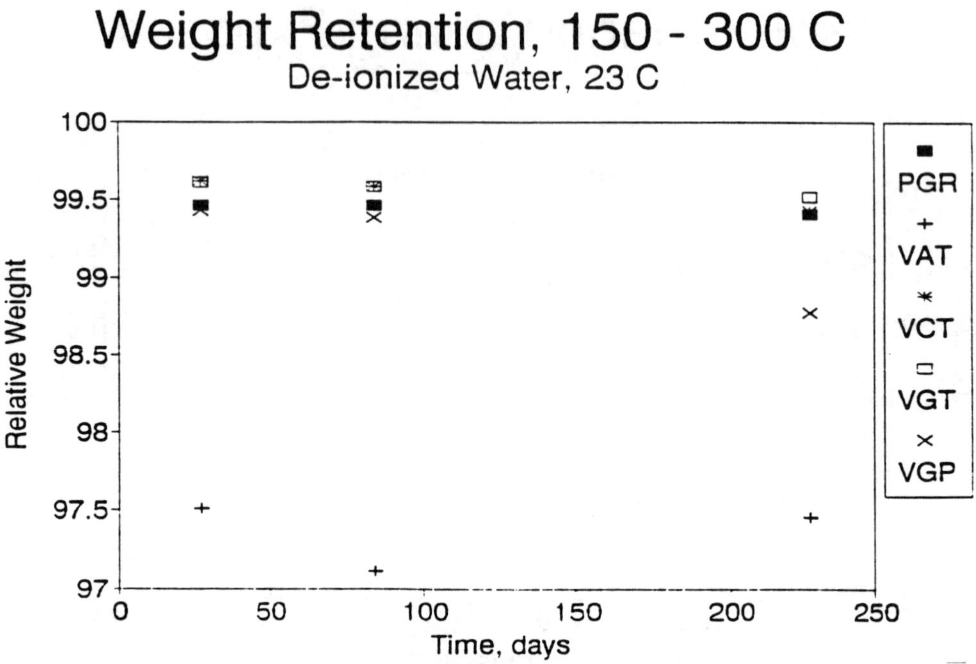

FIG. 12—*Relative weight retention between 150 and 300°C of various types of FRPs as a result of exposure to de-ionized water at 23°C as a function of time over prolonged periods.*

during the manufacturing process. This assumption is supported by the observation that prolonged heating of the material at 80°C causes a decrease in T_g in all regions of the material. The decrease in T_g upon heating, like the increase in the amount of volatile components shown by the TGA measurements (see above) may be attributed to thermally induced

TABLE 6—*Effects of aqueous media (de-ionized water and 0.6% acetic acid buffer, pH 2.88) on the properties of various FRPs upon exposure at 80°C for 14 days. Values of properties (except WR) relative to unexposed samples, %.*

	De-ionized Water			0.6% HOAc, pH 2.88		
Property	VAT	VCT	VGT	VAT	VCT	VGT
WE	103.86	100.56	100.37	102.56	101.39	100.33
WR	97.77	99.08	99.29	97.67	99.08	99.25
TM	94.3	96.8	99.8	92.4	97.2	102.4
TS	125.6	110.6	99.9	121.1	109.3	83.9
FM	99.4	89.6	101.3	98.0	103.9	100.6
FS	82.7	87.5	97.3	79.3	104.2	96.9
SBS	59.0	160.3	116.5	60.0	160.0	110.7
DC	59.6	128.0	88.2	65.4	122.8	89.6

NOTE—WE = weight after exposure/weight before exposure; WR = weight at 300°C/weight at 150°C; TM = tensile modulus; TS = tensile strength; FM = flexural modulus; FS = flexural strength; SBS = short beam shear strength; DC = diametral compression strength.

TABLE 7—*Effect of thermal exposure at 80°C for 28 days in air on the glass transition temperature and on the associated change in heat capacity.*

Specimen	T_g, °C	ΔC_p, J·K^{-1}·g^{-1}
Unheated, surface	119.2 ± 1.9	0.13 ± 0.01
Unheated, intermediate zone	113.8 ± 2.4	0.05 ± 0.02
Unheated, core	114.4 ± 1.7	0.08 ± 0.02
Heated, surface	115.8 ± 0.5	0.12 + 0.01
Heated, intermediate zone	112.4 ± 2.4	0.05 ± 0.02
Heated, core	112.7 ± 1.3	0.08 ± 0.01

depolymerization of the resin matrix. In general, plasticization and hydrolysis of the matrix facilitate chemical degradation, thus accounting for both the DSC and the TGA observations [*13*]. The magnitude of the change in heat capacity during the glass transition, ΔC_p, is also useful as it is indicative of the volume fraction of the polymeric resin undergoing the glass transition, and thus of the volume fraction of fibers. In the case of VGP, ΔC_p is highest in the case of the surface region, which probably contains close to 100% of resin. ΔC_p is lower in the core, which appears to contain, based on the DSC results, approximately 30% of fibers, and is lowest in the intermediate region, which appears to contain about 60% of fibers.

In summary, the thermochemical measurements reported here appear to be quite useful in shedding light on the nature of the degradation processes and in laying the basis for the development of accelerated tests of FRP degradation using accelerating factors such as elevation of the temperature. Further tests are currently in progress with a view to validating this conclusion, establishing more quantitative relationships between changes in thermochemical properties and in mechanical properties in the course of environmental exposure, and formulating the conditions for accelerated testing. More extensive and more accurate studies are necessary in order to establish a quantitative relation between accelerated and real-time aging. More multiplicate measurements will be carried out in order to quantify standard deviations. In particular, more quantitative control of moisture content before and after environmental exposure and determination of the equilibrium water content would improve the precision and usefulness of the weight measurements.

Acknowledgments

The authors are grateful to the Federal Highway Administration for support under Contract DTFH61-93-C-00012. Funding for the DSC and accessory equipment for the TGA was provided by the National Science Foundation. Valuable support for the investigating group was provided by Consolidated Edison Company of New York, Inc. Invaluable assistance and encouragement was provided by E. Munley (FHWA), A. H. Zurieck (Georgia Institute of Technology), E. Bryan (NSF), Br. P. Ellis (CUA), G. Cherepakhov (Con Edison), and R. D. Varrin, Jr. (Dominion Engineering).

References

[*1*] Barkatt, A., Bank L. C., Gentry, T. R., Prian, L., Shan, R., Sang, J. C., and Pollard, R., "Environmental Degradation of Fiber Reinforced Plastic Materials in Neutral, Acidic, and Basic Aqueous Solutions," Paper 138, *Corrosion/95*, National Association of Corrosion Engineers, Houston, TX, 1995.

[2] Springer, G. S., "Environmental Effects," in *Environmental Effects on Composite Materials*, Vol. 3, G. S. Springer, Ed., Technomic, Lancaster, PA, 1988, pp. 1–34.

[*3*] Guetta, B., Bunsell, A. R., Belinski, C., and Lalau Keraly, F., "Effect of Thermal Spiking on Carbon Toray 300B Reinforced Polystyrylpyridine Composites: A Kinetic, Spectroscopic and Mechanical Study," *Composites*, Vol. 20, 1989, pp. 461–465.

[*4*] Diamond, S., "Effects of Two Danish Flyashes on Alkali Contents of Pore Solutions of Cement-Flyash Pastes," *Cement and Concrete Research*, Vol. 11, 1981, pp. 383–394.

[*5*] Gerritse, A., "Durability Criteria for Non-Metallic Tendons in an Alkaline Environment," *Advanced Composite Materials in Bridges and Structures*, K. W. Neale and P. Labossiere, Eds., CSCE, Montreal, 1992, pp. 129–138.

[*6*] Mijovic, J., "Interplay of Physical and Chemical Aging in Graphite/Epoxy Composites," *Journal of Composite Materials*, Vol. 19, 1985, pp. 178–191.

[*7*] Brady, R. L. and Porter, R. S., "Interfacial Adsorption and Crystallization of Polycarbonate in Carbon Fiber Composites," *Journal of Applied Polymer Science*, Vol. 39, 1990, pp. 1873–1885.

[*8*] Hoa, S. V., Lin, S., and Chen, J. R., "Hygrothermal Effect on Mode II Interlaminar Fracture Toughness of a Carbon/Polyphenylene Sulfide Laminate," *Journal of Reinforced Plastics and Composites*, Vol. 11, 1992, pp. 3–31.

[*9*] "Method 1311 — Toxicity Characteristic Leaching Procedure (TCLP)," *Code of Federal Regulations*, Title 40 — Environmental Protection Agency, Part 261, Appendix II, 1990.

[*10*] Harper, J. F. and Naeem, M., "The Moisture Absorption of Glass Fibre Reinforced Vinylester and Polyester Composites," *Materials and Design*, Vol. 10, 1989, pp. 297–300.

[*11*] Rana, M. A. and Douglas, R. W., "The Reaction Between Glass and Water. Part I. Experimental Methods and Observations," *Physics and Chemistry of Glasses*, Vol. 2, 1961, pp. 179–205.

[*12*] Wang, J. Z., Dillard, D. A., Wolcott, M. P., Kamke, F. A., and Wilkes, G. L., "Transient Moisture Effects in Fibers and Composite Materials," *Journal of Composite Materials*, Vol. 14, 1990, pp. 994–1009.

[*13*] Ghorbel, I. and Valentin, D., "Hydrothermal Effects on the Physico-Chemical Properties of Pure and Glass Fiber Reinforced Polyester and Vinylester Resins," *Polymer Composites*, Vol. 14, 1993, pp. 324–334.

Jon Shively[1] Charles Miglionico,[2] Robert Roybal,[2] Thomas King,[2] Robert Robertson,[2] James Baird,[3] Steven Davis,[2] and Charles Stein[2]

Combined Effects of the Low-Earth-Orbit Environment on Polymeric Materials

REFERENCE: Shively, J., Miglionico, C., Roybal, R., King, T., Robertson, R., Baird, J., Davis, S., and Stein, C., **"Combined Effects of the Low-Earth-Orbit Environment on Polymeric Materials,"** *High Temperature and Environmental Effects on Polymeric Composites: 2nd Volume, ASTM STP 1302*, Thomas S. Gates and Abdul-Hamid Zureick, Eds., American Society for Testing and Materials, 1997, pp. 223–242.

ABSTRACT: Understanding the phenomenology of the interaction of the space environment with organic composite spacecraft surfaces is both interesting science and important to the longevity of our space assets. The importance of this understanding has been made even more critical by the desire to use new, organic composite materials whose lower weight and higher stiffness will make it possible to orbit larger payloads. Inconsistencies between ground simulation and space test data clearly indicate that our understanding of the interaction phenomenology is inadequate. Preliminary research carried out by the Phillips Laboratory has demonstrated that the *simultaneous* interaction of all of the space environmental factors is often required to obtain agreement between ground- and space-test data. This paper describes the test equipment designed by Phillips Laboratory to subject composite materials to a simulated low-earth-orbit environment consisting of simultaneous fluxes of hyperthermal neutral oxygen atoms, energetic electrons, vacuum ultraviolet light, and hypervelocity-particle impacts characteristic of space debris and micrometeorites. Preliminary results of studies of the combined effects of fluxes of atomic oxygen and electrons on polysulfone resins and composites, and an evaluation of impact damage to polysulfone resins and composites from small aluminum particles traveling at velocities up to 6 km/s, are presented. The need for testing standards and methods for measuring space environmental effects is assessed.

KEY WORDS: polymeric composites, space environmental effects, atomic oxygen, erosion damage, electron damage

The Space Environment

The environment experienced by spacecraft in orbit depends on the distance from the earth. Low-earth orbit pertains to orbits in the range of 300 to 1000 km from the surface of the earth while geosynchronous orbits are approximately 35 000 km. The environments at these two orbital locations have similarities and essential differences. For the purposes of this paper we will focus on low-earth orbit (LEO), and when it is appropriate, point out where the results may apply to geosynchronous orbits.

The environment in LEO is not benign, especially when polymer-based materials are considered for use. While the exact details of the environment depend upon the specific

[1] , CIAM Department, California State University at Northridge, Northridge, CA 91330.
[2] , , , , , , respectively, Phillips Laboratory, Kirtland Air Force Base, NM 87117.
[3] , Chemistry Department, University of Alabama-Huntsville, Huntsville, AL.

orbit, distances, shape, and orientation, the environment can be generally characterized by several major factors:

- Atomic oxygen particles
- Electromagnetic radiation, especially ultraviolet
- Charged particle radiation, electrons and protons
- Collisions with space debris and micrometeorites
- Temperature cycles
- Magnetic fields
- Partial vacuum (10^{-2} Pa or lower)

While the effects of individual components of the LEO environment have been studied, only recently have researchers focused on the possibility of synergism between components that may be different than the individual effects studied separately. The purposes of this paper is:

1. to review the environment associated with LEO including the results of the spacecraft: the Long Duration Exposure Facility (LDEF) and terrestrial studies on spacecraft materials,
2. to identify mechanisms that potentially cause chemical, mechanical, or physical changes,
3. to describe the Phillips Laboratory Space Environmental Interactions Facility, and
4. to present results of an initial study, conducted using JPL facilities, of the combined effects of atomic oxygen and electrons.

The conditions in the environment in LEO are complex and change daily. Several excellent references present the details of an LEO environment and the reader is referred to one of these for further information [*1*]. For purposes of assessing effects of the space environment in the laboratory, average values of the conditions are frequently used. These conditions are presented in Table 1.

Experience with Polymers in a Space Environment

Information about the performance of materials in LEO comes from two major sources, the Long Duration Experiment Facility (LDEF) and from ground-based testing. LDEF spent

TABLE 1—*Space environment and interactions in a typical LEO.*

Environmental Factor	Values	Interaction with Materials
Atomic oxygen	$E = 5$ eV $\Phi = 10^{-14}/\text{cm}^2/\text{s}$	erosion of surfaces
Vacuum ultraviolet light	1 sun	chemical changes
Thermal cycles	213 to 288 °K 16 cycles/day	materials degradation and cracking
Electrons and ions		polarization and charging
Vacuum	10^{-2} Pa or lower	outgassing
Space debris impacts	7 to 20 km/s; wide range of masses.	physical damage, contamination

69 months in space before it was retrieved in January 1990. Since then, many major research laboratories have carefully studied specimens and hardware components from LDEF for evidence of space environmental interactions. The results of their findings are summarized in two NASA reports [*2,3*]. Ground-based testing has not been able to simultaneously simulate all aspects of the space environment. However, testing has definitely contributed to a better understanding of the mechanisms that operate in space [*4–8*].

The LDEF environment is summarized in Table 2. The conditions for LDEF are not constant because the orbit decayed from 482 to 340 km when it was retrieved by the Space Shuttle. Because the conditions change with orbits, the values in Table 2 reflect averages over the 69 months in orbit. Observations about the behavior of organic composite materials are summarized in Refs *9* and *10*.

The observations of polymers retrieved from LEO strongly indicate that the space environment interacts with polymeric materials [*3,8*]. Unprotected polymers exhibit erosion damage from the flux of atomic oxygen particles, (AO), which react chemically with the polymer species. The reaction leads to gaseous products that are expelled from the surface. As atoms are extracted from the surface, the surface recedes. In the case of very thin materials, the erosion could lead to their complete disappearance. In addition, the reaction of polymeric materials with oxygen atoms is expected to be enhanced by ultraviolet (UV) radiation [*5*]. The atomic oxygen reactions can be minimized if a protective coating is present on the outside surface of the polymer [*11*]. However, if the coating is breached by a collision with a hypervelocity particle, then erosion can proceed locally.

The mechanical properties of polymers have been observed to be degraded by exposure to the environment in LEO [*3*]. The degradation leads to a reduction in tensile and bending strengths and embrittlement. The loss, while slight, is attributed to microcracking and surface damage from the exposure to space in LEO.

There is considerable evidence found on LDEF for contamination of surfaces [*2*]. While the source of major contamination was from oils and adhesive resins, some polymeric materials are expected to outgas in the vacuum of space. In addition, the source of contamination can also come from hypervelocity collisions by small particles of space debris and/or micrometeorites that will produce vapor plumes in the process of forming craters on polymeric surfaces. The plumes contaminate by spreading material on the satellite surfaces nearby [*12*].

Another source of concern is collisions by space debris and micrometeorites, which produce impact craters on the surfaces of materials. The craters may puncture any protective coating that is present. In addition to the ejecta produced by the cratering processes, the collisions generate shock waves that travel into the material and produce subsurface damage [*12*].

Another observation seen in materials retrieved from space are microcracks in the matrix of composite and in polymer materials. The cracks are thought to be the result of vacuum

TABLE 2—*LDEF space environment.*

Condition	Parameter
Orbit range	482 to 340 km
Time	April 1984 to Jan. 1990
Atomic oxygen fluence	9×10^{21} atoms/cm^2
Solar UV	11 000 h
Thermal cycles	30 000 90 min cycles
Impacts	35 000 craters
Vacuum	0.1 − 0.01 Pa

ultraviolet (VUV) radiation attack together with thermal cycling [*9*]. Also, thermal control coatings have exhibited degraded emissivity, due to contamination in the form of straining. The staining is a direct result of VUV, but could be enhanced by photochemical reactions or by interactions with AO.

What are the Potential Effects on Materials in LEO

AO Reactions

The reaction of atomic oxygen with polymers is complex. In the first place, the reaction involves a combination of chemical bonding and physical removal processes. In addition, VUV and/or electrons may play a role in breaking bonds in the polymer chain by affecting the activation energy for the process.

Terrestrial testing has provided some insights into the process of erosion by AO. The rate of removal of atoms appears to be linear, for a constant AO flux [*13*]. This result was also borne out by observation from LDEF in which identical specimens had erosion rates proportional to the AO flux, which depended on location on LDEF. Thus it seems that the amount of material removed from a polymer is directly proportional to the product of the flux of AO particles and the time, or fluence. Laboratory tests also show that AO erosion rates depend on the incident energy of the AO particles, up to an energy of about 5 eV where the energy effect appears to saturate [*8*]. This is not too surprising, considering the size of the bond energies involved with most mers. However, there is also some evidence that shows that the rate of removal varies with the flux when the incident collision rate is very high, 10^{18} $cm^{-2}s^{-1}$ [*14*].

Photochemical Reactions

Polymers are susceptible to photodegradation that can lead to chain scissions, crosslinking of chains, and secondary oxidation processes. The photooxidative behavior of polymers uses a free radical mechanism in which a chromophoric group absorbs electromagnetic energy and produces low molecular weight radical and/or polymeric radical. The initiation step involves the combination of oxygen with the low molecular weight radical or with the polymer radical. The propagation then results in a chain scissions or crosslinks [*15*].

The space environment in LEO contains solar radiant energy in the 100 to 200 nm wavelength region that is essentially absent on the surface of the earth as a result of absorption by the earth's atmosphere. Thus, UV-resistant polymers on earth may not be resistant in space. For example, VUV radiation of wavelengths near 180 nm are absorbed in a thin layer on the surface of the polymer. Because the bonds break quickly and the products are ejected because their specific volume exceeds that of the mer. The ejected radicals combine with the oxygen [*15*]. In addition to the VUV wavelength, the processes of degradation also depend on the intensity of the electromagnetic radiation.

Charged Particle Radiation

Charged particles, largely emitted by the sun, are trapped by the earth's magnetic field and may subsequently collide with the satellites. The collision by the particles leads to a charging of a satellite. This mechanism operates in geosynchronous orbits far outside the earth's atmosphere. In LEO, the presence of a thin atmosphere limits the flux of charged particles, except in polar orbits. In LEO, the photoelectric effect can produce a charge build-up on a satellite.

Phillips Laboratory Space Simulation Facility

Phillips Laboratory has established a facility to systematically and simultaneously study the effects of AO, VUV, charging, protons, thermal cycling, and hypervelocity particle impacts on the behavior of spacecraft materials. The facility consists of a vacuum chamber that has a source of AO particles with an energy of 5 eV, which corresponds to the energy (orbital velocity) a satellite would encounter AO particles in LEO. Depending on the orbit radius, the AO collisions occur at a rate of 10^{-12} to 10^{-14} cm^2/s. Simultaneously, UV radiation can be directed towards the specimen in the chamber. An electron beam can be directed onto the specimen, or the specimen may be charged by applying a dc voltage. Phillips Laboratory has also developed a technique to bombard the specimen with milligram particles of aluminum at velocities approaching 10 km/s. The chamber is equipped with a time-of-flight mass spectrometer, that is used to sample the chemical species generated by the bombarded specimen.

Atomic Oxygen Source

Phillips Laboratory uses a recently developed AO source that produces a constant flux of neutral oxygen atoms of 4.5×10^{13} cm^{-2}/s^{-1} in the (^3P) state with a mean kinetic energy of 5 eV and a range of 4 to 6 eV at half maximum [*16*].

The principle of the AO source is based on the high permeability of oxygen atoms through silver and the extremely low permeability of O_2. Oxygen gas is placed at a small pressure on one side of a pure silver membrane that is heated to ~ 550°C. The oxygen gas dissociates oxygen gas at the silver surface and the absorbed atoms dissolve in the silver. This sets up a diffusion gradient that moves the atoms through the membrane. On the opposite surface, the oxygen atoms are desorbed by means of electron-stimulation created by a low flux of 2 keV electrons bombarding the exit surface of the silver membrane. By adjusting the flux of electrons, the desorption rate of oxygen atoms and, consequently, the flow of oxygen atoms through the membrane is controlled.

In selecting a source for AO particles, Phillips Laboratory reviewed the other possible sources for atomic oxygen. Current AO sources fall into two categories; hyperthermal AO and thermal AO beams. Thermalized beams, 0.1 to 0.4 eV, contain multiple contaminant species such as ozone and oxygen ions. For example, plasma ashers have been used extensively to study AO erosion of polymers. However, besides containing a plasma of ions and atomic oxygen, an asher emits extensive amounts of UV radiation of unknown radiance and spectral distribution. Therefore, systematically separating the effects of AO and VUV would be difficult. Also, the incident AO energy is too low relative to LEO, and the presence of ion species contaminants renders this source of limited value for our objectives. Hyperthermal beams are produced in several ways. The AO beams may be produced by a laser discharging into oxygen gas that generates an instantaneous plasma that contains AO. The laser can be operated continuously or may be pulsed. When the continuous mode is used, the energy is around 2.5 eV, at a flux of 10^{14} cm^{-2}/s^{-1}. In the pulsed mode, the average energy is near 5 eV but the instantaneous flux is now four orders of magnitude higher. Because the laser recovery time is a few milliseconds, the time-averaged flux equals that in LEO. However, if the peak flux is too high, then there is competition for reaction sites on the polymer that may affect the observed rate of erosion. The competition for sites give lower erosion rates at equal fluences, when high and low flux data are compared. Hyperthermal beams produced by laser discharging contain unknown amounts of VUV radiation and, in some cases, contaminant species such a 0^- and 0_2 as well. However, the VUV can be chopped out of the AO beam, the ions can be removed by magnetic lens systems and 0_2 content is usually small.

UV Radiation

In LEO, the electromagnetic spectrum is characterized by the presence of VUV radiation, wavelengths from 100 to 200 nm, in addition to the terrestrial electromagnetic spectrum [*1*]. VUV poses a problem for laboratory testing using conventional UV light sources because optical systems, including electric discharge bulbs, use conventional glass or quartz envelopes and lenses, that have very high absorptivities at wavelengths below 200 nm. Thus, to raise the radiant energy in the range of 180 to 200 nm requires that the overall spectrum be raised to twice that in LEO at wavelengths <250 nm. Thus, spectral energy differences in the photon interactions with polymers are distorted by the differences in the two spectra.

Our approach is to examine the spectral dependence of chemical reactivities with and without AO, using a monochrometer to give us a specific range of wavelengths before integrating the combined effects of the whole solar spectrum in LEO without accelerating the effects by raising the intensities. Alternatively, one could use an *in-situ* source of VUV in the chamber. The monochrometer approach will enable us to separate the role of UV in photoinitiation of free radicals and/or chain scissions at irradiances comparable to LEO. One might expect a strong dependence on wavelength of the incident radiation. In addition, the presence of AO would assist the oxidation steps in polymer degradation initiated by the UV radiation in a range of wavelengths.

Charging Effects

The presence of charging effects on satellite materials is well documented, particularly for geosynchronous orbits [*17*]. Charging also occurs in LEO, principally due to the photoelectric effect. However, our interest is in the role of a charge distribution on the chemical reactivity in the presence of AO and/or UV. Since polymers are generally considered dielectric materials, a charge distribution results in polarization of the molecules. The polarization may reorient side groups and stretch primary bonds. As a result, the chemical reactivity would be expected to change. A more detailed evaluation of the role of charging on the effects of AO will be presented in section 4, below.

To study charging effects, we have introduced an electron gun into our apparatus that produces a flux of electrons of sufficient magnitude to charge an ungrounded specimen. Alternatively, we can apply a voltage to the specimen during bombardment by atomic oxygen.

Combined Effects of AO and Charging on Chemical Reactions

As previously noted, spacecraft surfaces facing the direction of motion are exposed to a flux of AO having translational energy of about 5 eV, and thermal energies (that is, an energy variation with respect to the 5 eV mean energy) corresponding to a temperature of roughly 1000 K. The translational energy is sufficient to break many types of polymer bonds and exceeds the activation energy required for many chemical reactions with atomic oxygen. Reaction rates of oxygen at the surface of polymeric materials depend on the nature of the reacting oxygen. Differences are observed, depending on whether molecular oxygen, thermal atomic oxygen, hyperthermal atomic oxygen in the ground state, or excited-state atomic oxygen is reacting.

The simultaneous presence of energetic charged particles, photons, and/or electric fields can significantly affect the rates of reaction leading to specific final states. For example, Tennyson [5] observed that the rate of oxidation of a Teflon™ surface by 2.4 eV atomic oxygen more than doubled, when the surface was simultaneously illuminated with 1.8 equivalent suns of UV radiation. This result contrasts with the measured factor-of-two *decrease* in reaction rate observed when hyperthermal atomic oxygen interacts with poly(imide-siloxane) copolymer in the presence of ultraviolet radiation [*18*]. In our own work, we have observed an order-of-magnitude increase in reaction rates when polysulfone resin is simultaneously bombarded by 5 eV atomic oxygen and 20 keV electrons, as will be discussed. Chemical reactions at the spacecraft surface have two principal effects: they lead to material removal through production of volatile reaction products and they change the properties (most importantly, conductivity and spectral absorptance and emittance) of the spacecraft surface.

In addition to AO, spacecraft in LEO are bombarded by ultraviolet radiation (UV) and energetic charged particles (primarily electrons), which can also produce chemical and mechanical changes in the materials of which the spacecraft surface is composed. When electron bombardment is combined with AO bombardment a number of things can happen. First, because the spacecraft is usually thicker than the electrons' range, the electrons will stop in the spacecraft, leading to an accumulation of charge. This charging mechanism can be offset by emission of photoelectron from the spacecraft surface, but, in general, spacecraft carry a net charge. Consequences of surface charging will be discussed in section 4.3. The bombarding electrons also break chemical bonds and excite and ionize atoms in the spacecraft surface, causing both transient and permanent changes in the species populations available for reaction with AO; the associated phenomenology is analogous to that of UV photochemistry.

The presence of vacuum ultraviolet radiation modifies the potential chemical reactions that may occur when a given polymer is bombarded by atomic oxygen and energetic electrons [*15*]. In general, photooxidative degradation is described by three steps: photoinitiation, propagation, and termination. Photoinitiation has three possible absorption paths: absorption by chromophoric groups forming low-molecular-weight radicals, by chromophoric groups transferring energy to another group, and by accumulated absorption at a bond by energy migration. The propagation step can take place via six different chemical reaction mechanisms: reaction of low-molecular-weight radicals with polymer alkyl radicals, reaction of polymer alkyl radicals with oxygen, abstraction of hydrogen by a hydroperoxyl group, photodecomposition of a hydroperoxide group to form polymer radicals, abstraction of hydrogen by a polymer alkyloxy radical to form a hydroxyl group, disproportionation reaction of polymer alkyloxy radicals to form end-aldehyde groups and end-polymer alkyl radicals, and rearrangement and addition reactions. Termination occurs as a result of polymer free-radical crosslinking, either inter-chain or with low-molecular-weight radicals, leading to termination reactions.

The phenomena of degradation of polymers involves several possible mechanisms. The main importance of singlet oxygen reactions is the formation of hydroperoxide groups, which are sensitive to UV, to form free radicals. Impurities present can act as photoinitiation sites. Extreme UV, wavelengths from 180 to 200 nm, is known to cause ablation of polymers, through disintegration of the polymers at the surface; because the polymers are so absorptive, less than 300 nm penetration, each photon is capable of breaking a bond. The broken bonds are free to combine with oxygen. Finally, there are mechanisms that involve charge transfer combined with light and depolymerization by photoinitiated cations.

In view of the brief discussion of polymer photochemical processes given above, it is not surprising that UV, atomic oxygen and energetic electrons, when simultaneously present, may result in chemical-reaction pathways that are not evident from studying their separate effects. While much work has been performed to characterize photodegradation of polymers, little or no work has focused on how the chemical reactions at the surface of a solid exposed to UV can be changed when the surface is bombarded with atomic oxygen and/or energetic electrons.

In view of the above, we expect that reaction cross sections for AO interacting with polymeric materials will, in general, be changed when the materials are simultaneously bombarded by energetic electrons; similar considerations for UV lead to similar conclusions when UV is present. In general, we find that accurate ground-based materials testing requires the capability to *simultaneously* impose the AO, UV, and electron environments found in space. Furthermore, it is important to pay attention to the fidelity with which the energy spectra of the different environments are reproduced. (This has been a particular problem for the AO environment.)

In the next section, we describe measurements we have made on polysulfone resin and graphite/polysulfone composite materials, to illustrate the complexity of the phenomena. The measurements described below support the hypothesis that it is the effect of surface electric fields, due to surface charging, that are most important in changing the cross-sections for the oxidation reactions we studied in these materials.

Experiments

First, we exposed a target of polysulfone resin, spin-coated onto a silicon substate, to a pulsed AO beam at NASA's Jet Propulsion Laboratory. Simultaneously, we applied an electron beam, whose energy could be varied downward from 20 keV, and whose current density at the target could be varied downward from about 50 $\mu A/cm^2$. The resin layer was about 5 μm thick, so that electrons of energy greater than about 10 keV would pass through it, stopping in the silicon substrates. The silicon substrate could be connected to the vacuum chamber ground, to an external high-voltage supply, or could be electrically isolated (floated). A quadruple mass filter was used to select the mass/charge (m/e) of species emitted from the target surface, and time-of-flight (TOF) techniques were used to measure the velocity distributions of reaction products leaving the target surface. The apparatus is shown in elevation and plan views in Figs. 1 and 2, respectively.

The AO beam was generated by a laser source that employs a pulsed CO_2 laser to generate a plasma, from which neutrals effuse through a skimmer. The pulse width at half maximum FWHM was typically a bit less than 50 μs, which corresponds to a velocity spread of about 1 km/s about the mean of 8 km/s, with about 10^{14} AO per pulse. Ion contamination was less than a percent, however there was a significant O_2 component in the beam. Beam contamination is a significant source of uncertainty, because its effect on our results is difficult to estimate. Photons from the plasma flash were removed by a chopper, which blocked the light, but opened to allow the AO to pass; we could also chop out the first component of the AO velocity distribution, when desired.

We collected TOF spectra for scattered AO, and for CO, CO_2, SO, and SO_2 reaction products emitted from the target surface. The TOF spectra were converted to velocity and energy distributions, which are easier to relate to the underlying phenomena. Figures 3*a*, *b*, and *c* show the time, velocity, and energy distributions, respectively, of the incident AO beam. Note that the energy width of the beam is about 2.5 eV, much wider than would be encountered in the space environment, but the mean energy is about right.

The velocity spectrum measured for AO scattered at an angle of 90° from the target surface is shown in Fig. 4. Deconvolution of the velocity distribution of the incident AO beam reveals that there are two principal peaks in the scattered spectrum, with mean velocities of about 2.5 and 4.5 km/s. Elastic scattering from a rigid target would result in a single peak, at the incident velocity (about 8 km/s).

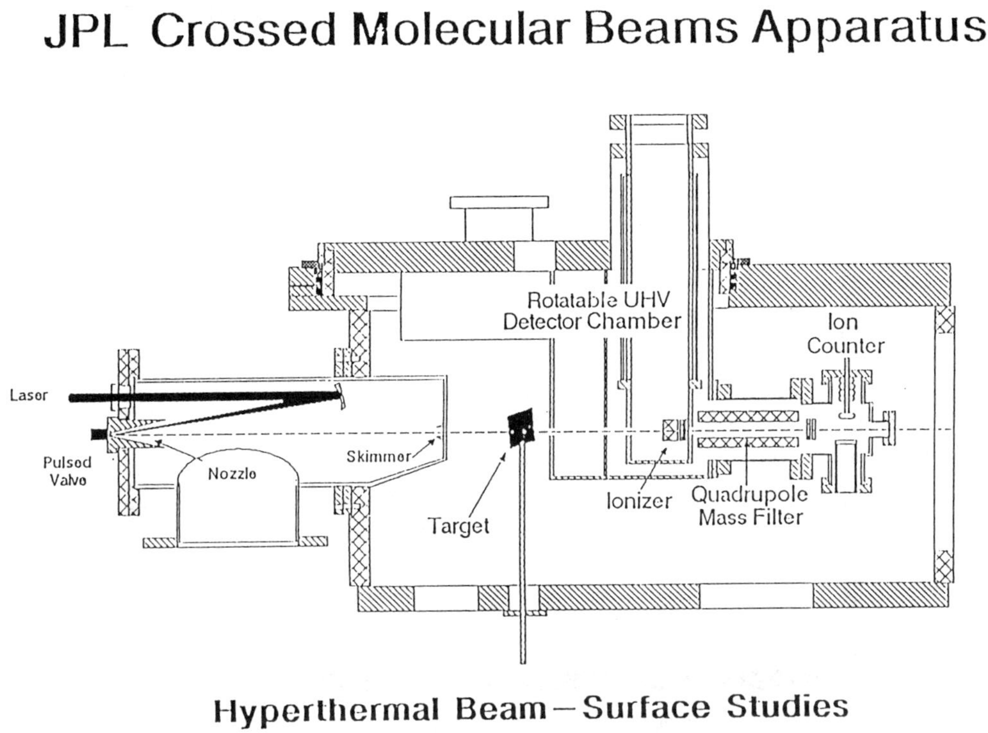

FIG. 1—*Elevation view of the JPL AO facility. The vacuum chamber is about 2 m wide by 0.5 m high.*

In the remainder of this section, we show data for rates of reactions of AO with polysulfone, leading to m/e = 44 (presumably CO_2) in the final state. This is done to avoid complicating the discussion with considerations of how the rates vary from one exit channel to another. The data we collected on reaction products other than CO_2 qualitatively exhibit the same behavior as discussed below for CO_2.

Velocity spectra of $m/e = 44$ reaction products emitted when the target is simultaneously bombarded with 20 keV electrons shown in Figs. 5*a*, *b*, and *c*. The electron beam had an area of about 1 cm^2 at the target, about the same as the AO beam on which it was superimposed. Figure 5*a* shows the spectrum obtained with the electron current turned off, and the target floating (electrically isolated from the ground); Figure 5*b* shows the spectrum obtained with 5 μA on target, with the target grounded, and Fig. 5*c* shows the spectrum obtained with the same current, with the target floating. All three spectra are normalized to the same number of AO incident. The only significant difference among the three spectra is the appearance of a new peak at about 12 km/s in the spectrum taken with 5 μA on an electrically-isolated target. When we varied the electron-beam current, we found that the spectrum was not measurably affected by currents below about 1 μA. A maximum in the cross-section for the fast peak, and the peak at about 1.5 km/s, was observed for a beam current of about 10 μA.

The effect of applying a DC bias to the target is shown in Figs. 6*a*, *b*, and *c*. The electron beam was off for these measurements. The spectrum in Fig. 6*a*, taken with the target grounded, is similar to those in Figs. 5*a* and 5*b*. Figure 6*b*, taken with a DC bias of 1 kV

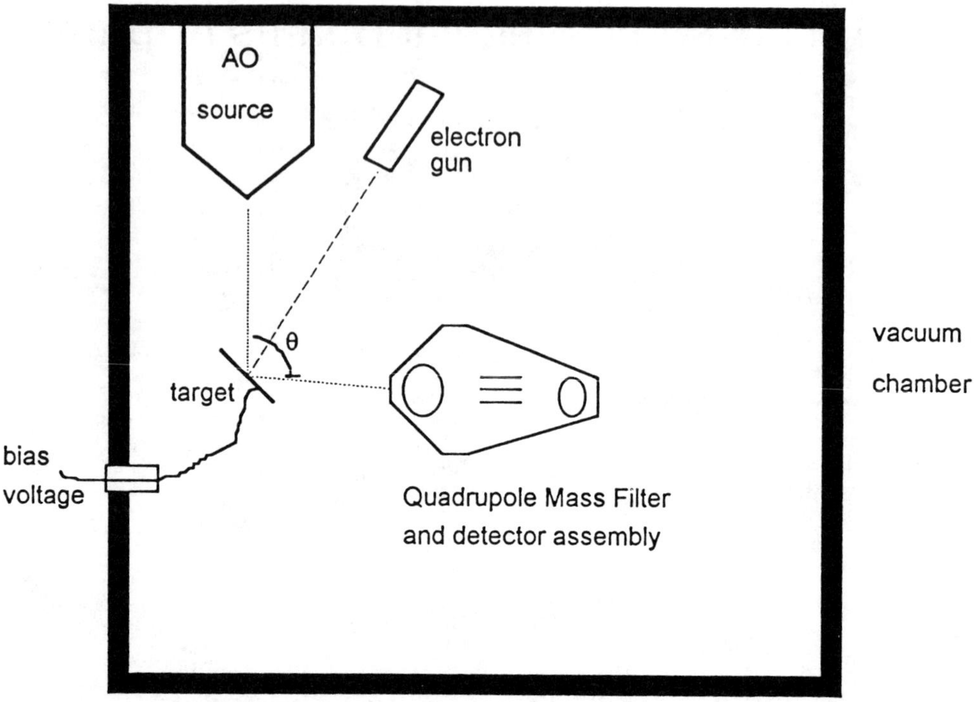

FIG. 2—*Plan view of the JPL AO facility. The vacuum chamber is about 2 m².*

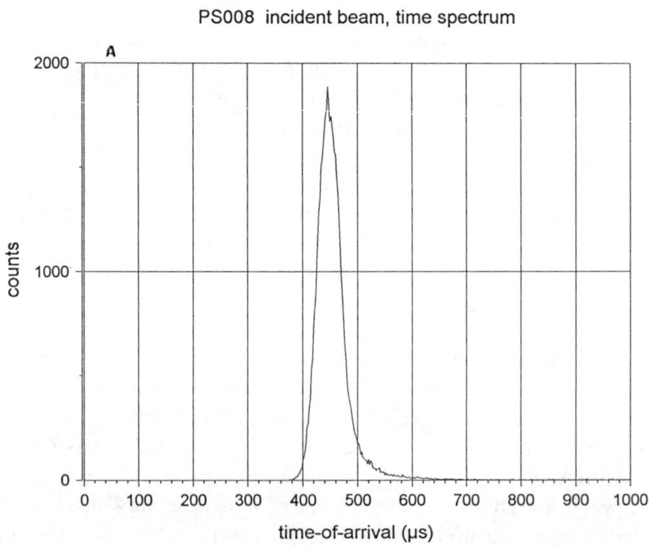

FIG. 3—*The incident AO beam. (a) arrival-time spectrum.*

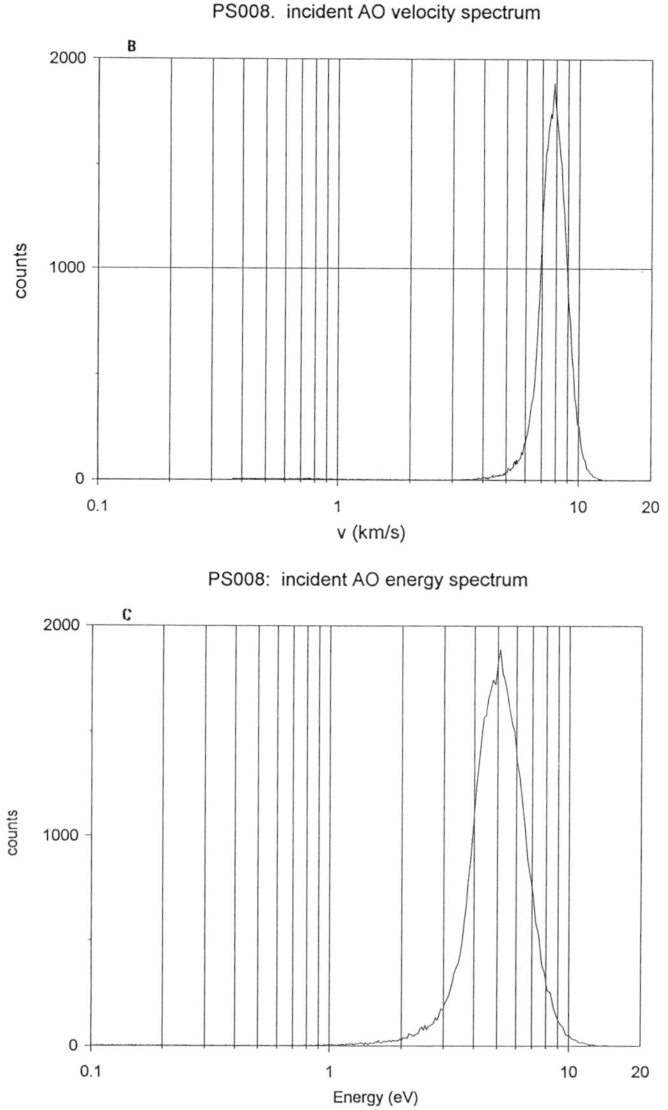

FIG. 3—*The incident AO beam. (b) velocity spectrum; (c) energy spectrum.*

applied to the target, has a new peak at 10 to 12 km/s, and exhibits a significant enhancement of the peak centered at about 1.5 km/s. Figure 6*c*, taken with a DC bias of +1 kV, does not have the fast peak, but the peak at 1.5 km/s is twice as high as in Fig. 6*a*. As for electron-beam current, we found that the variation of the reaction cross-sections exhibited a threshold and a saturation as the DC bias voltage was varied.

Figures 7*a*, *b*, and *c* show the results of having both electron bombardment and DC bias present when measuring the interaction of the AO beam. In Fig. 7*a*, the target is grounded and bombarded with 5 μA of 20 keV electrons; the spectrum is much the same as in Figs.

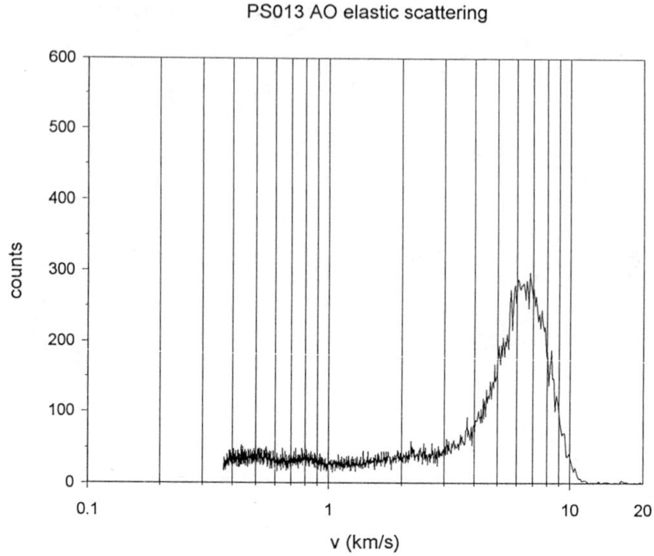

FIG. 4—*Velocity spectrum of AO scattered at 90° with respect to the incident beam direction. Note that the spectrum is complex, and the target does not behave as a simple rigid wall.*

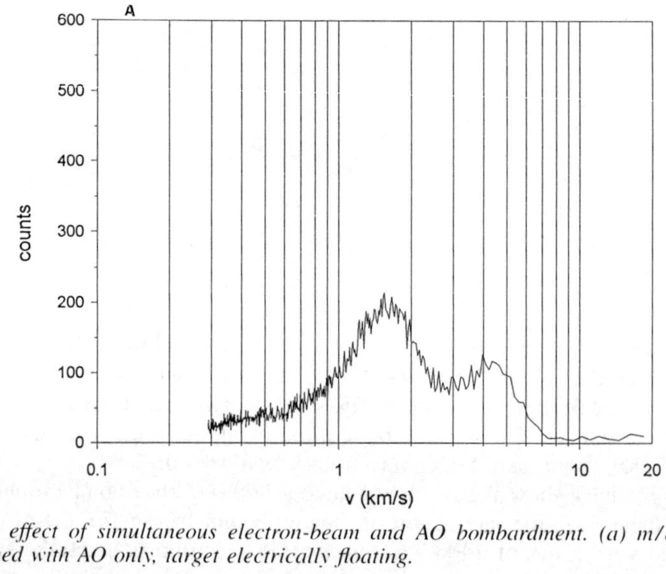

FIG. 5—*The effect of simultaneous electron-beam and AO bombardment. (a) m/e = 44 velocity spectrum obtained with AO only, target electrically floating.*

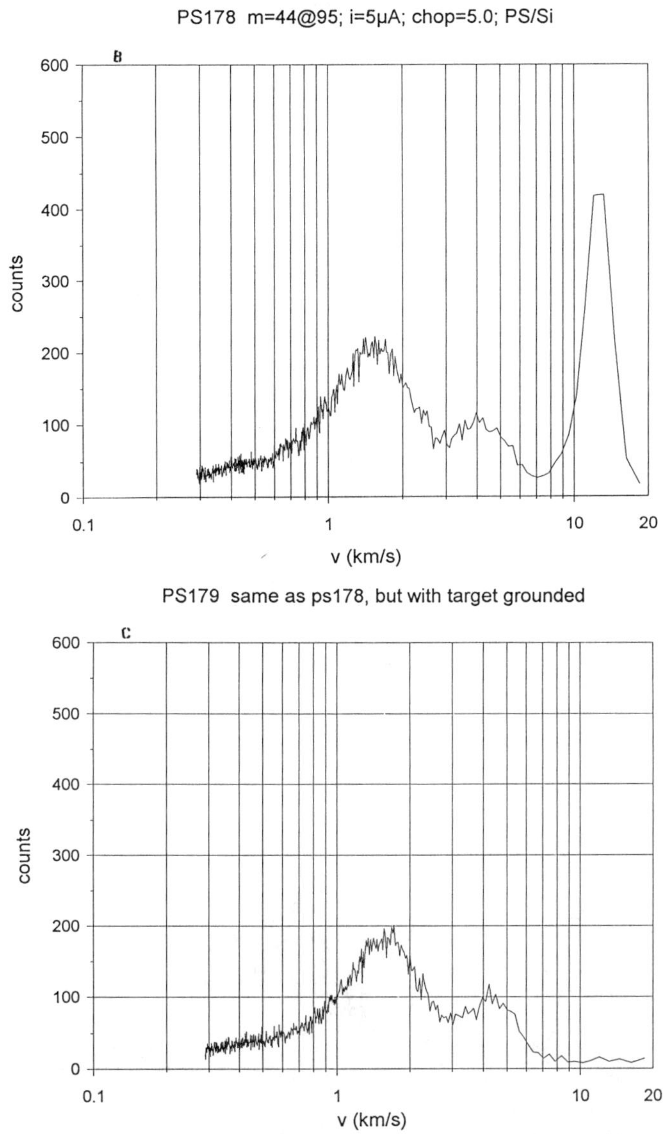

FIG. 5—*The effect of simultaneous electron-beam and AO bombardment. (b) target grounded and bombarded with 5 μA of 20 kV electrons during AO bombardment; (c) target floating, and bombarded by both electrons and AO. The etiology of the fast peak at about 12 km/s is at present unknown.*

5*a* and 6*a*. In Fig. 7*b*, the target has been biased to −1 kV, in addition to the electron bombardment; the peak at 10 to 12 km/s has appeared, and the peak at 1.5 km/s is enhanced. Finally, in Fig. 7*c*, the bias was changed to +1 kV; the fast peak is absent, and the 1.5 km/s peak is greatly enhanced.

The results of our measurements are summarized in Table 3. The simplest interpretation of the results is that the presence of an electric field at the target surface modifies the cross-

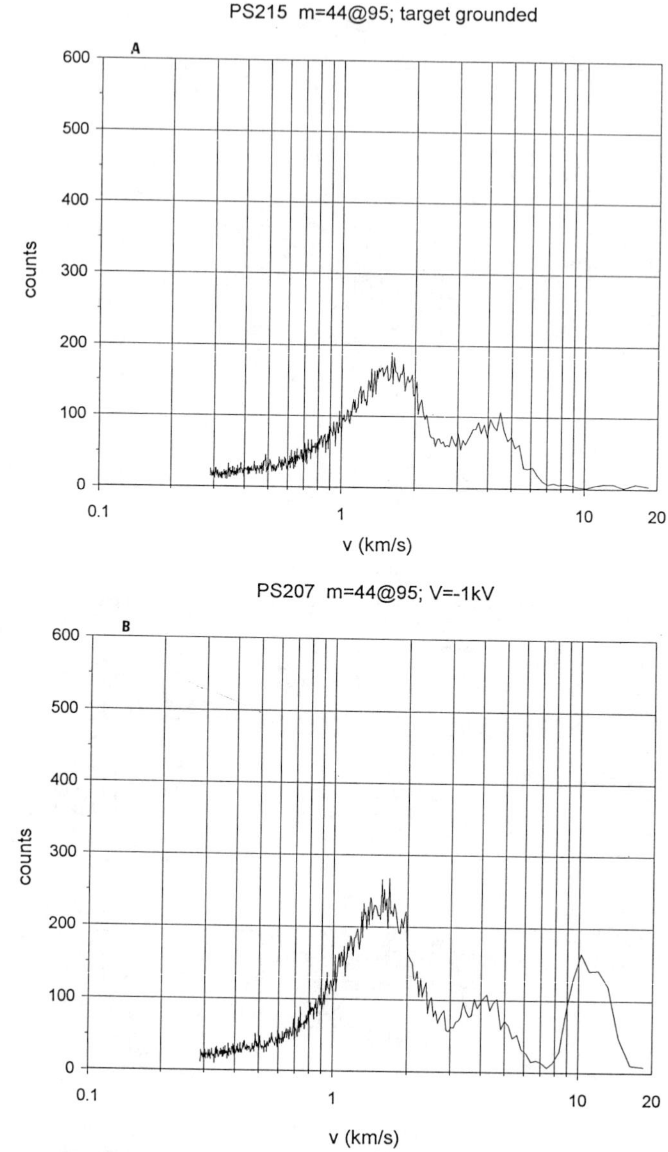

FIG. 6—*The effect of applying a DC bias during AO bombardment, with the electron beam turned off. (a) m/e = 44 velocity spectrum with* V = *0; (b) spectrum with target biased 1 kV negative with respect to the vacuum chamber wall.*

sections for reactions of AO with the target constituents at the surface. In the space environment, the field would be due to spacecraft charging. Some of the polymeric materials exposed on LDEF exhibited more erosion than was expected on the basis of ground tests; our results suggest one mechanism by which the discrepancy may be resolved. Unfortunately, detailed, quantitative comparison of our results with LDEF is not possible, for at least two

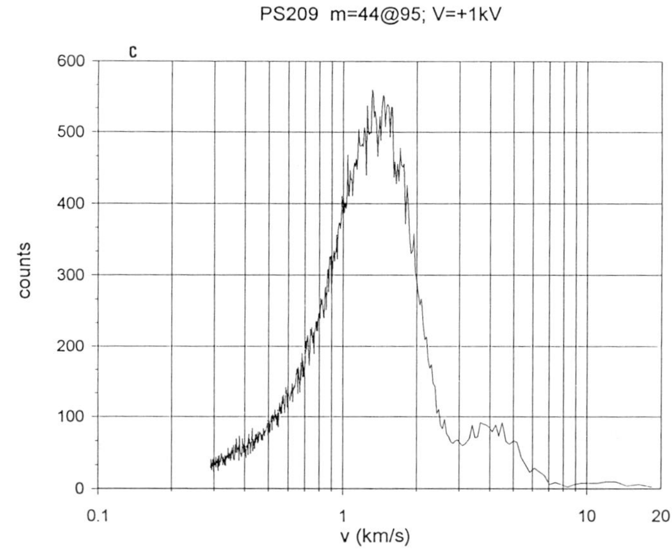

FIG. 6—*The effect of applying a DC bias during AO bombardment, with the electron beam turned off. (c) spectrum with target biased at +1 kV. In addition to a general increase in cross section as the magnitude of the bias voltage increases, there are obvious differences in the shapes of the spectra taken with positive bias and those taken with negative bias.*

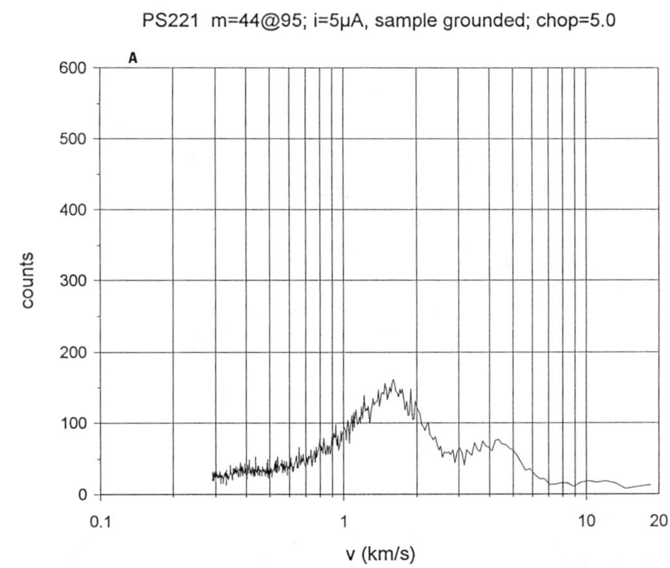

FIG. 7—*The effect of simultaneously applying DC bias and bombarding with 5 µA of 20 kV electrons, in addition to AO. (a) m/e = 44 velocity spectrum, with the target grounded while bombarding with electrons and AO.*

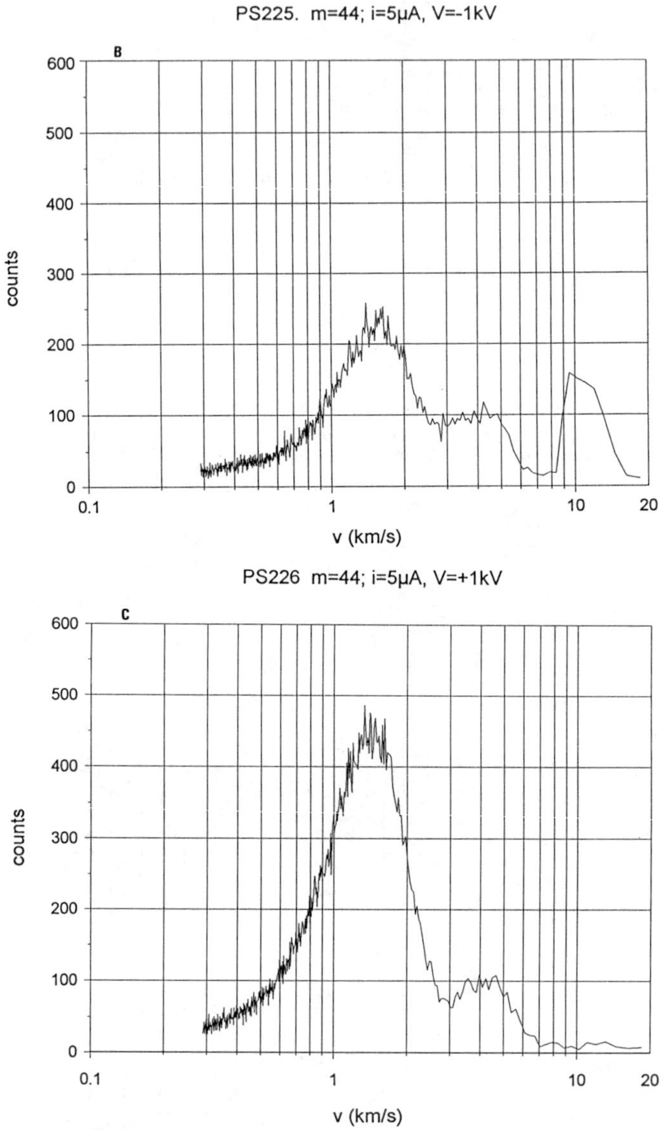

FIG. 7—*The effect of simultaneously applying DC bias and bombarding with 5 μA of 20 kV electrons, in addition to AO. (b) target biased at −1 kV, while bombarding with AO and electrons; (c) target biased at +1 kV, while bombarding with AO and electrons. Comparison of Figs. 5, 6, and 7 suggests that the radiation-chemistry effects of the electron beam are not as important as effects arising due to electric fields at the target surface.*

TABLE 3—*Summary of results.*

Target Voltage	Electron Beam	Target Response
Target floating	OFF	*Baseline* rate; no fast peak.
	ON	*Increased* rate, once current threshold is exceeded; fast peak.
Target grounded	OFF	*Baseline* rate; no fast peak.
	ON	*Baseline* rate; no fast peak.
Positive bias	OFF	*Increased* rate, once voltage threshold is exceeded; no fast peak.
	ON	Same as beam OFF.
Negative bias	OFF	*Increased* rate, once voltage threshold is exceeded; fast peak.
	ON	Same as beam OFF.

reasons. First, the total environment at the different sample locations is not known; usually only the total integrated AO flux is given. In particular, the charging and AO flux histories at each sample location are not known. Secondly, our measurements only studied two of the possible reaction products resulting in erosion: CO and CO_2; other oxidation products, such as SO and more-complicated species, might be expected to contribute as well.

Theory

In an electric field, the dipole component of the bonding orbitals' charge distributions will cause the orbitals to try to align with the applied field. This tendency will be opposed by the thermal motion of the atoms; at any non-zero temperature, the degree of alignment (polarization) will saturate at its maximum value when a certain field strength, whose magnitude depends on the temperature, is reached. The reorientation of bonds in the polymeric target material may be expected to result in changes in the interaction cross sections for AO. Qualitatively, one would expect that there would be a range of applied field where no effect would be seen, because the energy of interaction of the field was smaller than the energy of thermal motion; a region where the cross-section would change (in general, nonlinearly) with increasing field; and a saturation region, where the cross-section would not change, because all the bonds would be aligned. An elementary theory describing the situation is sketched below.

The energy $U(\theta)$ of a bond dipole in an external electric field is given by

$$U(\theta) = -pE\cos\theta \qquad (1)$$

where θ is the angle the bond direction makes with the field and p is the dipole moment.

The probability, $w(\theta,T)$, that p makes an angle θ with E, when the temperature is T, is

$$w(\theta,T) = \frac{1}{Z(T)}e^{-\beta U(\theta)} \qquad (2)$$

where

$$Z(T) = \int_{0}^{\pi} e^{-\beta U(\theta)} \sin\theta d\theta \tag{3}$$

and $\beta = (kT)^{-1}$, where k is Boltzmann's constant, and T is the absolute temperature.

We suppose that the reaction cross section σ depends on θ, among other things, and can be expanded in Legendre polynomials, in the customary fashion

$$\sigma = \sum_{l=0} \sigma_l P_l (\cos \theta) \tag{4}$$

where the partial cross sections $\sigma_0, \sigma_1, \ldots \sigma_n$ may be functions of the atom energy, surface temperature, and so forth, but not functions of θ.

The angle-averaged value $\langle\sigma\rangle$ of the cross section is obtained from Eq 4 by computing the angle-averaged Legendere polynomials $\langle P_n(\cos\theta)\rangle$ defined by

$$\langle P_n (\cos\theta)\rangle = \frac{1}{Z(T)} \int_{0}^{p} P_n (\cos \theta) e^{-\beta U(q)} \sin \theta d\theta \tag{5}$$

For simplicity, we ignore polynomials of order $n = 3$ and higher in averaging Eq 4. After evaluation of Eq 5, we obtain

$$\langle\sigma\rangle = \sigma_0 + \sigma_1 L_1(E) + \sigma_2 L_2(E) + \ldots \tag{6}$$

where we have kept only the leading terms. $L_1(E)$ and $L_2(E)$, which contain the temperature and electric field dependence of the cross section, are given by

$$L_1 (E) = \coth (\beta pE) - \frac{1}{\beta pE} \tag{7}$$

and

$$L_2 (E) = 1 - \frac{3}{\beta pE} \coth (\beta pE) + \frac{3}{(\beta pE)^2} \tag{8}$$

Equations 6 through 8 contain all of the properties of the polysulfone + AO erosion reaction we have observed so far. First, while $L_1 (E)$ is an odd function of E, $L_2(E)$ is an even function; hence, the term $s_1 L_1(E)$ in Eq 6 determines the dependence of $\langle\sigma\rangle$ upon the sign of the applied voltage. Second, as $E \rightarrow \infty$, both $L_1(E)$ and $L_2(E)$ approach unity and $\langle\sigma\rangle$ attains its saturation value, given by $\langle\sigma\rangle = \sigma_0 + \sigma_1 + \sigma_2$.

Although we have yet to measure the detailed behavior of $\langle\sigma\rangle$ at low fields, it is useful to note that when $\beta pE << 1$, Eq 6 follows the approximate polynomial form

$$\langle\alpha\rangle \approx s_0 + \frac{\sigma_1}{3} bpE + \frac{\sigma_2}{24} (\beta pE)^2 \tag{9}$$

This indicates that the target temperature (represented by the factor β) is important in determining the reaction rates that will be observed in a given experiment.

Conclusions

Ground-based tests to simulate the effects of the LEO environment on spacecraft materials must be designed with more care than heretofore, in the case of polymeric materials. Our measurements suggest that the effects of spacecraft charging on AO reaction cross-sections must be simulated in ground-based materials tests, in addition to the UV simulation already included at some test facilities. Furthermore, all three environments (AO, e^-, UV) must be simultaneously applied in order to ensure fidelity, and the target surface must be maintained at an appropriate temperature. Finally, our measurements provide tentative evidence for an energy dependence of the reaction mechanisms that lead to given final-state products (for example, for the production of CO), which we expect to confirm and quantify in future experiments. Thus, one will probably have to reproduce the LEO AO energy spectrum in order to correctly reproduce the details of the interactions in LEO.

Several laboratories have developed facilities to study one or more of the primary polymer degradation mechanisms, AO, VUV, charging, and hypervelocity impacts. However, we have a concern about the comparability of the results from these laboratories because there is a lack of control of experimental variables. These concerns relate to the techniques used to expose specimens to AO or VUV with little or no control of charging. For example, considerable experimental effort has deployed sources of AO which contain large unknown amounts of VUV. Some of the sources use AO with energies far below the 5 eV estimates for LEO. Some of the AO sources are pulsed with time-averaged AO fluxes that match LEO but with peak values that lie orders of magnitude higher than in LEO. Finally, the results of erosion test have used Kapton(TM) as a standard based upon the erosion rates calculated for LDEF as a means of calibration. However, other polymers do not exhibit a standard behavior in alternative AO facilities.

In addition to the problem of salient differences in the AO sources used in laboratories, there is a lack of an absolute standard for AO flux measurements. Several methods are used for measuring the flux of AO based upon behaviors or materials. A partial list of phenomena is: changes in resistivity, weight gain, heat expelled, and light transmission. All of these techniques require assumption about the phenomena of the reaction of hyperthermal AO with the detector material. Alternatively, the AO flux has been measured directly with a mass spectrometer by assuming an ionization efficiency.

Finally, the effects of VUV in LEO are not duplicated in today' laboratories because most sources of VUV do not match LEO. This is because oxygen absorbs UV in the range of 100 to 200 nm readily. Therefore, the presence of air or oxide containing optical systems filters out 90% of the short wavelength radiation compared to LEO. It is precisely the 100 to 200 nm radiation that is expected to contribute heavily to the combined effects of AO, charging, and photons. Of course, there are possible effects of UV in the range of 200 to 350 nm because many of the bond energies for polymers lie in this range. However, ablative photodegradation requires wavelengths less than 200 nm.

Acknowledgments

The authors wish to thank the Phillips Laboratory for its support for this project. In addition, JS and TK express their appreciation of the Air Force Office of Scientific Research for their support as URRP and NRC fellows, respectively. The authors thank Dr. Tim Minton and JPL for their help in conducting the experiments on polysulfone.

References

[1] Anderson, B. J. and Smith, R. E., "Natural Orbital Environment Guidelines for Use in Aerospace Vehicle Development," Marshall Space Flight Center, NASA TM 4527, June 1995.

[2] Levine, A. S., Ed., Proceedings of a Symposium Sponsored by NASA and AIAA, *LDEF 69 Months in Space, Third Post-Retrieval Symposium*, NASA Conference Publication 3275, Part 3, 8–12 Nov. 1993, Washington, D.C.

[3] Whitaker, A. F. and Gregory, J., Ed., Proceedings of a Conference Sponsored by NASA, The University of Alabama-Huntsville, and SAMPE, *LDEF Materials Results for Spacecraft Applications*, NASA Conference Publication 3257, 27–28 Oct. 1992, Huntsville, AL.

[4] Whitaker, A. F. and Jang, B. Z., "The Mass Loss Mechanisms of Polymers in a Radio Frequency Induced Atomic Oxygen Environment," *Journal of Applied Polymer Sciences*, Vol. 48, 1993, pp. 1341–1367.

[5] Tennyson, R. C., "Atomic Oxygen and its Effect on Materials," *The Behavior of Systems in the Space Environment*, NATO ASI Series E, Vol. 245, R. N. DeWitt et al., Eds., Kluwer Academic Publishers, Dordrecht, 1993, p. 233.

[6] George, P. E. and Dursch, H. W., "Low-Earth Orbit Effects on Organic Composites Flown on the Long Duration Exposure Facility," *Journal of Advanced Materials*, Vol. 25, No. 3, April 1995, pp. 10–19.

[7] Koontz, S., Leger, L., Albyn, K., and Cross, J., "Vacuum Ultraviolet Radiation/Atomic Oxygen Synergism in Materials Reactivity," *Journal of Spacecraft*, Vol. 27, No. 3, May-June 1990, pp. 346–348.

[8] Dever, J. A., "Low Earth Orbital Atomic Oxygen and Ultraviolet Radiation Effects on Polymers," NASA Technical Memorandum 103711, February 1991, N91-19294.

[9] George, P. E. and Dursch, H. W., "Low-Earth Orbits Effects on Organic Composite Materials," in Proceedings of a Conference Sponsored by NASA, The University of Alabama-Huntsville, and SAMPE, A. F. Whitaker and J. Gregory, Eds., *LDEF Materials Results for Spacecraft Applications*, NASA Conference Publication 3257, 27–28 Oct. 1992, Huntsville, AL.

[10] Young, R. P., Slemp, W. S., Whitley, K. S., Kalil, C. R., Siochi, E. J., Shen, J. Y., and Chang, A. C., "LDEF Polymeric Materials: A Summary of Langely Characterization in Proceedings of a Symposium Sponsored by NASA and AIAA, *LDEF 69 Months in Space, Third Post-Retrieval Symposium*, A. S. Levine, Ed., NASA Conference Publication 3275, Part 3, 8–12 Nov. 1993, Washington, D.C.

[11] Banks, B. A., Auer, B. M., Rutledge, S. K., de Groh, K. K., and Gebauer, L., "The Use of Plasma Ashers and Monte Carlo Modeling for the Projection of Atomic Oxygen Durability of Protected Polymers in Low Earth Orbit," in Proceedings of the 17th Space Simulation Conference-Terrestrial Test for Space Success, J. L. Strecher, Ed., NASA Conference Publication 3181, 1992, pp. 37–48.

[12] Roybal, R., Shively, J., Stein, C., and Miglionico, C., "Laboratory Simulation of Hypervelocity Debris," *The International Journal of Impact Engineering*, to be published in Vol. 17.

[13] Vaugh, J. A., Linton, R. C., Carruth, M. R., Whitaker, A. F., Cuthbertson, J. W., Langer, W. D., and Motley, R. W., Characterization of a 5 eV Neutral Atomic Oxygen Beam, NASA, Jan. 1991, n91-320736.

[14] Tagawa, M., Tomita, M., Umeno, M., and Ohmae, N., "Atomic Oxygen Generators for Surface Studies in Low Earth Orbit," *AIAA Journal*, Vol. 32, No. 1, Jan. 1994; p. 98.

[15] *Mechanisms of Photophysical Processes and Photochemical Reactions in Polymers*, Jan F. Rabek, John Wiley & Sons, New York, 1987.

[16] Outlaw, R. A. and Davidson, M. R., "Small Ultrahigh Vacuum Compatible Hyperthermal Oxygen Atom Generator, *Journal of Vacuum Science Technology* Vol. 12, No. 3, May/June 1994, pp. 854–860.

[17] Wrenn, G. L. and Sims, A. J., "Surface Charging of Spacecraft in Geosynchronous Orbit," in *The Behavior of Systems in the Space Environment*, R. N. De Witt, D. Duston, and A. K. Hyder, Eds., NATO ASI Series E, Vol. 245, R. N. DeWitt et al., Eds., Kluwer Academic Publishers, Dordrecht, 1993, p. 491.

[18] Gilman, J. W., et al. "Low Earth Orbit Resistant Materials," 6th Annual Workshop on Surface Reactions in the Space Environment, Northwestern University, 1991.

John E. Kamvouris,[1] Gary D. Roberts,[1] J. Michael Pereira,[1] and Christopher Rabzak[2]

Physical and Chemical Aging Effects in PMR-15 Neat Resin

REFERENCE: Kamvouris, J. E., Roberts, G. D., Pereira, J. M., and Rabzak, C., **"Physical and Chemical Aging Effects in PMR-15 Neat Resin,"** *High Temperature and Environmental Effects on Polymeric Composites: 2nd Volume, ASTM STP 1302*, Thomas S. Gates and Abdul-Hamid Zureick, Eds., American Society for Testing and Materials, 1997, pp. 243–258.

ABSTRACT: The objective of this study was to determine whether a shifting approach can be used to account for both chemical and physical effects during long-term aging of high temperature polyimide thermoset polymers, such as the commercial material PMR-15. Stress relaxation curves measured after aging at 316°C for times up to 800 h could be shifted to form a master curve. The horizontal shift factors were found to increase with increasing aging time, consistent with a reduction in molecular mobility that occurs as a result of either physical aging or increased cross-linking. There was a correlation between the vertical shift factors and cure related chemical changes that cause T_g to increase and mass to decrease. Aging effects were partially reversed by heating the aged specimen above its glass transition temperature, indicating that a large portion of the aging is physical in nature. Specimens aged in nitrogen had a larger reversible component than specimens aged in air.

KEY WORDS: stress relaxation, viscoelastic, polyimide, physical aging, chemical aging

Polymers undergo both physical and chemical change during aging. Changes in chemical structure occur if the polymer is exposed to a sufficiently high temperature or a reactive environment. These irreversible changes are called "chemical aging." Reversible changes in physical properties occur during aging of amorphous, glassy polymers at temperatures below the glass transition temperature T_g. Above T_g molecular mobility is high enough to allow the equilibrium volume to be maintained during any change in temperature. Below T_g, molecular mobility is sharply reduced, and the molecular rearrangements required to achieve the equilibrium volume do not occur instantaneously. As a result, the volume of a specimen cooled from a temperature above T_g to a temperature below T_g will be greater than the equilibrium volume and will slowly decrease to approach the equilibrium volume. This gradual approach to equilibrium without chemical change is called "physical aging." During physical aging, the change in volume and the associated change in elastic properties is small (typically on the order of 1% or less). However, physical aging can have a large effect on time dependent mechanical properties (for example, creep and stress relaxation) and on rate dependent failure properties (for example, tensile yield stress).

[1] National Research Council research associate, materials engineer, and materials engineer, respectively, NASA Lewis Research Center, Cleveland, OH 44135.

[2] Aerospace engineer, NYMA Inc., Cleveland, OH 44135.

When creep data are plotted in the form of log (compliance) versus log (time), the primary effect of physical aging is to shift the creep curve horizontally to longer times on the log (time) axis. The same is true for stress relaxation data plotted as log (modulus) versus log (time). Details of the physical aging process and the use of creep measurements for determining aging effects has been discussed in the monograph by Struik [*1*]. Struik suggests that physical aging is qualitatively similar for a wide variety of materials, including glassy polymers. The similarity lies in viewing the fundamental aspect of physical aging as a reduction in atomic or molecular mobility. Physical aging is described as a "self retarding" process since the molecular packing that occurs during aging causes a decrease in molecular mobility, which in turn causes a decrease in the rate at which further packing can occur. Such a process will continually slow down and approach equilibrium only asymptotically. The kinetics of the aging process is determined primarily by this self retarding effect rather than the nature of the material. As an extension of this view, Struik suggests that other forms of aging (such as chemical aging) may appear to be similar to the physical aging if the primary effect of the aging process is to reduce mobility. An example is cited in which a controlled amount of chemical crosslinking introduced into a polymeric material causes shifting of creep curves similar to that produced by physical aging.

Physical aging has been observed in a wide variety of polymers [*1*] and polymer matrix composites, including thermoplastics [*2,3*] and thermosets [*4*]. In most of these investigations, the polymer is assumed to be sufficiently stable that chemical effects are negligible for the aging and test temperatures used. However, chemical aging of polyimide composites is unavoidable when the composites are used in aircraft engines where temperatures can exceed 300°C. The purpose of this work was to determine if the methods described by Struik could be used to evaluate the relative importance of physical and chemical effects during 316°C aging of the polyimide resin PMR-15. In a previous report [*5*], test methods were reported for performing shear stress relaxation measurements on unaged PMR-15 specimens at temperatures ranging from 288 to 348°C. In this report, stress relaxation curves are measured at 316°C, and shift factors are determined for specimens aged up to 800 h at 316°C. Both air and nitrogen environments are used during aging to assess the effect of oxidative degradation. After specimens have been aged and tested, they are exposed to a temperature above T_g and re-tested. The thermal cycle erases the physical portion of the aging, so that the shift factor measured after the thermal cycle is an indication of the extent of chemical aging. Specimen mass, dimensions, and T_g were measured before and after aging as an additional indication of the extent of chemical aging.

Experimental Procedures

PMR-15 resin plates were made by compression molding of PMR-15 powder (HyComp 100 from HyComp, Inc.) in a 6- by 6-in. mold. Temperature and pressure were gradually increased, using an automated cure cycle, until the temperature reached 316°C (600°F). The temperature was then held at 316°C for 2 h while the pressure was held at 1350 psi (9.31 MPa). Plates were then cut into pieces 60-mm long by 5.0-mm wide. The nominal thickness of all pieces was 2.5 mm. The cut specimens were post-cured in an oven at 316°C for 16 h. When the specimens were removed from the oven they were placed immediately into a desiccator and cooled. The weight and dimensions of the cooled specimens were recorded and checked prior to performing any subsequent tests to ensure that no moisture had been absorbed. In order to eliminate the effect of process variability, specimens used for measuring glass transition temperatures and stress relaxation curves after aging were cut from a single plate. A separate set of plates was made to determine the effect of cure and postcure conditions on the glass transition temperature. Specimens were aged in a stainless steel tube furnace at 316°C with pre-heated air or nitrogen flowing through the tube. The dynamic shear modulus and the relaxation shear modulus were measured using an RMS 800 instrument (manufactured by Rheometrics Scientific™). Dynamic shear moduli were measured by deforming the test specimens in torsion at a frequency of 1 Hz with a maximum strain of 0.2%. The storage modulus G' was recorded as the specimen was heated at a rate of 5°C/min. The glass transition temperature T_g was determined using the intercept method illustrated in Fig. 1. Relaxation shear moduli were determined by deforming the test specimens in torsion to a strain of 0.15% over a time interval of 0.375 s. The relaxation modulus was then recorded as the specimen was held at constant strain for 1 h. In order to obtain stable measurements, the instrument had to be pre-heated and held at 316°C for 10 min prior to applying the strain. Without this step, thermal expansion of the test fixtures caused an unacceptable normal force to develop during the test. At the end of the 1-h test, the specimen was kept in the test fixtures, heated to a temperature of T_g + 10°C (about 375°C) and held for 10 m. The specimen was then cooled to 316°C and re-tested for 1 h. The purpose of this thermal cycle was to erase the effects of physical aging. Mass loss was measured using an electronic balance accurate to 0.0001 g. Dimension measurements were made using digital calipers accurate to 0.01 mm.

Dynamic Mechanical Analysis

A typical curve illustrating the dynamic storage modulus G' as a function of temperature is shown in Fig. 1. T_g is the glass transition temperature measured by the intercept method, and T_d is the temperature at which rapid chemical degradation begins to induce blistering in the test specimen. The regions designated as A, B, and C represent temperature intervals in which the behavior of the resin is primarily elastic (Region A), viscoelastic with the possibility of slow chemical changes (Region B), and viscoelastic with rapid chemical degradation (Region C). In Region A, thermal expansion causes an increase in intermolecular spacing

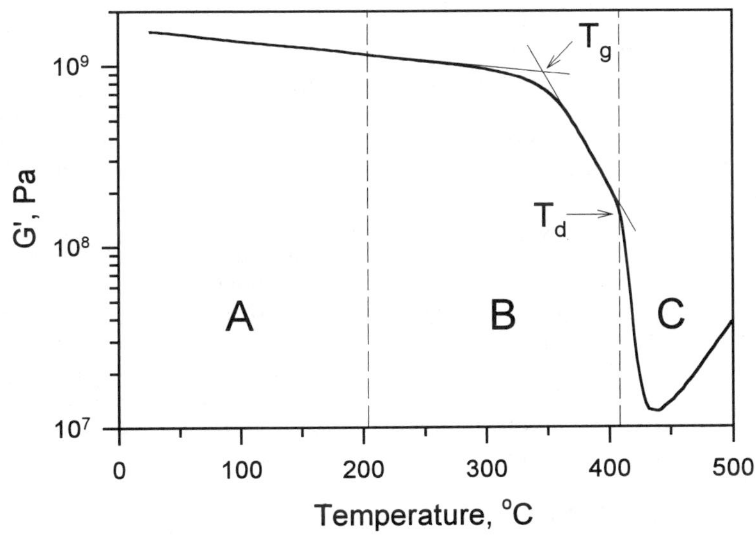

FIG. 1—*Variation in relaxation shear modulus* G′ *with temperature. Typical curve for PMR-15 showing glass transition temperature* T_g *and decomposition temperature* T_d.

and an associated decrease in modulus. The linear decrease of log (G') with temperature is common for polymers at temperatures well below T_g. The transition from Region A to B is determined by the temperature at which mechanical properties become time-dependent. In order to establish some bounds on this temperature, some simple creep tests on PMR-15 resin were done in our lab at room temperature and at 204°C with loads up to 50% of the failure load. The loading and unloading curves were linear, and no measurable creep was detected for times as long as one month. Therefore, the resin behavior can be described as linear elastic for temperatures up to 204°C, and possibly higher. In a previous report [*5*], stress relaxation was measured at 288°C. The transition from Region A to B therefore occurs somewhere between 204 and 288°C. In Region B, measurable creep and stress relaxation occur over times of 1 h or less. Some initial measurements have indicated that stress relaxation rates are reduced when specimens are aged and that the stress relaxation curves are shifted in a direction consistent with the expected effects for physical aging. However, as shown later in this report T_g increases and the mass of the specimen decreases during aging. This is an indication that chemical changes are occurring. According to the analysis by Struik [*1*], it is not possible to determine whether the shift in stress relaxation curves is a result of physical aging or chemical aging without further supporting evidence. One method for separating the effects of physical and chemical aging is to measure the stress relaxation a second time on the same specimen after it has been heated to a temperature above T_g in order to erase the effects of physical aging. For PMR-15 resin, this requires that the specimen be exposed to a temperature between T_g and T_d in Region B of Fig. 1. In a previous report [*5*], it was shown that exposing the PMR-15 resin to temperatures of T_g + 20°C or greater can cause an increase in modulus. This is an indication that chemical changes are occurring. When the resin was exposed to a temperature of T_g + 10°C for 30 min no increase in modulus was observed. For results presented in this report, a maximum temperature of T_g + 10°C was used in order to minimize chemical effects during thermal cycling. In Region C, the PMR-15 resin remains viscoelastic in nature, but the behavior of the resin is dominated by chemical effects. At the thermal decomposition temperature, T_d volatiles generated by thermal decomposition plasticize the resin causing G' to decrease. In addition, the internal pressure of the volatile species causes the softened resin to blister. As the temperature increases beyond T_d, thermal degradation leads to further cross-linking causing G' to increase.

Changes that occur during long-term aging at temperatures below T_g in Region B are the primary subject of this report. One complication in performing these studies is that plate-to-plate variations can obscure some of the subtle effects of aging on the stress relaxation curves. To avoid this complication, all specimens used in the aging portion of this work were cut from the same plate. Since this limits the number of specimens available for testing, aging conditions were limited to 316°C with either air or nitrogen as the aging environment. Aging at lower temperatures and evaluation of the effects of plate to plate variations will be considered in future work. Additional plates were molded and used for investigating the evolution of T_g as a function of cure and post-cure conditions.

An interesting artifact of using the intercept method for calculating T_g became apparent during this investigation. The slope of the curve between T_g and T_d decreases as specimens are aged. The effect of this change in slope on the measured T_g is shown in Fig. 2. From the shapes of the G' curves, it does not appear that the glass transition has shifted to a lower temperature between 50 and 400 h. However, the values of T_g measured using the intercept method indicate that a decrease in T_g has occurred.

Cure and Post-Cure of PMR-15 Resin

In order to investigate the effect of cure conditions on the state of cure of PMR-15 resin, a series of plates were molded for times ranging from 15 min to 16 h and at temperatures

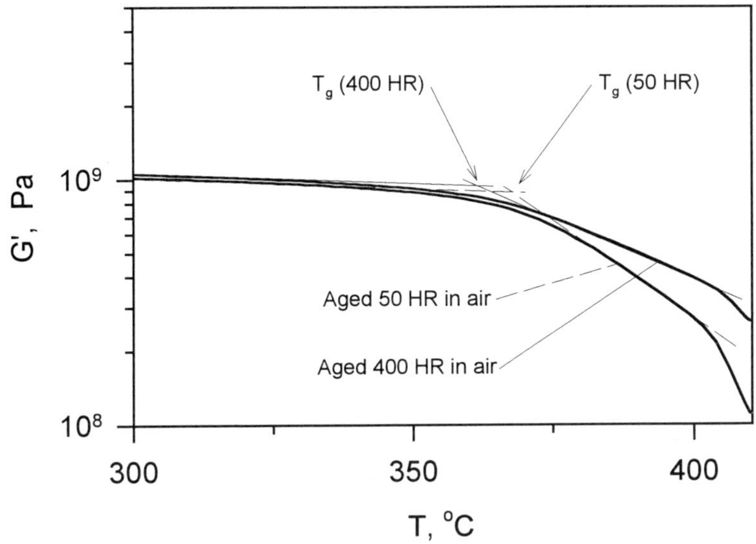

FIG. 2—*Variation in* T_g *as an artifact of using the intercept method.*

ranging from 260 to 343°C. The evolution of the glass transition temperature, which is one indicator of the state of cure, is shown in Figs. 3 and 4. In Fig. 3, specimens molded for 2 h appear to reach a plateau value of T_g at a cure temperature of 316°C. For lower temperature thermosets, such as epoxies, a constant T_g with increasing cure temperature indicates that the resin is fully cured (that is, cross-linking reactions are complete). If this were the case for PMR-15, there should be no further increase in T_g when plates are cured for longer

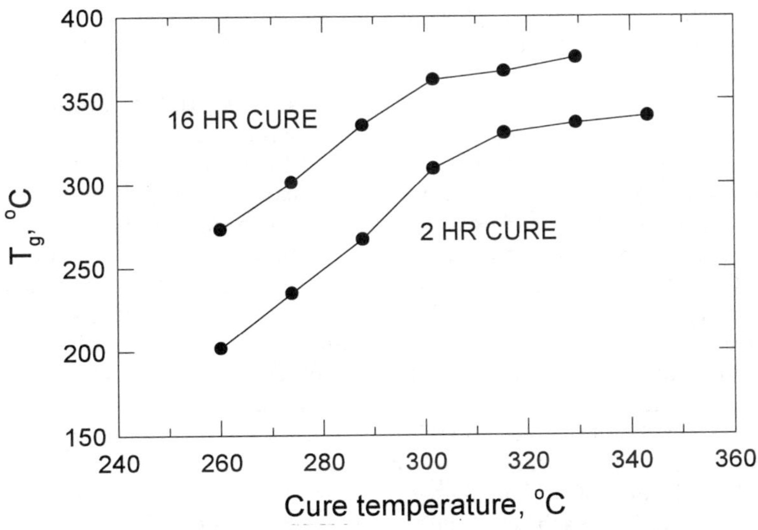

FIG. 3—*Variation in glass transition temperatures* T_g *with cure temperature.*

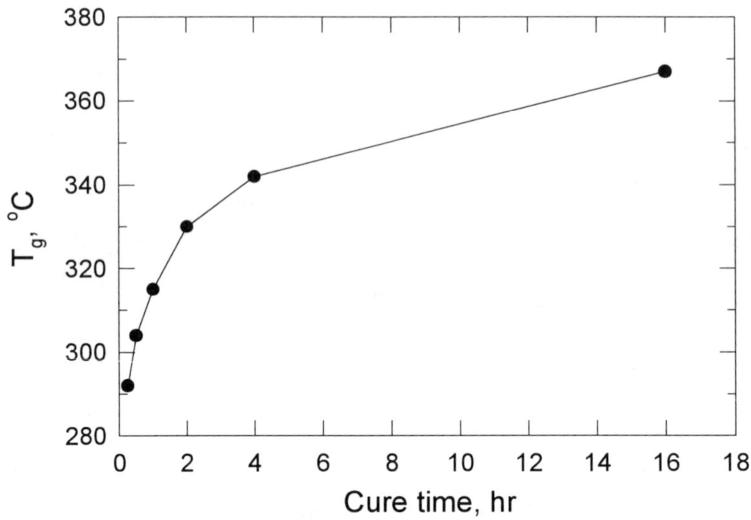

FIG. 4—*Variation in glass transition temperatures* T_g *with cure time at 316°C.*

times at temperatures above 316°C. This is clearly not the case for the 16-h cure data shown in Fig. 3. It appears that T_g can be increased to even higher values than those shown in Fig. 3 if higher cure temperatures or longer cure times are used. However, there is a practical limit on the maximum attainable T_g. In Fig. 1, T_d was shown to be about 410°C. Thermal exposures at temperatures approaching T_d result in excessive mass loss and embittlement of the resin. This leads to the conclusion that PMR-15 resin is never "fully cured" and that continuation of the cure reactions are likely to occur during aging along with other chemical effects such as oxidative degradation. Another way of stating this conclusion is that resin cure, physical aging, and chemical aging are simultaneously involved from the earliest stages of processing, but the relative importance of each of these is different during cure, post-cure, and aging.

For simplicity, cure, post-cure, and aging were all done at 316°C in this initial investigation. The evolution of T_g as a function of cure time during molding is shown in Fig. 4. T_g increases most rapidly during the first 2 h or cure. The usual practice with PMR-15 is to remove the resin from the mold after 2 h and finish the cure by applying a 16-h post-cure in an oven. Storage modulus curves are shown in Figs. 5 and 6 for specimens cured for 2 h in a compression mold then post-cured in a tube furnace with flowing nitrogen (Fig. 5) or air (Fig. 6) for times ranging from 0 to 32 h. The final T_g is determined primarily by the total time-temperature history of the specimen rather than the method of applying the thermal exposure (that is, cure versus post-cure or air versus nitrogen environment). However, there is a subtle effect of post-cure environment on the curves in Figs. 5 and 6. The specimens post-cured in nitrogen converge to a single curve at temperatures above 430°C while the specimens post-cured in air do not. This difference may be related to the development of an oxidized surface layer that is known to form on PMR-15 during long-term aging in air at 316°C [*6*]. Investigation of the initial stages of surface oxidation and associated void formation and cracking is the subject of an on-going investigation.

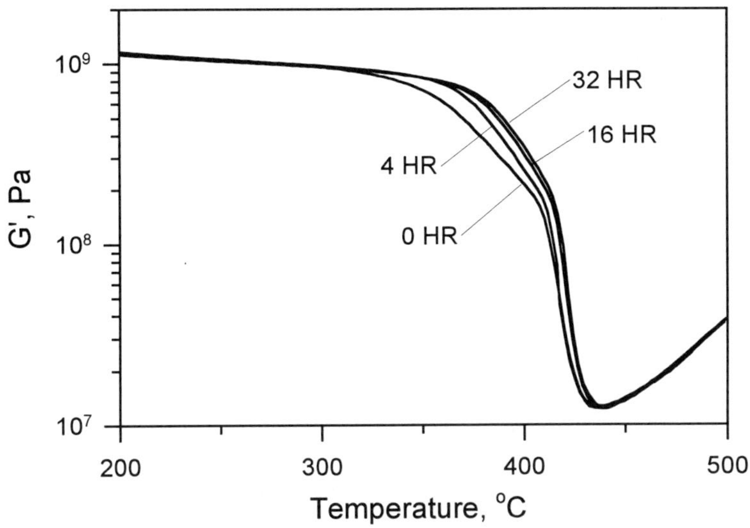

FIG. 5—*Variation in storage modulus* G′ *with temperature. Specimens postcured at 316°C in nitrogen for indicated time.*

Aging of PMR-15 Resin

Glass Transition Temperatures, Mass, and Dimension Changes during Aging

Specimens having a 2-h cure followed by a 16-h post-cure in air are the starting point for the aging studies. This is indicated as 0 h of aging in the figures. Changes in T_g for specimens

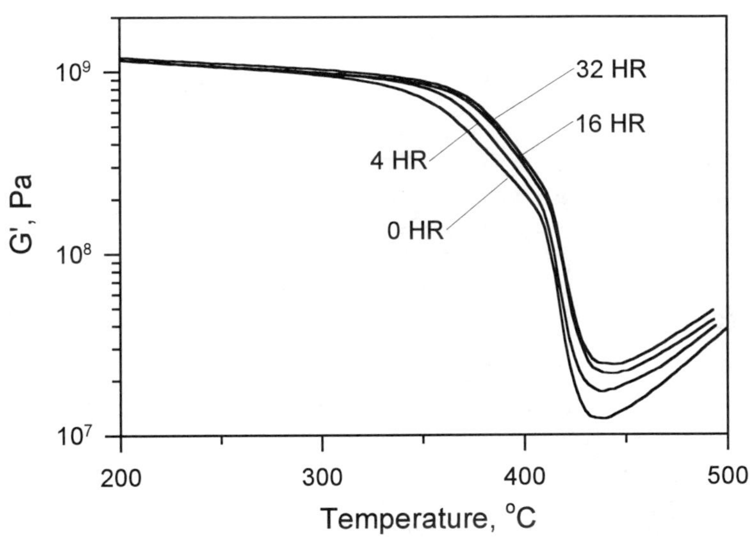

FIG. 6—*Variation in storage modulus* G′ *with temperature. Specimens postcured at 316°C in air for indicated time.*

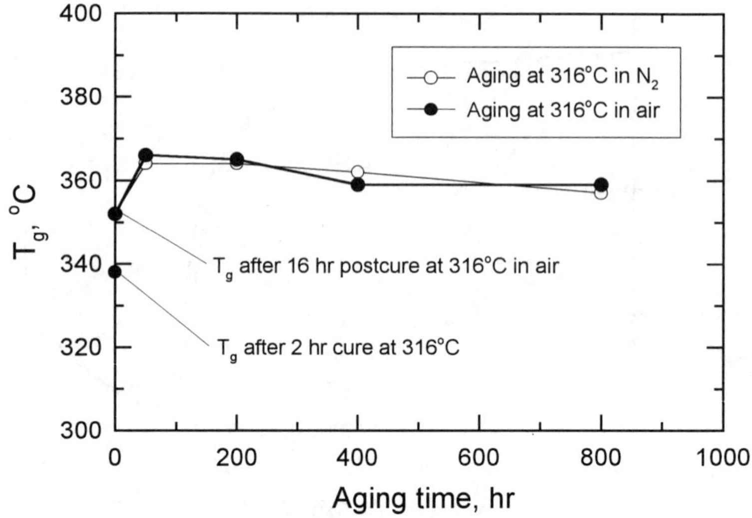

FIG. 7—*Variation in glass transition temperature* T_g *with aging time at 316°C in nitrogen and air environment.*

aged up to 800 h in nitrogen and in air are shown in Fig. 7. T_g after cure and post-cure are also indicated in Fig. 7. The mass loss during aging is shown in Fig. 8. About 0.5% mass loss typically occurs during postcure. T_g increases during the first 50 h of aging then remains nearly constant up to 800 h. The increase in T_g during the first 50 h is a result of a continuation of the cure reactions throughout the bulk region of the specimen. The gradual decrease in T_g shown in Fig. 7 is an artifact of using the intercept method for calculating T_g. (See

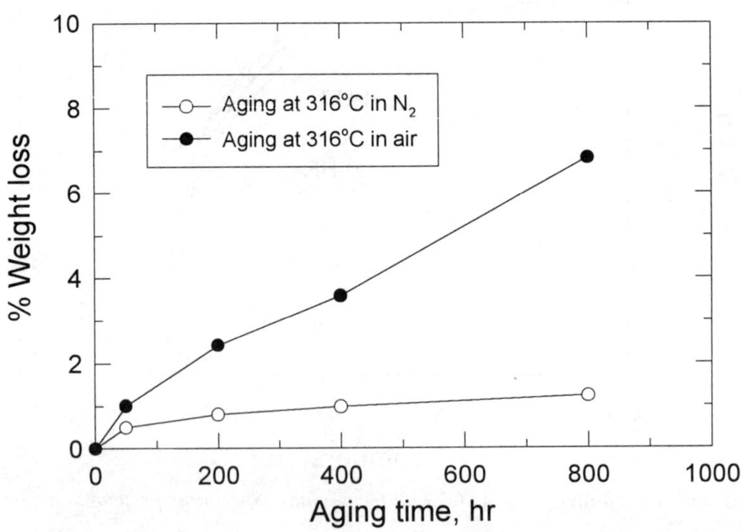

FIG. 8—*Variation in mass loss with aging time at 316°C in nitrogen and air environment.*

Fig. 2 and the related discussion.) The aging environment (O_2 versus N_2) has little effect on T_g. For specimens aged in nitrogen, mass loss during aging is correlated with the increase in T_g with most of the mass loss occurring during the first 50 h. Specimens aged in air lose more mass than specimens aged in nitrogen and the mass lose is not correlated with changes in T_g. This suggests that specimens aged in air loss mass from the surface in addition to the bulk mass loss experienced by specimens aged in nitrogen. This interpretation is supported by the changes in specimen dimensions after aging shown in Figs. 9 and 10. The open circles in Fig. 9 represent the percent decrease in length for specimens aged in nitrogen. The maximum decrease in length is about 0.3% after 800 h of aging. Since mass loss is expected to be a bulk rather than a surface effect when specimens are aged in an inert atmosphere like nitrogen, the width and thickness should also shrink by about 0.3%. However, the change in width and thickness caused by a 0.3% decrease would be too small to be detected by caliper measurements. There were no significant changes in specimen width or thickness for any specimens aged in nitrogen. If the width and thickness are assumed to decrease in the same proportion as the length, the percent volume decrease after 800 h of aging would be approximately three times 0.3%, which is about 0.9%. This is close to the percent mass loss indicated in Fig. 8. For shorter aging times, the dimension changes appear to lag behind the change in mass. A possible explanation for this is that volume relaxation following mass loss is a time-dependent property, similar to creep and stress relaxation. More accurate measurements of volume shrinkage are currently underway to examine the time-dependence of the volume relaxation. In constrast to the results for specimens aged in nitrogen, specimens aged in air showed measurable changes in all dimensions. The solid circles in Fig. 9 represent the percent decrease in length for specimens aged in air. The solid symbols in Fig. 10 represents the absolute decreases in length, width, and thickness for specimens aged in air. Since no changes in width or thickness occur for specimens aged in nitrogen, the changes in width and thickness shown in Fig. 10 must occur as a result of a surface effect such as oxidation. In a previous investigation of PMR-15 specimens that were aged in air [6], Bowles

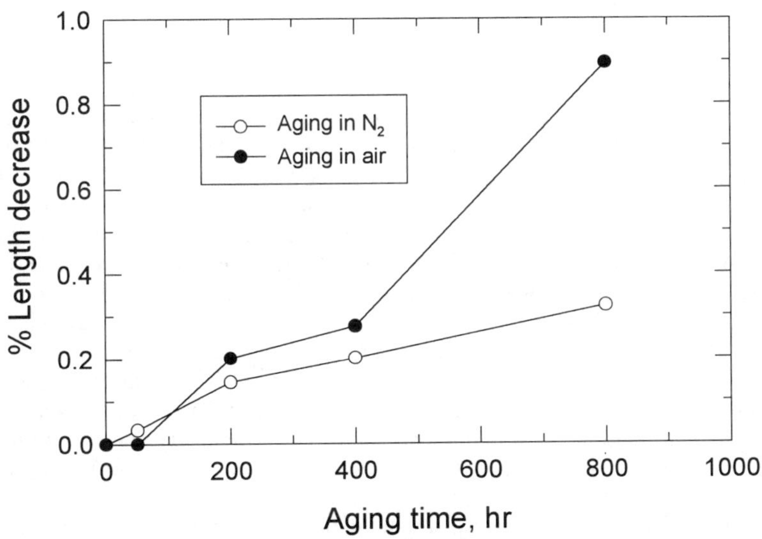

FIG. 9—*Variation in percent length decrease with aging time at 316°C in nitrogen and air environments.*

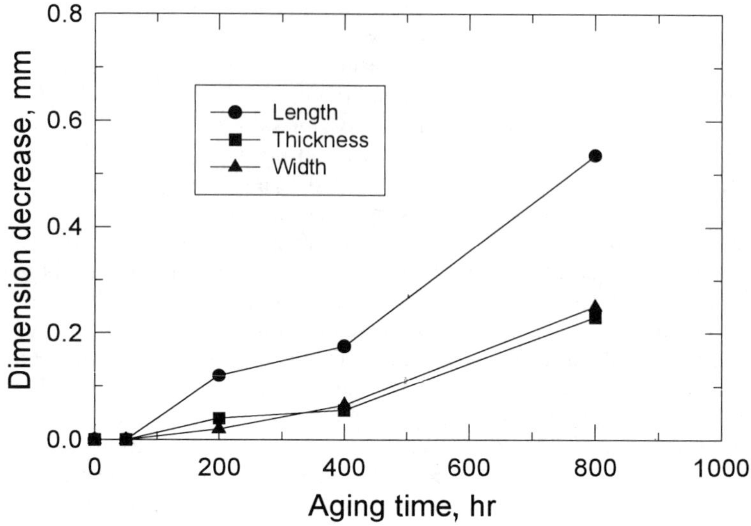

FIG. 10—*Variation in absolute decrease in length, width, and thickness with aging time at 316°C for specimens aged in air.*

has observed the formation of voids near the surface followed by cracking that appeared to be driven by localized shrinkage in the surface region. The dimension changes for times up to 400 h are probably caused by oxidative degradation and shrinkage in the surface region. The accelerated rate of decrease in length, width, and thickness between 400 and 800 h is probably a result of the on-set of cracking. A separate investigation is underway to examine the formation of surface damage as a function of aging temperature, aging time, aging environment, and applied mechanical load.

Effects of Aging on Stress Relaxation Curves

In this section, the feasibility of using shear stress relaxation measurements for determining the extent of physical aging in PMR-15 is examined. The data in Figs. 3–10 indicate that any attempt to measure physical aging in PMR-15 at 316°C must account for a continuously changing chemical structure during cure, post-cure and aging. Since the relative importance of physical and chemical aging was not known a priori, some assumptions had to be made. The first assumption is that chemical aging will be dominant during cure and post-cure, while physical aging will be dominant during aging. The data in Figs. 3 and 6 indicate that the rate of increase of T_g is considerably slowed after a cure time of 2 h and a post-cure time of 16 h. Therefore, processing conditions required for "complete cure" were arbitrarily defined as a 2-h cure followed by a 16-h post-cure. The slower increase in T_g which occurs after post-cure (Fig. 7) is considered to be an aging effect. The second assumption is that chemical aging is a small effect superimposed on physical aging. This assumption is required in order to proceed with an analysis of mechanical relaxation data according to the methods described by Struik. The final assumption is that surface effects (for example, surface oxidation) are small enough to be ignored in calculation of the relaxation shear modulus.

Shear stress relaxation curves are shown in Fig. 11 for specimens aged from 0 to 800 h in nitrogen. Both horizontal and vertical shifts were required to form a master curve. The

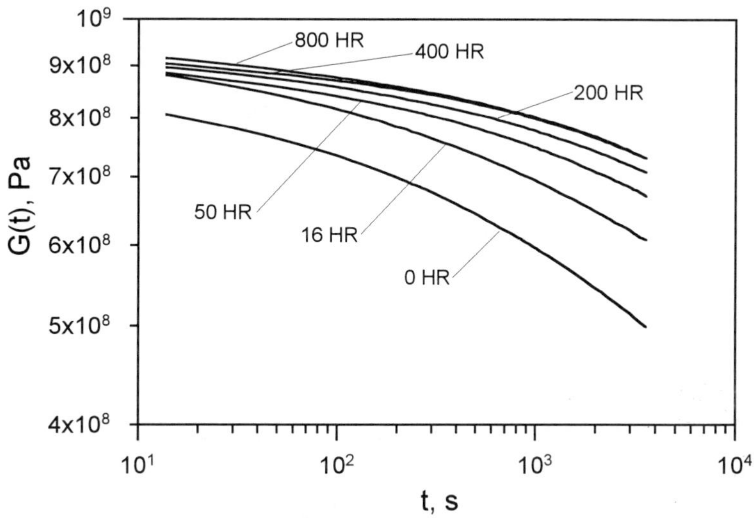

FIG. 11—*Relaxation shear stress modulus curves* G(t) *for specimens aged at 316°C in nitrogen. Effect of aging time.*

shift factors and master curve formed using the unaged condition as the reference state are shown in Fig. 12. A small vertical shift is required for all aging times. Vertical shifts are often required when forming a master curve from tests conducted at different temperatures. Since these measurements were made at the same temperature (but different aging times), the vertical shift may occur as a result of the increase in T_g which occurs at short aging times (Fig. 7). The horizontal shift factors increase with increasing aging time. This is con-

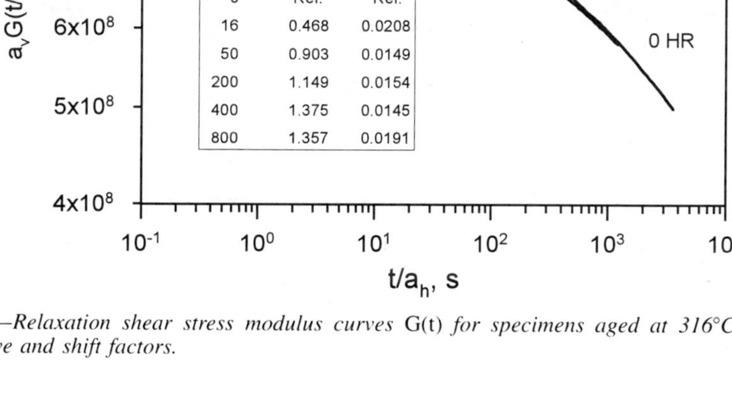

t, hr	$-Log(a_h)$	$-Log(a_v)$
0	Ref.	Ref.
16	0.468	0.0208
50	0.903	0.0149
200	1.149	0.0154
400	1.375	0.0145
800	1.357	0.0191

FIG. 12—*Relaxation shear stress modulus curves* G(t) *for specimens aged at 316°C in nitrogen. Master curve and shift factors.*

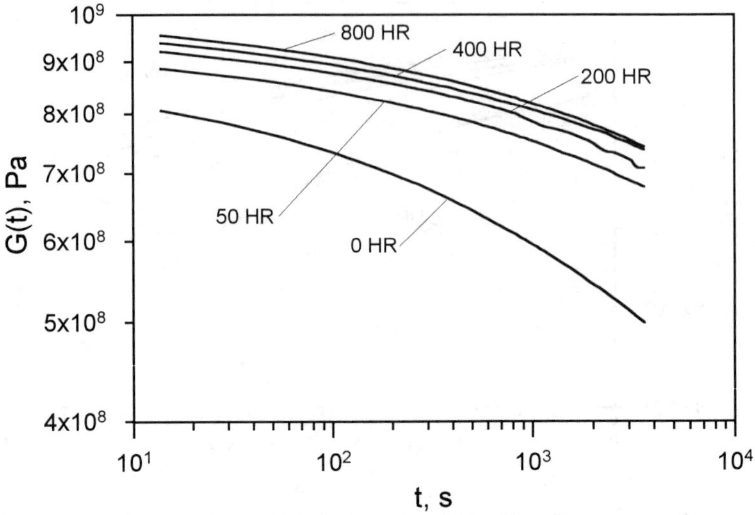

FIG. 13—*Relaxation shear stress modulus curves* G(t) *for specimens aged at 316°C in air. Effect of aging time.*

sistent with a reduction in molecular mobility that would occur as a result of either physical aging or increased cross-linking. Shear stress relaxation curves are shown in Fig. 13 for specimens aged in air. The master curve and shift factors are shown in Fig. 14. Since the curve for the specimen aged 800 h could not be fit to the master curve, the best fit is shown

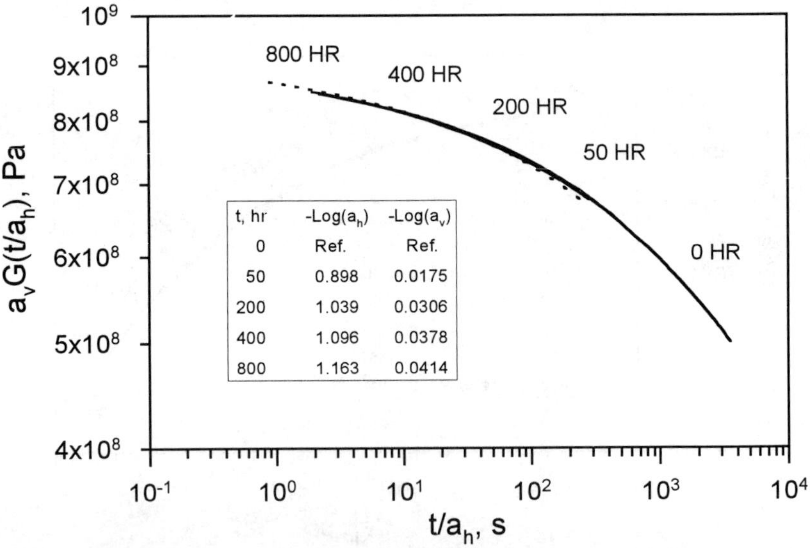

t, hr	$-\text{Log}(a_h)$	$-\text{Log}(a_v)$
0	Ref.	Ref.
50	0.898	0.0175
200	1.039	0.0306
400	1.096	0.0378
800	1.163	0.0414

FIG. 14—*Relaxation shear stress modulus curves* G(t) *for specimens aged at 316°C in air. Master curve and shift factors.*

as a dashed line in Fig. 14. It is not surprising that the 800 h curve does not fit the master curve since about 7% mass loss and substantial surface damage has occurred during aging (Figs. 8 and 10). Interpretation of the shift factors is complicated by the possibility that mass loss and oxidation are concentrated near the surface. This non-uniformity does not appear to be significant after 50 h since the horizontal and vertical shift factors are similar for specimens aged in air and in nitrogen. For longer aging times, the vertical shift factors increase with increasing aging time to values higher than those for specimens aged in nitrogen. The cause of vertical shifting at temperatures below T_g is not fully understood. However, there is a correlation between the time scale for the increase in the vertical shift factors and the time scale for changes in T_g and mass shown in Figs. 7 and 8. For specimens aged in nitrogen, a vertical shift occurs after 16 h of aging and does not increase at longer aging times. This indicates that the vertical shift may be associated with the same cure-related chemical changes which cause T_g to increase and mass to decrease at short aging times. For specimens aged in air, the vertical shift increases continuously with increasing aging time. This may be related to the same chemical degradation that causes a continuous increase in mass loss and the formation of an oxidized surface region when specimens are aged in air. This increase could be associated with the growth of an oxidized surface region. The horizontal shift factors increase with increasing aging time, but the rate of increase is less than that for specimens aged in nitrogen. It is not possible to determine if this difference is related to a real difference in molecular mobility or if it is a consequence of excessive, nonuniform degradation of specimens aged in air.

Thermal Reversibility of Aging Effects

Thermal reversibility of aging effects is an indication that the aging is physical, rather than chemical, in nature. As discussed previously, the thermal cycle used to erase physical aging in PMR-15 must be limited to short times at a maximum temperature of T_g + 10°C to minimize chemical changes induced by the thermal cycle rather than by prior aging. In Fig. 15, stress relaxation curves measured after 16 h of aging in nitrogen are shown along with curves for the same specimen after exposure to three separate thermal cycles. The exposure times at T_g + 10°C were 10, 20, and 30 minutes, resulting in accumulated times of 10, 30, and 60 minutes. The first thermal cycle (labeled 10 min) causes a significant decrease in modulus, which is consistent with erasing of prior physical aging. Subsequent exposures cause a slight increase in modulus, indicating that some chemical changes occur even when the maximum temperature is kept within 10°C of T_g. The first thermal cycle causes the curve to be shifted below the unaged curve in Fig. 11. This suggests that some aging occurred during cure and post-cure and that this aging was also erased by thermal cycling. In order to check for this effect more directly, an unaged specimen was exposed to a thermal cycle with a 10-min hold and maximum temperature of T_g + 10°C. Results are shown in Fig. 16. The downward shift indicates that a significant amount of physical aging has occurred during processing. All of the curves in Figs. 15 and 16 can be shifted onto the master curve in Fig. 12 with suitable choice of shift factors. In order to test for thermal reversibility after longer aging times, the specimens used to obtain the results in Figs. 11 and 13 were exposed to a thermal cycle. To illustrate the results, curves corresponding to 0, 50, and 800 h of aging in Figs. 11 and 13 are reproduced in Fig. 17. Results after exposure to the thermal cycle are shown in Fig. 18. All curves are shifted down after exposure to the thermal cycle, and specimens aged in air are shifted less than specimens aged in nitrogen. This is consistent with the idea that permanent chemical changes will be greater when specimens are aged in air. There is one inconsistency in the data of Fig. 18. After thermal cycling, the curves corresponding to 0 h aging and 50 h aging in nitrogen are nearly the same. This

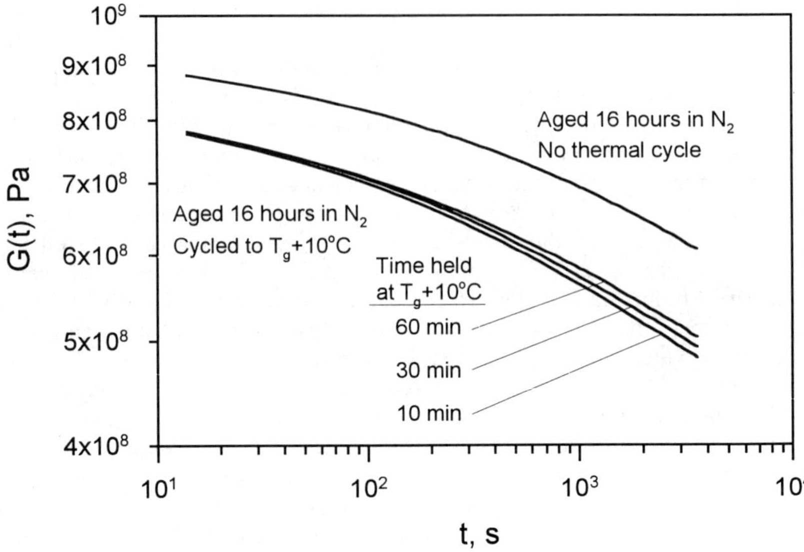

FIG. 15—*Thermal reversibility of aging effects. Effect of time exposed at maximum temperature in thermal cycle.*

result is not consistent with the large difference in T_g shown in Fig. 7 for these specimens. The unaged specimen has a lower T_g and should therefore have a lower initial modulus and a faster relaxation rate than the specimen aged for 50 h. In spite of this inconsistency, these results demonstrate qualitatively that a large part of the aging induced during long-term aging is physical in nature. Before the aging shift factors can be used as a reliable, quantitative

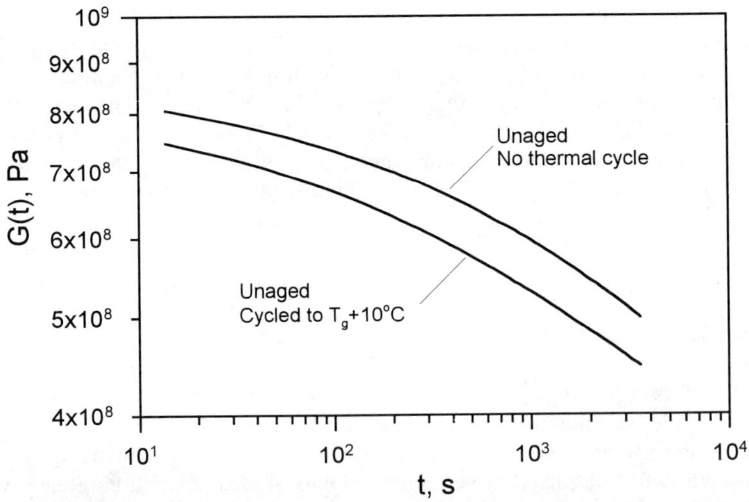

FIG. 16—*Thermal reversibility of aging effects. Aging induced during cure and post-cure (unaged specimen).*

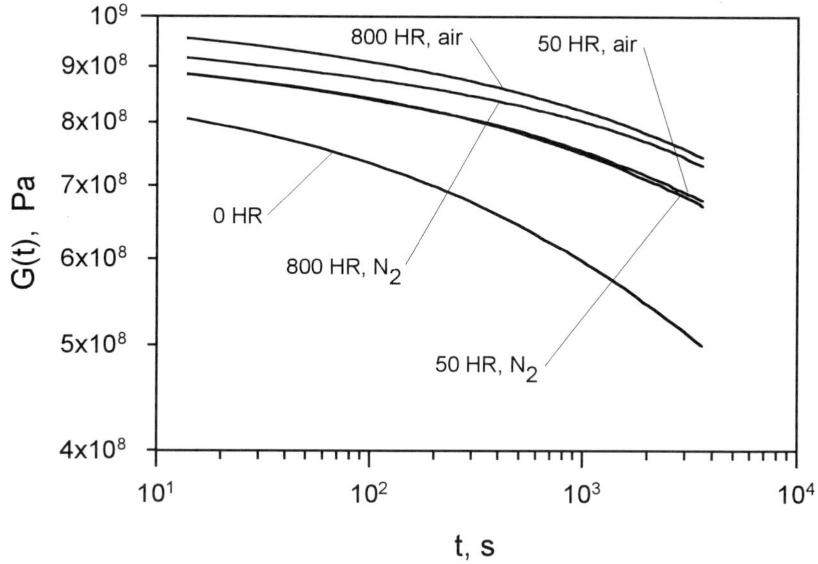

FIG. 17—*Thermal reversibility of aging effects in specimens aged at 316°C for 0, 50, and 800 h in nitrogen and air. Curves after aging.*

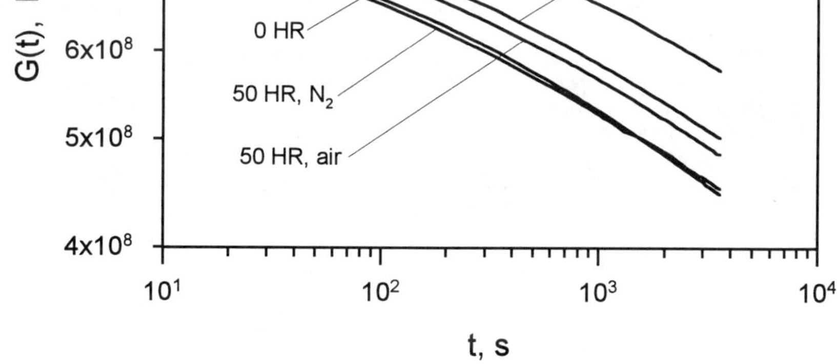

FIG. 18—*Thermal reversibility of aging effects in specimens aged and at 316°C for 0, 50, and 800 h in nitrogen and air. Curves after aging plus a thermal cycle to reverse aging.*

indicator of the extent of physical and chemical aging, further work is needed to optimize the cycle used to measure thermal reversibility. It may be possible to use a lower maximum temperature in the thermal cycle to further reduce the possibility of inducing chemical changes during the thermal cycle. In addition, the effect of cooling rate on the magnitude of the shift needs to be examined. Cooling rate effects have been observed [2] to have an effect on shift factors measured after thermal cycling.

Summary and Conclusions

Shear stress relaxation curves measured using PMR-15 specimens aged in nitrogen at 316°C for times up to 800 h can be superimposed by horizontal and vertical shifts to form a master curve. Curves measured using specimens aged in air can be shifted onto the same master curve if the aging time is less than 400 h. For specimens aged in nitrogen, the vertical shift may be associated with the same cure-related chemical changes which cause T_g to increase and mass to decrease at short aging times. For specimens aged in air, the vertical shift may be related to the same chemical degradation that causes a continuous increase in mass loss and the formation of an oxidized surface region when specimens are aged in air. Thermal reversibility of the aging effects indicates that a large portion of the aging is physical in nature. Specimens aged in nitrogen had a larger reversible component than specimens aged in air.

The results of this investigation indicated that the shift factors and the reversibility of aging effects can provide quantitative information about the extent of physical and chemical aging in PMR-15 for temperatures up to 316°C. However, more data is needed for aging temperatures less than 316°C, and the thermal cycle used to reverse physical aging needs to be optimized. In addition, the effects of non-uniform oxidation of specimens aged in air needs further consideration.

References

[1] Struik, L. C. E., *Physical Aging in Amorphous Polymers and Other Materials*, Elsevier/North-Holland Inc., New York, NY, 1978.

[2] Hastie, R. L. and Morris, D. H. "The Effects of Physical Aging on the Creep Response of a Thermoplastic Composite," *High Temperature and Environmental Effects on Polymeric Composites, ASTM STP 1174*, American Society for Testing and Materials, West Conshohocken, PA, pp. 163–185.

[3] Gates, T. S., "Effects of Elevated Temperature on the Viscoplastic Modeling of Graphite/Polymeric Composites," *High Temperature and Environmental Effects on Polymeric Composites, ASTM STP 1174*, American Society for Testing and Materials, West Conshohocken, PA, pp. 201–221.

[4] Sullivan, J. L., "Creep and Physical Aging in Composites," *Composites Science and Technology*, Vol. 39, 1990, pp. 207–232.

[5] Roberts, G. D., Malarik, D. C., and Robaidek, J. O., "Viscoelastic Properties of Addition-Cured Polyimides Used in High Temperature Polymer Matrix Composites," Composites Design, Manufacturing, and Application; *Proceedings of the Eighth International Conference on Composite Materials*, S. W. Tsai and G. S. Springer, Eds., Society for Advanced Materials and Process Engineering, Covina, CA, 1991, pp. 12-H-1 to 12-H-10.

[6] Bowles, K. J., Jayne, D., and Leonhardt, T. A., "Isothermal Aging Effects on PMR-15 Resin," *SAMPE Quarterly*, Vol. 14, No. 2, Jan. 1993, pp. 2–9.